Lecture Notes in Computer Scienc

Commenced Publication in 1973
Founding and Former Series Editors:
Gerhard Goos, Juris Hartmanis, and Jan van Leeuwen

Wendy MacCaull   Michael Winter
Ivo Düntsch (Eds.)

Relational Methods in Computer Science

8th International Seminar on Relational Methods in Computer Science
3rd International Workshop on Applications of Kleene Algebra
and Workshop of COST Action 274: TARSKI
St. Catharines, ON, Canada, February 22-26, 2005
Selected Revised Papers

Springer

Volume Editors

Wendy MacCaull
St. Francis Xavier University
Department of Mathematics, Statistics and Computer Science
P.O. Box 5000, Antigonish, Nova Scotia, Canada B2G 2W5
E-mail: wmaccaul@stfx.ca

Michael Winter
Ivo Düntsch
Brock University, Computer Science Department
St. Catharines, Ontario, Canada L2S 3A1
E-mail: {winter, duentsch}@brocku.ca

Library of Congress Control Number: 2006924115

CR Subject Classification (1998): F.4, D.2.4, F.3, I.1, I.2.3, G.2

LNCS Sublibrary: SL 1 – Theoretical Computer Science and General Issues

ISSN 0302-9743
ISBN-10 3-540-33339-8 Springer Berlin Heidelberg New York
ISBN-13 978-3-540-33339-5 Springer Berlin Heidelberg New York

Springer is a part of Springer Science+Business Media

springer.com

Printed in Germany

Typesetting: Camera-ready by author, data conversion by Scientific Publishing Services, Chennai, India
Printed on acid-free paper SPIN: 11734673 06/3142 5 4 3 2 1 0

Preface

This volume is the post conference proceedings of the 8th International Seminar on Relational Methods in Computer Science (RelMiCS 8), held in conjunction with the 3rd International Workshop on Applications of Kleene Algebra and a COST Action 274 (TARSKI) Workshop. This combined meeting took place in St. Catharines, Ontario, Canada, from February 22 to February 26, 2005. The purpose of this meeting was to bring together researchers from various subdisciplines of computer science and mathematics who use the calculus of relations and/or Kleene algebra as methodological and conceptual tools in their work.

The meeting was a continuation of three different series of meetings. Previous RelMiCS meetings were held in:

- Schloß Dagstuhl, Germany, January 1994
- Parati (near Rio de Janeiro), Brazil, September 1995
- Hammamet, Tunisia, January 1997
- Stefan Banach Center, Warsaw, Poland, September 1998
- Valcartier (near Québec City), Canada, January 2000
- Osterwijk (near Tilburg), The Netherlands, October 2001
- Malente (near Kiel), Germany, May 2003

The first two workshops on applications of Kleene algebra were held Schloß Dagstuhl, Germany, February 2001 and Malente (near Kiel), Germany, May 2003. COST Workshops were held four times per year in various locations in the European Union since 2001. The substantial common interests and overlap of these communities motivated the joint meeting. Proceedings, edited by Düntsch and Winter and containing extended abstracts of the 34 accepted papers and abstracts of the 4 invited talks, were available at the conference. After the conference, a Call for Papers was issued.

This volume contains the 17 (full) papers accepted from the submissions at that time and three invited papers, "Topological Representation of Precontact Algebras" by Georgi Dimov and Dimiter Vakarelov, "Relational Semantics through Duality" by Ewa Orłowska, Ingrid Rewitzky and Ivo Düntsch and "Duality Theory for Projective Algebras" by Alasdair Urquhart. The contributed papers underwent a thorough refereeing process, in which each paper was sent to several referees. Contributed papers include such topics as static analysis of programs, representation theory, theories of programming, evolutionary algorithms, verification and quantifier elimination.

We are grateful to the members of the Program Committee and to the many people who acted as external referees and who must remain anonymous. Their efforts have ensured the high quality of the papers in this volume.

January 2006

Wendy MacCaull
Michael Winter
Ivo Düntsch

Organization

Organizing Committee

Ivo Düntsch, Brock University, Canada
Wendy MacCaull, St. Francis Xavier University, Canada
Michael Winter, Brock University, Canada

Program Committee

Roland Backhouse, University of Nottingham, UK
Rudolf Berghammer, Christian Albrechts University of Kiel, Germany
Jules Desharnais, Université Laval, Canada
Ivo Düntsch, Brock University, Canada
Marc Frappier, Université de Sherbrooke, Canada
Marcelo Frias, University of Buenos Aires, Argentina
Peter Jipsen, Chapman University, USA
Wolfram Kahl, McMaster, Canada
Yasuo Kawahara, Kyushu University, Japan
Wendy MacCaull, St. Francis Xavier University, Canada
Bernhard Möller, University of Augsburg, Germany
Ewa Orłowska, Institute of Telecommunications, Poland
Ivo Rosenberg, Université de Montréal, Canada
Gunther Schmidt, UniBw Munich, Germany
Georg Struth, University of Augsburg, Germany
Burhan Türksen, University of Toronto, Canada
Michael Winter, Brock University, Canada

Sponsoring Institutions

Financial support from the following is gratefully acknowledged:
Computer Science Department, Brock University, St. Catharines, Canada
European Union COST Action 274 (TARSKI)

Table of Contents

Invited Papers

Contributed Papers

Topological Representation of Precontact Algebras*

Georgi Dimov and Dimiter Vakarelov

Faculty of Mathematics and Computer Science, Sofia University
gdimov@fmi.uni-sofia.bg, dvak@fmi.uni-sofia.bg

Abstract. The notions of *2-precontact* and *2-contact spaces* as well as of *extensional* (and other kinds) *3-precontact* and *3-contact spaces* are introduced. Using them, new representation theorems for precontact and contact algebras (satisfying some additional axioms) are proved. They incorporate and strengthen both the discrete and topological representation theorems from [3, 1, 2, 4, 10]. It is shown that there are bijective correspondences between such kinds of algebras and such kinds of spaces. In particular, such a bijective correspondence for the RCC systems of [8] is obtained, strengthening in this way the previous representation theorems from [4, 1].

1 Introduction

In this paper we present a common approach both to the discrete and to the non-discrete region-based theory of space. It is a continuation of the line of investigation started in [10] and continued in [1, 2].

Standard models of non-discrete theories of space are the contact algebras of regular closed subsets of some topological spaces ([10, 1, 2, 4]). In a sense these topological models reflect the continuous nature of the space. However, in the "real-world" applications, where digital methods of modeling are used, the continuous models of space are not so much suitable. This motivates a search for good "discrete" versions of the theory of space. One kind of discrete models are the so called *adjacency spaces*, introduced by Galton [6] and generalized by Düntsch and Vakarelov in [3]. Based on the Galton's approach, Li and Ying [7] presented a "discrete" generalization of the Region Connection Calculus (RCC). The latter, introduced in [8], is one of the main systems in the non-discrete region-based theory of space. A natural class of Boolean algebras related to adjacency spaces are the *precontact algebras*, introduced in [3] under the name of *proximity algebras*. The notion of precontact algebra is a generalization of the notion of contact algebra. Each adjacency space generates canonically a precontact algebra. It is proved in [3] (using another terminology) that each precontact algebra can be embedded in the precontact algebra of an adjacency space. In [1]

* This paper was supported by the project NIP-123 "Applied Logics and Topological Structures" of the Bulgarian Ministry of Education and Science.

W. MacCaull et al. (Eds.): RelMiCS 2005, LNCS 3929, pp. 1–16, 2006.

we prove that each contact algebra can be embedded in the standard contact algebra of a semiregular T_0-space, answering the question of Düntsch and Winter, posed in [4], whether the contact algebras have a topological representation. This shows that contact algebras possess both a discrete and a non-discrete (topological) representation. In this paper we extend the representation techniques developed in [1, 2] to precontact algebras, proving that each precontact algebra can be embedded in a special topological object, called a *2-precontact space.* We also establish a bijective correspondence between all precontact algebras and all 2-precontact spaces. This result is new even in the special case of contact algebras: introducing the notion of *2-contact space* as a specialization of 2-precontact space, we show that there is a bijective correspondence between all contact algebras and all 2-contact spaces. Similar representation theorems hold also for precontact and contact algebras satisfying some additional axioms, namely, for *extensional* (resp., *N-regular; regular; normal*) *precontact* and *contact algebras.* The topological objects that correspond to these algebras are introduced here under the names of *extensional* (resp., *N-regular; regular; normal*) *3-precontact* and *3-contact spaces.*

The paper is organized as follows. In Section 1 we introduce the notions of precontact and contact algebra and give the two main examples of them: the precontact algebras on adjacency spaces, and the contact algebras on topological spaces. In Section 2 we introduce different kinds of points in precontact algebras: ultrafilters, clans, maximal clans, clusters, co-ends. The notions of *topological adjacency space* and *Stone adjacency space* are introduced and our first representation theorem for precontact algebras is proved there. In Section 3 we introduce the notions of *2-precontact space* and *canonical precontact algebra of a 2-precontact space.* In Section 4 we associate with each precontact algebra $\mathbf{B}$ a 2-precontact space, called the *canonical 2-precontact space of* $\mathbf{B}$. In Section 5 we present the main theorem of the paper, the representation theorem for precontact algebras. In Section 6 we introduce the notion of *2-contact space* and we prove that there exists a bijective correspondence between all (up to isomorphism) contact algebras and all (up to isomorphism) 2-contact spaces. This is a generalization of the similar result about complete contact algebras obtained in [1]. In Section 7 we introduce the axiom of extensionality for precontact algebras which generalizes the well-known axiom of extensionality for contact algebras. In this way we obtain a class of precontact algebras containing as a subclass some well-known systems as, for example, RCC systems from [8]. We modify the representation constructions in order to obtain topological representation theorems for extensional precontact algebras. The notions of *extensional 3-precontact space* and *extensional 3-contact space* are introduced there and it is proved that there exists a bijective correspondence between all (up to isomorphism) extensional precontact algebras (resp., extensional contact algebras) and all (up to isomorphism) extensional 3-precontact spaces (resp., extensional 3-contact spaces). This is a generalization of a similar result about complete extensional contact algebras obtained in [1]. In Section 8 we extend the results for the extensional precontact and contact algebras to many other kinds of precontact and contact algebras.

The main reference book for all undefined in the paper topological notions is [5]. In this paper we present only the scheme of the proofs (i.e., all lemmas and propositions which are used in the proofs). The detailed proofs will be given in the full version of this paper.

2 Precontact Algebras

Definition 1. *An algebraic system* $\mathbf{B} = (B, C)$ *is called a* **precontact algebra** *([3]) (abbreviated as PCA) if the following holds:*

- $B = (B, 0, 1, +, ., *)$ *is a Boolean algebra (where the complement is denoted by "$*$");*
- C *is a binary relation on* B*, called the* **precontact relation***, which satisfies the following axioms:*

$(C0)$ *If* aCb *then* $a, b \neq 0$*;*
$(C+)$ $aC(b+c)$ *iff* aCb *or* aCc*;* $(a+b)Cc$ *iff* aCc *or* bCc*.*

A precontact algebra (B, C) *is said to be* **complete** *if the Boolean algebra* B *is complete. Two precontact algebras* $\mathbf{B} = (B, C)$ *and* $\mathbf{B_1} = (B_1, C_1)$ *are said to be* **PCA-isomorphic** *(or, simply,* **isomorphic***) if there exists a Boolean isomorphism* $\varphi : B \longrightarrow B_1$ *such that, for every* $a, b \in B$*,* aCb *iff* $\varphi(a)C_1\varphi(b)$*.*

The negation of the relation C *is denoted by* $(-C)$*.*

Let us define the **non-tangential inclusion** "$\ll_C$" *by* $a \ll_C b$ *iff* $a(-C)b^*$*.*

We will also consider precontact algebras satisfying some additional axioms:

$(Cref)$ *If* $a \neq 0$ *then* aCa *(reflexivity axiom);*
$(Csym)$ *If* aCb *then* bCa *(symmetry axiom);*
(Ctr) *If* $a \ll_C c$ *then* $(\exists b)(a \ll_C b \ll_C c)$ *(transitivity axiom);*
$(Ccon)$ *If* $a \neq 0, 1$ *then* aCa^* *or* a^*Ca *(connectedness axiom).*

A precontact algebra (B, C) *is called a* **contact algebra** *([1]) (and* C *is called a* **contact relation***) if it satisfies the axioms* $(Cref)$ *and* $(Csym)$*. A precontact algebra* (B, C) *is called* **connected** *if it satisfies the axiom* $(Ccon)$*.*

The following lemma says that in every precontact algebra we can define a contact relation.

Lemma 1. *Let* (B, C) *be a precontact algebra and define* $aC^{\#}b$ *iff* aCb *or* bCa *or* $a.b \neq 0$*. Then* $C^{\#}$ *is a contact relation on* B *and hence* $(B, C^{\#})$ *is a contact algebra.*

Remark 1. We will also consider precontact algebras satisfying the following variant of the transitivity axiom (Ctr):
$(Ctr\#)$ If $a \ll_{C^\#} c$ then $(\exists b)(a \ll_{C^\#} b \ll_{C^\#} c)$.

Examples of Precontact and Contact Algebras

1. Precontact algebras on adjacency spaces. (Galton [6], Düntsch and Vakarelov [3])

By an **adjacency space** we mean a relational system (W, R) where W is a non-empty set whose elements are called **cells**, and R is a binary relation on W called the **adjacency relation**; the subsets of W are called **regions**.

The reflexive and symmetric closure $R^\flat$ of R is defined as follows: $xR^\flat y$ iff xRy or yRx or $x = y$.

A precontact relation C_R between the regions of an adjacency space (W, R) is defined as follows: for every $a, b \subseteq W$,

$$aC_Rb \text{ iff } (\exists x \in a)(\exists y \in b)(xRy). \tag{1}$$

Proposition 1. ([3]) *Let (W, R) be an adjacency space and let 2^W be the Boolean algebra of all subsets of W. Then:*

(a) $(2^W, C_R)$ is a precontact algebra;
(b) $(2^W, C_R)$ is a contact algebra iff R is a reflexive and symmetric relation on W. If R is a reflexive and symmetric relation on W then C_R coincides with $(C_R)^\#$ and $C_{R^\flat}$;
(c) C_R satisfies the axiom (Ctr) iff R is a transitive relation on W;
(d) C_R satisfies the axiom (Ccon) iff R is a connected relation on W (which means that if $x, y \in W$ and $x \neq y$ then there is an R-path from x to y or from y to x).

Theorem 1. ([3]) *Each precontact algebra (B, C) can be isomorphically embedded in the precontact algebra $(2^W, C_R)$ of some adjacency space (W, R). Moreover if (B, C) satisfies some of the axioms (Cref), (Csym), (Ctr) then the relation R is, respectively, reflexive, symmetric, transitive.*

2. Contact algebras on topological spaces. Let X be a topological space and let $RC(X)$ be the set of all regular closed subsets of X (recall that a subset F of X is said to be *regular closed* if $F = cl(int(F))$). Let us equip $RC(X)$ with the following Boolean operations and contact relation C_X:

- $a + b = a \cup b$;
- $a^* = cl(X \setminus a)$;
- $a.b = (a^* \cup b^*)^*$;
- $0 = \emptyset$, $1 = X$;
- aC_Xb iff $a \cap b \neq \emptyset$.

The following lemma is a well-known fact.

Lemma 2. $(RC(X), C_X) = (RC(X), 0, 1, +, ., *, C_X)$ *is a contact algebra.*

Recall that a space X is said to be **semiregular** if $RC(X)$ is a closed base for X.

The following theorem answers the question, posed by Düntsch and Winter in [4], whether contact algebras have a topological representation:

Theorem 2. ([1]) *Each contact algebra (B, C) can be isomorphically embedded in the contact algebra $(RC(X), C_X)$ of some semiregular T_0-space X. The algebra (B, C) is connected iff the space X is connected.*

The aim of this work is to generalize Theorem 1 and Theorem 2 in several ways: to find a topological representation of precontact algebras which incorporates both the "discrete" and the "continuous" nature of the space; to find representation theorems in the style of the Stone representation of Boolean algebras instead of embedding theorems; to establish, again as in the Stone theory, a bijective correspondence between precontact algebras and the corresponding topological objects; to extend this new representation theory to different classes of precontact algebras, satisfying some natural additional axioms.

3 Points in Precontact Algebras

In this section we will introduce different kinds of abstract points in precontact algebras: ultrafilters, clans, maximal clans, clusters and co-ends. This is done by analogy with the case of contact algebras (see, e.g., [1, 10]). We assume that the notions of a filter and ultrafilter in a Boolean algebra are familiar. Clans were introduced by Thron [9] in proximity theory. Our definition is a lattice generalization of Thron's definition.

The set of all ultrafilters of a Boolean algebra B is denoted by $Ult(B)$.

Definition 2. *Let $\mathbf{B} = (B, C)$ be a precontact algebra. A non-empty subset Γ of B is called a* **clan** *if it satisfies the following conditions:*

(*Clan*1) $0 \notin \Gamma$;
(*Clan*2) *If* $a \in \Gamma$ *and* $a \leq b$ *then* $b \in \Gamma$;
(*Clan*3) *If* $a + b \in \Gamma$ *then* $a \in \Gamma$ *or* $b \in \Gamma$;
(*Clan*4) *If* $a, b \in \Gamma$ *then* $aC^{\#}b$.

A clan Γ in $\mathbf{B}$ is called a **maximal clan** *in $\mathbf{B}$ if it is maximal among all clans in $\mathbf{B}$ with respect to set-inclusion.*

The set of all clans (resp., maximal clans) of a precontact algebra $\mathbf{B}$ is denoted by $Clans(\mathbf{B})$ (resp., $MClans(\mathbf{B})$).

The following lemma is obvious:

Lemma 3. *Each ultrafilter is a clan and hence $Ult(B) \subseteq Clans(\mathbf{B})$.*

We will define a binary relation R between ultrafilters in B which will make the set $Ult(B)$ an adjacency space.

Definition 3. *Let $\mathbf{B} = (B, C)$ be a precontact algebra and let U_1, U_2 be ultrafilters. We set*

$$U_1 R U_2 \text{ iff } (\forall a \in U_1)(\forall b \in U_2)(aCb)(\text{ i.e., iff } U_1 \times U_2 \subseteq C). \quad (2)$$

The relational system $(Ult(B), R)$ is called the **canonical adjacency space of B**.

We say that U_1, U_2 are **connected** *iff $U_1 R^\flat U_2$, where $R^\flat$ is the reflexive and symmetric closure of R.*

Lemma 4. *Let I be a set of connected ultrafilters. Then the union $\bigcup\{U \mid U \in I\}$ is a clan and every clan can be obtained in this way.*

Lemma 5. ([1, 3]) **Ultrafilter and clan characterizations of precontact and contact relations.**

Let $\mathbf{B} = (B, C)$ be a precontact algebra and $(Ult(B), R)$ be the canonical adjacency space on $\mathbf{B}$. Then the following is true for any $a, b \in B$:

(a) aCb *iff* $(\exists U_1, U_2 \in Ult(B))(U_1 R U_2)$;
(b) $aC^{\#}b$ *iff* $(\exists U_1, U_2 \in Ult(B))(U_1 R^\flat U_2)$;
(c) $aC^{\#}b$ *iff* $(\exists \Gamma \in Clans(\mathbf{B})(a, b \in \Gamma)$ *iff* $(\exists \Gamma \in MClans(\mathbf{B})(a, b \in \Gamma)$;
(d) *R is a reflexive relation iff $\mathbf{B}$ satisfies the axiom $(Cref)$.;*
(e) *R is a symmetric relation iff $\mathbf{B}$ satisfies the axiom $(Csym)$;*
(f) *R is a transitive relation iff $\mathbf{B}$ satisfies the axiom (Ctr).*

Definition 4. Clusters and co-ends. *Let $\mathbf{B} = (B, C)$ be a precontact algebra.*
• *A clan Γ in $\mathbf{B}$ is called a* **cluster** *in $\mathbf{B}$ if it satisfies the following condition:*
(Clust) *If for every $x \in \Gamma$ we have $xC^{\#}y$ then $y \in \Gamma$.*
• *A clan Γ in $\mathbf{B}$ is called a* **co-end** *in $\mathbf{B}$ if it satisfies the following condition:*
(Coend) *If $x \notin \Gamma$ then there exists a $y \notin \Gamma$ such that $x(-C^{\#})y^*$.*

The set of all clusters (resp., co-ends) in $\mathbf{B}$ is denoted by $Clust(\mathbf{B})$ (resp., $Coend(\mathbf{B})$).

Stone Adjacency Spaces and Representation of Precontact Algebras

Definition 5. *An adjacency space (X, R) is called a* **topological adjacency space** *(abbreviated as TAS) if X is a topological space and R is a closed relation on X. When X is a compact Hausdorff zero-dimensional space (i.e. X is a* Stone space*), we say that the topological adjacency space (X, R) is a* **Stone adjacency space**.

Two topological adjacency spaces (X, R) and (X_1, R_1) are said to be **TAS-isomorphic** *if there exists a homeomorphism $f : X \longrightarrow X_1$ such that, for every $x, y \in X$, xRy iff $f(x)R_1 f(y)$.*

If X is a topological space, we denote by $Clopen(X)$ the set of all clopen subsets of X.

Now we can obtain the following strengthening of Theorem 1:

Theorem 3. *(a) Each precontact algebra (B, C) is isomorphic to the precontact algebra $(Clopen(X), C_R)$ of a Stone adjacency space (X, R), where X is the Stone space of the Boolean algebra B and the isomorphism is just the Stone map $s : B \longrightarrow 2^{Ult(B)}$, $a \longrightarrow \{U \in Ult(B) \mid a \in U\}$. Moreover, the relation C satisfies the axiom (Cref) (resp., (Csym); (Ctr)) iff the relation R is reflexive (resp., symmetric; transitive).*

(b) There exists a bijective correspondence between the class of all (up to PCA-isomorphism) precontact algebras and the class of all (up to TAS-isomorphism) Stone adjacency spaces (X, R).

As it is shown in [3], there is no bijective correspondence between precontact algebras and adjacency spaces (up to isomorphisms). Hence, the role of the topology in Theorem 3 is essential. However, Theorem 3 is not completely satisfactory because the representation of the precontact algebras (B, C) obtained here does not give a topological representation of the contact algebras $(B, C^{\#})$ generated by (B, C): we would like to have an isomorphism f such that, for every $a, b \in B$, aCb iff $f(a)C_R f(b)$, and $aC^{\#}b$ iff $f(a) \cap f(b) \neq \emptyset$ (see (1) for C_R). The isomorphism s in Theorem 3 is not of this type. Indeed, there are many examples of contact algebras (B, C) where $a.b = 0$ (and hence $s(a) \cap s(b) = \emptyset$) but aCb (note that C and $C^{\#}$ coincide for contact algebras). We now construct some natural topological objects which correspond bijectively to the precontact algebras and satisfy the above requirement. In the case when (B, C) is a contact algebra (satisfying some additional axioms), we will show that these topological objects are just topological pairs (resp., topological triples) satisfying some natural conditions. In such a way we will obtain new representation theorems for the contact algebras (satisfying some additional axioms), completely different from those given in [10, 1, 2, 4]. In particular, we will describe the topological triples which correspond bijectively to the connected extensional contact algebras, i.e. to the RCC systems of Randel, Cui and Cohn [8], strengthening in such a way the representation theorems for RCC systems given in [4, 1].

4 2-Precontact Spaces

Definition 6. *(a) Let X be a topological space and X_0, X_1 be dense subspaces of X. Then the pair (X, X_0) is called a* **topological pair** *and the triple (X, X_0, X_1) is called a* **topological triple**.

(b) Let (X, X_0) be a topological pair. Then we set

$$RC(X, X_0) = \{cl_X(A) \mid A \in Clopen(X_0)\}. \tag{3}$$

(c) Let (X, X_0, X_1) be a topological triple. Then we set

$$RC(X, X_0, X_1) = \{X_1 \cap cl_X(A) \mid A \in Clopen(X_0)\}. \tag{4}$$

Lemma 6. *(a) Let (X, X_0) be a topological pair. Then $RC(X, X_0) \subseteq RC(X)$; the set $RC(X, X_0)$ with the standard Boolean operations on the regular closed subsets of X is a Boolean subalgebra of $RC(X)$; $RC(X, X_0)$ is isomorphic to the Boolean algebra $Clopen(X_0)$; the sets $RC(X)$ and $RC(X, X_0)$ coincide iff X_0 is an extremally disconnected space. If $C_{(X,X_0)}$ is the restriction of the contact relation C_X (see Lemma 2) to $RC(X, X_0)$ then $(RC(X, X_0), C_{(X,X_0)})$ is a contact subalgebra of $(RC(X), C_X)$.*

(b) Let (X, X_0, X_1) be a topological triple. Then $RC(X, X_0, X_1) \subseteq RC(X_1)$; the set $RC(X, X_0, X_1)$ with the standard Boolean operations on the regular closed subsets of X_1 is a Boolean subalgebra of the Boolean algebra $RC(X_1)$; the Boolean algebra $Clopen(X_0)$ is isomorphic to $RC(X, X_0, X_1)$; the sets $RC(X, X_0, X_1)$ and $RC(X_1)$ coincide iff X_0 is an extremally disconnected space. Let us denote

by $C_{(X,X_0,X_1)}$ *the restriction of the contact relation* C_{X_1} *to* $RC(X,X_0,X_1)$*. Then* $(RC(X,X_0,X_1),C_{(X,X_0,X_1)})$ *is a contact subalgebra of* $(RC(X_1),C_{X_1})$*.*

Definition 7. Let (X,τ) be a topological space, X_0 be a subspace of X and $x \in X$. We put

$$\sigma_x = \{F \in RC(X) \mid x \in F\};\ \Gamma_{x,X_0} = \{F \in Clopen(X_0) \mid x \in cl_X(F)\}. \quad (5)$$

Definition 8. *Let* X_0 *be a topological space and* R *be a binary relation on* X_0*. Then the pair* $(Clopen(X_0),C_R)$ *(see (1) for* C_R*) is a precontact algebra, called the* **precontact algebra of the relational system** (X_0,R) **determined by the clopen subsets of** X_0**.**

Definition 9. 2-Precontact spaces. *(a) A triple* $\mathbf{X} = (X,X_0,R)$ *is called a* **2-precontact space** *(abbreviated as PCS) if the following conditions are satisfied:*

$(PCS1)$ (X,X_0) *is a topological pair and* X *is a* T_0*-space;*
$(PCS2)$ (X_0,R) *is a Stone adjacency space;*
$(PCS3)$ $RC(X,X_0)$ *is a closed base for* X*;*
$(PCS4)$ *For every* $F,G \in Clopen(X_0)$*,* $cl_X(F) \cap cl_X(G) \neq \emptyset$ *implies that* $F(C_R)^\# G$ *(see (1) for* C_R*);*
$(PCS5)$ *If* $\Gamma \in Clans(Clopen(X_0),C_R)$ *then there exists a point* $x \in X$ *such that* $\Gamma = \Gamma_{x,X_0}$ *(see (5) for* Γ_{x,X_0}*).*

(b) Let $\mathbf{X} = (X,X_0,R)$ *be a 2-precontact space. Define, for every* $F,G \in RC(X,X_0)$*,* $FC_R^X G$ *iff there exist* $x \in F \cap X_0$ *and* $y \in G \cap X_0$ *such that* xRy*. Then the precontact algebra* $\mathbf{B}(\mathbf{X}) = (RC(X,X_0),C_R^X)$ *is said to be the* **canonical precontact algebra of X**.

(c) A 2-precontact space $\mathbf{X} = (X,X_0,R)$ *is called* **reflexive** *(resp.,* **symmetric; transitive***) if the relation* R *is reflexive (resp., symmetric; transitive);* $\mathbf{X}$ *is called* **connected** *if the space* X *is connected.*

(d) Let $\mathbf{X} = (X,X_0,R)$ *and* $\widehat{\mathbf{X}} = (\widehat{X},\widehat{X}_0,\widehat{R})$ *be two 2-precontact spaces. We say that* $\mathbf{X}$ *and* $\widehat{\mathbf{X}}$ *are* **PCS-isomorphic** *(or, simply,* **isomorphic***) if there exists a homeomorphism* $f : X \longrightarrow \widehat{X}$ *such that:*

(1) $f(X_0) = \widehat{X}_0$*; and*
(2) $(\forall x,y \in X_0)(xRy \leftrightarrow f(x)\widehat{R}f(y))$*.*

Proposition 2. *(a) Let* (X,X_0,R) *be a 2-precontact space. Then* X *is a semiregular space and, for every* $F,G \in Clopen(X_0)$*,*

$$cl_X(F) \cap cl_X(G) \neq \emptyset \text{ iff } F(C_R)^\# G. \quad (6)$$

(b) Let $\mathbf{X}$ *and* $\widehat{\mathbf{X}}$ *be two isomorphic 2-precontact spaces. Then the corresponding canonical precontact algebras* $\mathbf{B}(\mathbf{X})$ *and* $\mathbf{B}(\widehat{\mathbf{X}})$ *are isomorphic.*

Lemma 7. Correspondence Lemma. *Let* $\mathbf{X} = (X,X_0,R)$ *be a 2-precontact space and let* $\mathbf{B}(\mathbf{X}) = (RC(X,X_0),C_R^X)$ *be the canonical precontact algebra of* $\mathbf{X}$*. Then the following equivalences hold:*

(a) *The space* **X** *is reflexive iff the algebra* **B**(**X**) *satisfies the axiom* $(Cref)$;
(b) *The space* **X** *is symmetric iff* **B**(**X**) *satisfies the axiom* $(Csym)$.;
(c) *The space* **X** *is transitive iff* **B**(**X**) *satisfies the axiom* (Ctr);
(d) *The space* **X** *is connected iff* **B**(**X**) *is connected.*

5 The Canonical 2-Precontact Space of a Precontact Algebra

In this section we will associate with each precontact algebra a 2-precontact space.

Definition 10. *Let* $\mathbf{B} = (B, C)$ *be a precontact algebra. We associate with* **B** *a 2-precontact space* $\mathbf{X}(\mathbf{B}) = (X, X_0, R)$, *called the* **canonical 2-precontact space of B**, *as follows:*

- $X = Clans(\mathbf{B})$ *and* $X_0 = Ult(B)$;
- *A topology on the set* X *is defined in the following way: for any* $a \in B$, *let*

$$g(a) = \{\Gamma \in X \mid a \in \Gamma\}; \tag{7}$$

  then the family $\{g(a) \mid a \in B\}$ *is a closed base for a topology* τ *on* X. *If, for every* $a \in B$, $g_0(a) = g(a) \cap X_0$ *then the family* $\{g_0(a) \mid a \in B\}$ *is a closed base of* X_0 *and* $g_0(a) = s(a)$, *where* $s : B \longrightarrow 2^{X_0}$ *is the Stone map;*
- (X_0, R) *is the canonical adjacency space of* **B** *(see Definition 3).*

Proposition 3. *Let* $\mathbf{B} = (B, C)$ *be a precontact algebra. Then the system* $\mathbf{X}(\mathbf{B}) = (X, X_0, R)$ *defined above is a 2-precontact space.*

Proposition 4. *Let* $\mathbf{B_1}$ *and* $\mathbf{B_2}$ *be two isomorphic precontact algebras. Then the corresponding canonical 2-precontact spaces* $\mathbf{X}(\mathbf{B_1})$ *and* $\mathbf{X}(\mathbf{B_2})$ *are isomorphic.*

Proposition 5. *Let* **B** *be a precontact algebra, let* **X**(**B**) *be the canonical 2-precontact space of* **B** *and let* **B**′ *be the canonical precontact algebra of the 2-precontact space* **X**(**B**). *Then* **B** *and* **B**′ *are PCA-isomorphic.*

Lemma 8. Topological characterization of the connectedness of B. *Let* **B** *be a precontact algebra and* **X**(**B**) *be the canonical 2-precontact space of* **B**. *Then* **B** *is connected iff* **X**(**B**) *is connected.*

6 The Main Theorem

Theorem 4. Representation theorem for precontact algebras.

(a) *Let* $\mathbf{B} = (B, C)$ *be a precontact algebra and let* $\mathbf{X}(\mathbf{B}) = (X, X_0, R)$ *be the canonical 2-precontact space of* **B**. *Then the function* g, *defined in (7), is a PCA-isomorphism from* (B, C) *onto the canonical precontact algebra* $(RC(X, X_0), C_R^X)$ *of* **X**(**B**). *The same function* g *is a PCA-isomorphism* *between the contact algebras* $(B, C^{\#})$ *and* $(RC(X, X_0), C_{(X,X_0)})$. *The sets* $RC(X, X_0)$ *and* $RC(X)$ *coincide iff the precontact algebra* $\mathbf{B}$ *is complete. The algebra* $\mathbf{B}$ *satisfies the axiom* $(Cref)$ *(resp.,* $(Csym)$; (Ctr)*) iff the 2-precontact space* $\mathbf{X}(\mathbf{B})$ *is reflexive (resp., symmetric; transitive). The algebra* $\mathbf{B}$ *is connected iff* $\mathbf{X}$ *is connected.*

(b) *There exists a bijective correspondence between all (up to PCA-isomorphism) (connected) precontact algebras and all (up to PCS-isomorphism) (connected) 2-precontact spaces.*

7 2-Contact Spaces and a New Representation Theorem for Contact Algebras

Proposition 6. *Let* X_0 *be a subspace of a topological space* X. *For every* $F, G \in Clopen(X_0)$ *set*

$$F\delta_{(X,X_0)}G \text{ iff } cl_X(F) \cap cl_X(G) \neq \emptyset. \tag{8}$$

Then $(Clopen(X_0), \delta_{(X,X_0)})$ *is a contact algebra.*

Definition 11. 2-Contact spaces. *(a) A topological pair* (X, X_0) *is called a* **2-contact space** *(abbreviated as CS) if the following conditions are satisfied:*

$(CS1)$ X *is a* T_0*-space;*
$(CS2)$ X_0 *is a Stone space;*
$(CS3)$ $RC(X, X_0)$ *is a closed base for* X*;*
$(CS4)$ *If* $\Gamma \in Clans(Clopen(X_0), \delta_{(X,X_0)})$ *then there exists a point* $x \in X$ *such that* $\Gamma = \Gamma_{x,X_0}$ *(see (5) for* Γ_{x,X_0}*).*

A 2-contact space (X, X_0) *is called* **connected** *if the space* X *is connected.*

(b) Let (X, X_0) *be a 2-contact space. Then the contact algebra* $\mathbf{B}(\mathbf{X}, \mathbf{X_0}) = (RC(X, X_0), C_{(X,X_0)})$ *is said to be the* **canonical contact algebra of the 2-contact space** (X, X_0).

(c) Let (B, C) *be a contact algebra,* $X = Clans(B, C)$, $X_0 = Ult(B)$ *and* τ *be the topology on* X *described in Definition 10. Take the subspace topology on* X_0. *Then the pair* (X, X_0) *is called the* **canonical 2-contact space of the contact algebra** (B, C).

(d) Let (X, X_0) *and* $(\widehat{X}, \widehat{X}_0)$ *be two 2-contact spaces. We say that* (X, X_0) *and* $(\widehat{X}, \widehat{X}_0)$ *are* **CS-isomorphic** *(or, simply,* **isomorphic***) if there exists a homeomorphism* $f : X \longrightarrow \widehat{X}$ *such that* $f(X_0) = \widehat{X}_0$.

Lemma 9. *For every 2-contact space* (X, X_0) *there exists a unique reflexive and symmetric binary relation* R *on* X_0 *such that* (X, X_0, R) *is a 2-precontact space.*

Theorem 5. New representation theorem for contact algebras.

(a) *Let* (B, C) *be a contact algebra and let* (X, X_0) *be the canonical 2-contact space of* (B, C) *(see Definition 11(c)). Then the function* g*, defined in (7), is a PCA-isomorphism from the algebra* (B, C) *onto the canonical contact* *algebra* $(RC(X, X_0), C_{(X,X_0)})$ *of* (X, X_0)*. The sets* $RC(X, X_0)$ *and* $RC(X)$ *coincide iff the contact algebra* (B, C) *is complete. The contact algebra* (B, C) *is connected iff the 2-contact space* (X, X_0) *is connected.*

(b) *There exists a bijective correspondence between all (up to PCA-isomorphism) (connected) contact algebras and all (up to CS-isomorphism) (connected) 2-contact spaces.*

Definition 12. ([1]) *A semiregular* T_0*-space* (X, τ) *is* **C-semiregular** *if for every clan* Γ *in* $(RC(X), C_X)$ *there exists a point* $x \in X$ *such that* $\Gamma = \sigma_x$ *(see (5) for* σ_x*).*

Corollary 1. ([1]) *There exists a bijective correspondence between the class of all (up to PCA-isomorphism) complete (connected) contact algebras and the class of all (up to homeomorphism) (connected) C-semiregular spaces.*

8 Precontact Algebras with the Axiom of Extensionality

The axiom of extensionality is one of the most interesting axioms for contact algebras. Since $C^{\#}$ is a contact relation for every precontact relation C, we formulate it for the relation $C^{\#}$:

$(\forall c)(aC^{\#}c \leftrightarrow bC^{\#}c) \rightarrow (a = b)$.

The axiom of extensionality has several equivalent formulations (see, e.g., [1]); the shortest one is the following:

$(\forall a \neq 1)(\exists b \neq 0)(a(-C^{\#})b)$, or, equivalently,

(Cext) $(\forall a \neq 1)(\exists b \neq 0)(a(-C)b$ and $b(-C)a$ and $a.b = 0)$.

Note that the class of extensional precontact algebras contains some well-known systems as, for example, RCC system from [8].

In this section we will modify the representation theory for precontact algebras in order to obtain similar theory for extensional precontact algebras.

Definition 13. *A precontact algebra* (B, C) *which satisfies the axiom* $(Cext)$ *is said to be an* **extensional precontact algebra** *(abbreviated as EPA).*

Recall that a topological space (X, τ) is called **weakly regular** ([4]) if it is semiregular and for each non-empty $U \in \tau$ there exists a non-empty $V \in \tau$ such that $cl(V) \subseteq U$.

Definition 14. *A quadruple* $\mathbf{X} = (X, X_0, X_1, R)$ *is called an* **extensional 3-precontact space** *(abbreviated as EPS) if it satisfies the following conditions:*

$(EPS1)$ (X, X_0, R) *is a 2-precontact space;*
$(EPS2)$ X_1 *is a dense subspace of* X*;*
$(EPS3)$ X_1 *is a weakly regular* T_1*-space;*
$(EPS4)$ *If* $x \in X_1$ *then the set* Γ_{x,X_0} *(see (5)) is a maximal clan in the precontact algebra* $(Clopen(X_0), C_R)$*; conversely, for every maximal clan* Γ *in* $(Clopen(X_0), C_R)$ *there exists a point* $x \in X_1$ *such that* $\Gamma = \Gamma_{x,X_0}$*.*

(b) Let $\mathbf{X} = (X, X_0, X_1, R)$ *be an extensional 3-precontact space. Define, for every* $F, G \in RC(X, X_0, X_1)$, $FC_R^{X,X_1}G$ *iff there exist* $x \in cl_X(F) \cap X_0$ *and* $y \in cl_X(G) \cap X_0$ *such that* xRy*. Then the precontact algebra* $\mathbf{B}(\mathbf{X}) = (RC(X, X_0, X_1), C_R^{X,X_1})$ *is said to be the* **canonical extensional precontact algebra of the extensional 3-precontact space X**.

(c) An extensional 3-precontact space $\mathbf{X} = (X, X_0, X_1, R)$ *is called* **reflexive** *(resp.,* **symmetric**; **transitive***) if the relation* R *is reflexive (resp., symmetric; transitive);* $\mathbf{X}$ *is called* **connected** *if the space* X_1 *is connected.*

(d) Let $\mathbf{X} = (X, X_0, X_1, R)$ *and* $\widehat{\mathbf{X}} = (\widehat{X}, \widehat{X}_0, \widehat{X}_1, \widehat{R})$ *be two extensional 3-precontact spaces. We say that* $\mathbf{X}$ *and* $\widehat{\mathbf{X}}$ *are* **EPS-isomorphic** *(or, simply,* **isomorphic***) if there exists a homeomorphism* $f : X \longrightarrow \widehat{X}$ *such that:*

(1) $f(X_0) = \widehat{X}_0$;
(2) $f(X_1) = \widehat{X}_1$; *and*
(3) $(\forall x, y \in X_0)(xRy \leftrightarrow f(x)\widehat{R}f(y))$.

Proposition 7. *Let* (X, X_0, X_1, R) *be an extensional 3-precontact space. Then* X *is a semiregular space and, for every* $F, G \in Clopen(X_0)$*, we have that*

$$cl_X(F) \cap cl_X(G) \neq \emptyset \text{ iff } cl_X(F) \cap cl_X(G) \cap X_1 \neq \emptyset. \tag{9}$$

We will now associate with each extensional precontact algebra an extensional 3-precontact space.

Definition 15. *Let* $\mathbf{B} = (B, C)$ *be an extensional precontact algebra. We will associate with* $\mathbf{B}$ *an extensional 3-precontact space* $\mathbf{X}(\mathbf{B}) = (X, X_0, X_1, R)$ *called the* **canonical extensional 3-precontact space of B**. *Let* (X, X_0, R) *be the canonical 2-precontact space of the precontact algebra* (B, C)*. We set* $X_1 = MClans(B, C)$*. Then* $X_1 \subseteq X$ *and we take the subspace topology on* X_1.

Theorem 6. Representation theorem for extensional precontact algebras.

(a) *Let* $\mathbf{B} = (B, C)$ *be an extensional precontact algebra and let* $\mathbf{X}(\mathbf{B}) = (X, X_0, X_1, R)$ *be the canonical extensional 3-precontact space of* $\mathbf{B}$*. Then the function* g*, defined in (7), is a PCA-isomorphism from* (B, C) *onto the canonical extensional precontact algebra* $(RC(X, X_0, X_1), C_R^{X,X_1})$ *of* $\mathbf{X}(\mathbf{B})$*. The same function* g *is a PCA-isomorphism between the contact algebras* $(B, C^{\#})$ *and* $(RC(X, X_0, X_1), C_{(X,X_0,X_1)})$*. The algebra* $\mathbf{B}$ *is complete iff the sets* $RC(X, X_0, X_1)$ *and* $RC(X_1)$ *coincide. The algebra* $\mathbf{B}$ *satisfies the axiom* $(Cref)$ *(resp.,* $(Csym)$; (Ctr)*) iff the extensional 3-precontact space* $\mathbf{X}(\mathbf{B})$ *is reflexive (resp., symmetric; transitive). The algebra* $\mathbf{B}$ *is connected iff* $\mathbf{X}$ *is connected.*
(b) *There exists a bijective correspondence between all (up to PCA-isomorphism) extensional (connected) precontact algebras and all (up to EPS-isomor phism) extensional (connected) 3-precontact spaces.*

Definition 16. Extensional 3-Contact spaces.

(a) A topological triple (X, X_0, X_1) *is called an* **extensional 3-contact space** *(abbreviated as ECS) if the following conditions are satisfied:*

$(ECS1)$ (X, X_0) *is a 2-contact space;*
$(ECS2)$ X_1 *is a weakly regular* T_1*-space;*
$(ECS3)$ *If* $x \in X_1$ *then the set* Γ_{x,X_0} *(see (5)) is a maximal clan in the contact algebra* $(Clopen(X_0), \delta_{(X,X_0)})$*; conversely, for every maximal clan* Γ *in* $(Clopen(X_0), \delta_{(X,X_0)})$ *there exists a point* $x \in X_1$ *such that* $\Gamma = \Gamma_{x,X_0}$.

An extensional 3-contact space (X, X_0, X_1) *is called* **connected** *if the space* X_1 *is connected.*

(b) Let (X, X_0, X_1) *be an extensional 3-contact space. Then the contact algebra* $(RC(X, X_0, X_1), C_{(X,X_0,X_1)})$ *is said to be the* **canonical extensional contact algebra of the extensional 3-contact space** (X, X_0, X_1).

(c) Let (B, C) *be an extensional contact algebra and* (X, X_0) *be the canonical 2-contact space of* (B, C)*. Set* $X_1 = MClans(B, C)$*. Then* $X_1 \subseteq X$*. Take the subspace topology on* X_1*. Then the triple* (X, X_0, X_1) *is called the* **canonical extensional 3-contact space of the extensional contact algebra** (B, C).

(d) Let (X, X_0, X_1) *and* $(\widehat{X}, \widehat{X}_0, \widehat{X}_1)$ *be two extensional 3-contact spaces. We say that* (X, X_0, X_1) *and* $(\widehat{X}, \widehat{X}_0, \widehat{X}_1)$ *are* **ECS-isomorphic** *(or, simply,* **isomorphic***) if there exists a homeomorphism* $f : X \longrightarrow \widehat{X}$ *such that* $f(X_0) = \widehat{X}_0$ *and* $f(X_1) = \widehat{X}_1$.

Lemma 10. *For every extensional 3-contact space* (X, X_0, X_1) *there exists a unique reflexive and symmetric binary relation* R *on* X_0 *such that* (X, X_0, X_1, R) *is an extensional 3-precontact space.*

Theorem 7. New representation theorem for extensional contact algebras.

(a) *Let* (B, C) *be an extensional contact algebra and let* (X, X_0, X_1) *be the canonical extensional 3-contact space of* (B, C) *(see Definition 16(c)). Then the function* g*, defined in (7), is a PCA-isomorphism from* (B, C) *onto the canonical extensional contact algebra* $(RC(X, X_0, X_1), C_{(X,X_0,X_1)})$ *of the extensional 3-contact space* (X, X_0, X_1)*. The sets* $RC(X, X_0, X_1)$ *and* $RC(X_1)$ *coincide iff the algebra* (B, C) *is complete. The contact algebra* (B, C) *is connected iff the extensional 3-contact space* (X, X_0, X_1) *is connected.*

(b) *There exists a bijective correspondence between all (up to PCA-isomorphism) extensional (connected) contact algebras and all (up to ECS-isomorphism) extensional (connected) 3-contact spaces.*

Let X be a set and Γ be a family of subsets of X. Recall that the family Γ is said to be **fixed** if $\bigcap \Gamma \neq \emptyset$.

Definition 17. ([1]) *A weakly regular* T_1*-space* (X, τ) *is* **C-weakly regular** *if every maximal clan* Γ *in* $(RC(X), C_X)$ *is fixed.*

Corollary 2. ([1]) *There exists a bijective correspondence between the class of all (up to PCA-isomorphism) complete extensional (connected) contact algebras and the class of all (up to homeomorphism) (connected) C-weakly regular spaces.*

9 Precontact Algebras Satisfying Some Additional Axioms

Definition 18. Some new axioms. *Let (B, C) be a precontact algebra.*
N-regularity Axiom. *If xCy then there exists a cluster Γ containing x and y.*
Regularity Axiom. *If xCy then there exists a co-end Γ containing x and y.*

Definition 19. *Let (B, C) be an extensional precontact algebra. It is called an* **N-regular (resp., regular; normal) precontact algebra** *if it satisfies the N-regularity axiom (resp., Regularity axiom; the axiom (Ctr#)).*

Definition 20. ([1]) *A topological space (X, τ) is* **N-regular** *if it is semiregular and the family $RC(X)$ is a network for (X, τ) (i.e., for every point $x \in X$ and every open neighborhood U of x there exists an $F \in RC(X)$ such that $x \in F \subseteq U$).*

Definition 21. *A quadruple $\mathbf{X} = (X, X_0, X_1, R)$ is called an* **N-regular 3-precontact space** *(abbreviated as NPS) (resp.,* **regular 3-precontact space** *(abbreviated as RPS)) if it satisfies the following conditions:*

(1) *(X, X_0, R) is a 2-precontact space;*
(2) *X_1 is a dense subspace of X;*
(3) *X_1 is an N-regular T_1-space (resp., regular T_2-space);*
(4) *If $x \in X_1$ then the set Γ_{x,X_0} (see (5)) is a cluster (resp., co-end) in the precontact algebra $(Clopen(X_0), C_R)$; conversely, for every cluster (resp., co-end) Γ in $(Clopen(X_0), C_R)$ there exists a point $x \in X_1$ such that $\Gamma = \Gamma_{x,X_0}$.*

An N-regular 3-precontact space (X, X_0, X_1) is said to be a **normal 3-precontact space** *if X_1 is a compact Hausdorff space.*

Replacing in Definition 14(b)(c)(d) the word "extensional" with the word "N-regular" (resp., "regular", "normal"), we introduce the notions of **canonical N-regular (resp., regular; normal) precontact algebra of an N-regular (resp., regular; normal) 3-precontact space X** *as well as the notions of* **reflexive** *(or* **symmetric***;* **transitive***;* **connected***)* **N-regular** *(resp.,* **regular***;* **normal***)* **3-contact space***, and the notions of* **isomorphism** *between such spaces.*

Definition 22. *Let $\mathbf{B} = (B, C)$ be an N-regular (resp., regular; normal) precontact algebra. We will associate with $\mathbf{B}$ an N-regular (resp., regular; normal) 3-precontact space $\mathbf{X}(\mathbf{B}) = (X, X_0, X_1, R)$ called the* **canonical N-regular (resp., regular; normal) 3-precontact space of B***. Let (X, X_0, R) be the canonical 2-precontact space of the precontact algebra (B, C). We set $X_1 = Clust(B, C)$ (resp., $X_1 = Coend(B, C)$; $X_1 = Clust(B, C)$). Then $X_1 \subseteq X$ and we take the subspace topology on X_1.*

Theorem 8. Representation theorem for N-regular, regular and normal precontact algebras.

(a) *Let* $\mathbf{B} = (B, C)$ *be an N-regular (resp., regular; normal) precontact algebra and let* $\mathbf{X}(\mathbf{B}) = (X, X_0, X_1, R)$ *be the canonical N-regular (resp., regular; normal) 3-precontact space of* $\mathbf{B}$. *Then the function* g*, defined in (7), is a PCA-isomorphism from* (B, C) *onto the canonical N-regular (resp., regular; normal) precontact algebra* $(RC(X, X_0, X_1), C_R^{X, X_1})$ *of* $\mathbf{X}(\mathbf{B})$*. The same function* g *is a PCA-isomorphism between the contact algebras* $(B, C^{\#})$ *and* $(RC(X, X_0, X_1), C_{(X,X_0,X_1)})$*. The sets* $RC(X, X_0, X_1)$ *and* $RC(X_1)$ *coincide iff the algebra* $\mathbf{B}$ *is complete. The algebra* $\mathbf{B}$ *satisfies the axiom (Cref) (resp., (Csym); (Ctr)) iff the N-regular (resp., regular; normal) 3-precontact space* $\mathbf{X}(\mathbf{B})$ *is reflexive (resp., symmetric; transitive). When* $\mathbf{B}$ *is normal, the algebra* $\mathbf{B}$ *is connected iff* $\mathbf{X}$ *is connected.*

(b) *There exists a bijective correspondence between all (up to PCA-isomorphism) N-regular (resp., regular; (connected) normal) precontact algebras and all (up to isomorphism) N-regular (resp., regular; (connected) normal) 3-precontact spaces.*

Definition 23. N-regular, regular and normal 3-Contact spaces. *A topological triple* (X, X_0, X_1) *is called an* **N-regular (resp., regular) 3-contact space** *if the following conditions are satisfied:*

(1) (X, X_0) *is a 2-contact space;*
(2) X_1 *is an N-regular* T_1*-space (resp., regular* T_2*-space);*
(3) *If* $x \in X_1$ *then the set* Γ_{x,X_0} *(see (5)) is a cluster (resp., co-end) in the contact algebra* $(Clopen(X_0), \delta_{(X,X_0)})$*; conversely, for every cluster (resp., co-end)* Γ *in* $(Clopen(X_0), \delta_{(X,X_0)})$ *there exists a point* $x \in X_1$ *such that* $\Gamma = \Gamma_{x,X_0}$.

An N-regular 3-contact space (X, X_0, X_1) *is called a* **normal 3-contact space** *if* X_1 *is a compact Hausdorff space.*

Replacing in Definition 16(b)(c)(d) the word "extensional" with the word "N-regular" (resp., "regular", "normal"), we introduce the notions of **canonical N-regular (resp., regular; normal) contact algebra of an N-regular (resp., regular; normal) 3-contact space X** *as well as the notions of* **canonical N-regular (resp., regular; normal) 3-contact space of an N-regular (resp., regular; normal) contact algebra** (B, C)*, and the notions of* **isomorphism** *between such spaces. An N-regular (resp., regular; normal) 3-contact space* (X, X_0, X_1) *is called* **connected** *if the space* X_1 *is connected.*

Theorem 9. New representation theorem for N-regular, regular and normal contact algebras.

(a) *Let* (B, C) *be an N-regular (resp., regular; normal) contact algebra and let* (X, X_0, X_1) *be the canonical N-regular (resp., regular; normal) 3-contact space of* (B, C)*. Then the function* g*, defined in (7), is a PCA-isomorphism from* (B, C) *onto the canonical N-regular (resp., regular; normal) contact algebra* $(RC(X, X_0, X_1), C_{(X,X_0,X_1)})$ *of the N-regular (resp., regular; normal)*

3-contact space (X, X_0, X_1). *The sets* $RC(X, X_0, X_1)$ *and* $RC(X_1)$ *coincide iff the algebra* (B, C) *is complete. When* (B, C) *is normal, the contact algebra* (B, C) *is connected iff the normal 3-contact space* (X, X_0, X_1) *is connected.*

(b) *There exists a bijective correspondence between all (up to PCA-isomorphism) N-regular (resp., regular; (connected) normal) contact algebras and all (up to isomorphism) N-regular (resp., regular; (connected) normal) 3-contact spaces.*

Definition 24. ([1]) *An N-regular space X is said to be* **CN-regular** *if every cluster in* $(RC(X), C_X)$ *is fixed. A regular space* (X, τ) *is called* **C-regular** *if every co-end in* $(RC(X), C_X)$ *is fixed.*

Corollary 3. ([1]) *There exists a bijective correspondence between the class of all (up to PCA-isomorphism) N-regular (resp., regular; normal) complete (connected) contact algebras and the class of all (up to homeomorphism) (connected) CN-regular* T_1*-spaces (resp., C-regular* T_2*-spaces; compact* T_2*-spaces).*

The above results can be extended to the local contact algebras as well. This will be done in the full version of this paper.

Acknowledgements. We are very grateful to the referee for the useful remarks and suggestions.

References

1. DIMOV, G. AND VAKARELOV, D. Contact Algebras and Region-based Theory of Space: A Proximity Approach - I. 2004. (Submitted.)
2. DIMOV, G. AND VAKARELOV, D. Contact Algebras and Region-based Theory of Space: A Proximity Approach - II. 2004. (Submitted.)
3. DÜNTSCH, I. AND VAKARELOV, D. Region-based theory of discrete spaces: A proximity approach. In: Nadif, M., Napoli, A., SanJuan, E., and Sigayret, A. eds, *Proceedings of Fourth International Conference Journées de l'informatique Messine*, 123-129, Metz, France, 2003. (To appear in *Discrete Applied Mathematics.*)
4. DÜNTSCH, I. AND WINTER, M. A Representation theorem for Boolean Contact Algebras. 2003. (To appear in Theoretical Computer Science.)
5. ENGELKING, R. General Topology. PWN, Warszawa, 1977.
6. GALTON, A. The mereotopology of discrete spaces. In: Freksa, C. and Mark, D.M. eds, *Spatial Information Theory, Proceedings of the International Conference COSIT'99,* Lecture Notes in Computer Science, 251–266, Springer-Verlag, 1999.
7. SANJIANG LI, MINGSHENG YING. Generalized Connection Calculus. *Artificial Intelligence*, 2004 (in print).
8. RANDELL D., CUI Z., COHN A. G. A Spatial Logic based on Regions and Connection. In: NEBEL B., SWARTOUT W., RICH C. EDS, *Principles of Knowledge Representation and Reasoning: Proc. of the 3rd International Conference,* Cambridge MA, Oct. 1992, Morgan Caufmann, 165-176.
9. THRON, W. Proximity structures and grills. *Math. Ann.*, 206 (1973), 35-62.
10. VAKARELOV, D., DIMOV, G., DÜNTSCH, I. & BENNETT, B. A proximity approach to some region-based theory of space. *Journal of applied non-classical logics*, 12 (3-4) (2002), 527-559.

Relational Semantics Through Duality

Ewa Orłowska[1], Ingrid Rewitzky[2], and Ivo Düntsch[3]

[1] National Institute of Telecommunications, Warsaw
[2] University of Stellenbosch, South Africa
[3] Brock University, Canada

Abstract. In this paper we show how the classical duality results extended to a Duality via Truth contribute to development of a relational semantics for various modal-like logics. In particular, we present a Duality via Truth for some classes of information algebras and frames. We also show that the full categorical formulation of classical duality extends to a full Duality via Truth.

1 Introduction

In this paper we show how Stone-like or Priestley-like dualities can be split into two parts: one referring to the algebraic aspects and the other to the logical aspects. In this way a relationship between algebraic structures and relational structures (or frames, as they are called in non-classical logics) can be rooted in their common origin as semantic structures of formal languages. We will follow a method called Duality via Truth [OrR05b]. This aims to exhibit a relationship between a class of algebras and a class of frames based on their corresponding notions of truth for a formal language. We show that the appropriate elements of Stone-like or Priestley-like dualities can easily be extended to Duality via Truth. Usually, on an algebraic side, we are interested in representation theorems for a class of algebras which involves representing elements of those algebras as subsets of some universal set. On the logical side, we consider a class of frames and prove a completeness of the logic with respect to a class of models determined by those frames. We show that these two approaches can be put together and can be extended to a Duality via Truth which exhibits a principle according to which the two classes of structures are dual.

Given a formal language Lan, a class of frames Frm which determines a frame semantics for Lan, and a class Alg of algebras which determines its algebraic semantics, a Duality via Truth theorem says that these two kinds of semantics are equivalent in the following sense:

DvT A formula $\phi \in \mathsf{Lan}$ is true in every algebra of Alg iff is true in every frame of Frm.

In order to prove such a theorem we proceed as follows. From each algebra $L \in \mathsf{Alg}$ we form a canonical frame $\mathcal{X}(L)$, and from each frame $X \in \mathsf{Frm}$ we form a complex algebra $\mathcal{C}(X)$. Then we prove that $\mathcal{X}(L) \in \mathsf{Frm}$ and $\mathcal{C}(X) \in \mathsf{Alg}$.

Furthermore, we prove what is called a complex algebra theorem:

W. MacCaull et al. (Eds.): RelMiCS 2005, LNCS 3929, pp. 17–32, 2006.

CA For every frame $X \in \mathsf{Frm}$, a formula $\phi \in \mathsf{Lan}$ is true in X iff ϕ is true in $\mathcal{C}(X)$.

Finally, we prove a representation theorem:

R Every algebra $L \in \mathsf{Alg}$ is isomorphic to a subalgebra of the complex algebra of its canonical frame $\mathcal{C}(\mathcal{X}(L))$.

With a complex algebra theorem and a representation theorem we can prove a Duality via Truth theorem. The right-to-left implication of **DvT** follows from the left-to-right implication of **CA** and the left-to-right implication of **DvT** follows from right-to-left implication of **CA** and **R**.

In Sections 2 and 4 we present Duality via Truth results for modal algebras and modal frames, and for sufficiency algebras and sufficiency frames. In Sections 3 and 5 we consider a duality between some classes of information algebras and information frames. These frames have an indexed family of binary relations satisfying certain properties and were introduced (in [DeO02]) to capture intuitions about relations arising from information systems.

2 Modal Algebras and Frames

In this section we review Jónsson/Tarski duality for Boolean algebras with operators. This is then used as a case study for illustrating how the Duality via Truth approach extends this duality with dual notions of truth of formulae of a propositional language.

The class of algebras will consist of *modal algebras* (B, f) where B is a Boolean algebra $(B, \vee, \wedge, -, 0, 1)$ and f is an unary operator over B that is additive (i.e. $f(a \vee b) = f(a) \vee f(b)$) and normal (i.e. $f(0) = 0$). The class of frames will consist of *frames* (X, R) where X is a set endowed with a binary relation R over X. Let Alg_M denote the class of modal algebras, and Frm denote the class of frames.

First we show that any frame gives rise to a modal algebra. Let (X, R) be a frame. The binary relation R over X, induces monotone unary operators over 2^X, including, $f_R : 2^X \to 2^X$ defined by

$$f_R(A) = \{x \in X \mid R(x) \cap A \neq \emptyset\} \qquad \text{for } A \subseteq X,$$

and its dual, namely, $f_R^d : 2^X \to 2^X$ defined by

$$f_R^d(A) = \{x \in X \mid R(x) \subseteq A\} = -f_R(-A) \qquad \text{for } A \subseteq X.$$

It is a trivial exercise to show that the operator f_R is normal and (completely) additive, and its dual f_R^d is full (i.e. $f^d(1) = 1$) and (completely) multiplicative (i.e. $f^d(a \wedge b) = f^d(a) \wedge f^d(b)$). (See [BrR01] for further details.) So the powerset Boolean algebra 2^X endowed with the operator f_R is a modal algebra.

Next we show that any modal algebra in turn gives rise to a frame. In the case of a normal and completely additive operator f over a powerset Boolean algebra 2^X, a relation r_f over X may be defined, as in [BrR01, DuO01], by

$$x r_f y \quad \text{iff} \quad x \in f(\{y\}), \qquad \text{for } x, y \in X.$$

For the general case we invoke Stone's representation theorem — i.e., we represent the elements of the Boolean algebra as subsets of some universal set (namely the set of all prime filters), and then define a binary relation over this universe. Let (B, f) be a modal algebra, and let $\mathcal{X}(B)$ be the set of all prime filters in the Boolean algebra B. From the operator f define a binary relation R_f over $\mathcal{X}(B)$ by

$$FR_fG \quad \text{iff} \quad \forall y \in B,\ y \in G \Rightarrow f(y) \in F \quad \text{iff} \quad G \subseteq f^{-1}(F), \quad \text{for } F, G \in \mathcal{X}(B).$$

Note f^{-1} is the inverse image map given by $f^{-1}(F) = \{x \mid f(x) \in F\}$. An exercise in [BrR01] explains that the definition of r_f corresponds to the definition of R_f in the general case.

Lemma 1. *For any frame (X, R) and $F, G \in \mathcal{X}(B)$,*

$$G \subseteq f^{-1}(F) \quad \text{iff} \quad (f^{-1})^d(F) \subseteq G.$$

We now show how a modal algebra can be recovered from the frame it gave rise to. That is, if we start with a modal algebra (B, f), form its canonical frame $(\mathcal{X}(B), R_f)$ and form the complex algebra $(2^{\mathcal{X}(B)}, f_{R_f})$ of that, then this last modal algebra contains an isomorphic copy of the original modal algebra. For this it suffices to show that the Stone mapping $h : B \to 2^{\mathcal{X}(B)}$, which is an embedding of the Boolean algebra B into the Boolean algebra $2^{\mathcal{X}(B)}$, preserves operators over B. That is,

Theorem 1. *For any modal algebra (B, f) and $a \in B$,* $\quad h(f(a)) = f_{R_f}(h(a))$.

Proof: For any $a \in B$,

$$\begin{aligned} f_{R_f}(h(a)) &= \{F \in \mathcal{X}(B) \mid (\exists G \in h(a))[FR_fG]\} \\ &= \{F \in \mathcal{X}(B) \mid (\exists G \in \mathcal{X}(B))[a \in G \text{ and } G \subseteq f^{-1}(F)]\}. \end{aligned}$$

To show that this is equal to $h(f(a)) = \{F \in \mathcal{X}(B) \mid f(a) \in F\}$ we have to show that $f(a) \in F \quad \text{iff} \quad (\exists G \in \mathcal{X}(B))[a \in G \text{ and } G \subseteq f^{-1}(F)]$.

The right-to-left direction is easy, because if $a \in G$ and $G \subseteq f^{-1}(F)$ then $G \subseteq \{x \mid f(x) \in F\}$, and hence $f(a) \in F$. For the left-to-right direction consider the set $Z_f = \{b \in B \mid f^d(b) \in F\}$. Let F' be the filter generated by $Z_f \cup \{a\}$, that is, $F' = \{b \in B \mid \exists a_1, \ldots, a_n \in Z_f, \quad a_1 \wedge \ldots \wedge a_n \wedge a \leq b\}$. Then F' is proper. Suppose otherwise. Then for some $a_1, \ldots, a_n \in Z_f, \quad a_1 \wedge \ldots \wedge a_n \wedge a = 0$, i.e., $a_1 \wedge \ldots \wedge a_n \leq -a$. Since f^d is monotone, $f^d(a_1 \wedge \ldots \wedge a_n) \leq f^d(-a)$, that is, $f^d(a_1) \wedge \ldots \wedge f^d(a_n) \leq f^d(-a)$. By definition of Z_f we have $f^d(a_1), \ldots, f^d(a_n) \in F$ so, since F is a filter, $f^d(a_1) \wedge \ldots \wedge f^d(a_n) \in F$ and hence $f^d(-a) \in F$. Thus $-a \in Z_f$ which is a contradiction. So, by ([DaP90], p188), there is a prime filter G containing F'. Since $a \in F'$, $a \in G$ and hence $g \in h(a)$. Also $G \subseteq f^{-1}(F)$ since if $y \notin f^{-1}(F)$ then $f(y) \notin F$, i.e., $f^d(-y) \in F$, so $-y \in F' \subseteq G$ and hence $y \notin G$. □

With this result we can prove a representation theorem (see Theorem 3(a)) for modal algebras. For a representation theorem for frames we show how a frame can be recovered from the modal algebra it gave rise to. That is, if we start with a frame (X, R), form its complex algebra $(2^X, f_R)$ and form the canonical frame $(\mathcal{X}(2^X), R_{f_R})$ of that, then this last frame contains an isomorphic copy of the original frame. For this we invoke the one-one correspondence between the elements of X and certain prime filters of 2^X, namely the principal ones given by the mapping $k : X \to \mathcal{X}(2^X)$ where $k(x) = \{A \in 2^X \mid x \in A\}$. It is an easy exercise to show that $k(x)$ is a prime filter. We have to show that this mapping preserves structure. That is,

Theorem 2. *For any frame* (X, R) *and* $x, y \in X$, xRy iff $k(x)R_{f_R}k(y)$.

Proof: Note, for any $x, y \in X$,

$$\begin{aligned} k(x)R_{f_R}k(y) \quad &\text{iff} \quad k(y) \subseteq (f_R)^{-1}(k(x)) \\ &\text{iff} \quad \{Y \subseteq X \mid y \in Y\} \subseteq \{Z \subseteq X \mid x \in f_R(Z)\}. \end{aligned}$$

We now prove the desired double implication. For the left-to-right direction, suppose xRy. Take any $Y \subseteq X$ with $y \in Y$. Then $R(x) \cap Y \neq \emptyset$ and hence $Y \in \{Z \mid x \in f_R(Z)\}$. Thus $k(x)R_{f_R}k(y)$. For the right-to-left direction, suppose $k(x)R_{f_R}k(y)$. Since $y \in \{y\}$, by the above, $x \in f_R(\{y\})$, that is, xRy. □

A consequence of the above theorems is a Jónsson/Tarski duality between modal algebras and frames.

Theorem 3.

(a) *Every modal algebra* (B, F) *is isomorphic to a subalgebra of the complex algebra of its canonical frame* $(2^{\mathcal{X}(B)}, f_{R_f})$.
(b) *Every frame* (X, R) *is isomorphic to a substructure of the canonical frame of its complex algebra* $(\mathcal{X}(2^X), R_{f_R})$.

The final part of the duality consists of establishing a bijective correspondence between maps between modal algebras and maps between frames. In the case of modal algebras (B_1, f_1) and (B_2, f_2) the map is a Boolean algebra homomorphism $l : B_1 \to B_2$ and in the case of frames (X_1, R_1) and (X_2, R_2) the map is a bounded morphism $n : X_1 \to X_2$ (with the properties xR_1y implies $n(x)R_2n(y)$, and if $n(x)R_2y_2$ then for some $y_1 \in X_1$, xR_1y_1 and $f(y_1) = y_2$).

Theorem 4. *Let* (B_1, f_1) *and* (B_2, f_2) *be modal algebras and let* $l : B_1 \to B_2$ *be a homomorphism between them. Let* (X_1, R_1) *and* (X_2, R_2) *be frames and let* $n : X_1 \to X_2$ *be a bounded morphism between them. Then* $l^{-1} : \mathcal{X}(B_2) \to \mathcal{X}(B_1)$ *is a bounded morphism, and* $n^{-1} : 2^{X_2} \to 2^{X_1}$, *is a homomorphism.*

It is not difficult to extend Theorem 4 to show that injective/surjective homomorphisms correspond to surjective/injective bounded morphisms and vice versa. Let us use the category-theoretical device of denoting the function m^{-1} by $\mathcal{X}(m)$ (thus invoking a functorial notation), then the duality is finally completed by proving the following results.

Theorem 5. *Let l and n be as in* Theorem 4. *Suppose that the maps $h_{B_1} : B_1 \to 2^{\mathcal{X}(B_1)}$ and $h_{B_2} : B_2 \to 2^{\mathcal{X}(B_2)}$, and $k_{X_1} : X_1 \to \mathcal{X}(2^X)$ and $k_{X_2} : X_2 \to \mathcal{X}(2^Y)$ are the isomorphisms used in Theorem 3. Then*

$$(\mathcal{X}(l))^{-1} \circ h_{B_1} = h_{B_2} \circ l \quad \textit{and} \quad \mathcal{X}(n^{-1}) \circ k_{X_1} = k_{X_2} \circ n.$$

That is, the following diagrams commute:

$$\begin{array}{ccc} B_1 & \xrightarrow{h_{B_1}} & 2^{\mathcal{X}(B_1)} \\ \downarrow l & & \downarrow (\mathcal{X}(l))^{-1} \\ B_2 & \xrightarrow{h_{B_2}} & 2^{\mathcal{X}(B_2)} \end{array} \qquad \begin{array}{ccc} X_1 & \xrightarrow{k_{X_1}} & \mathcal{X}(2^{X_1}) \\ \downarrow n & & \downarrow \mathcal{X}(n^{-1}) \\ X_2 & \xrightarrow{k_{X_2}} & \mathcal{X}(2^{X_2}) \end{array}$$

Theorem 6. *Let l, h_{B_1}, h_{B_2} be as in Theorem 5. Then, for any $a \in B_1$,*

$$(\mathcal{X}(l))^{-1}(f_{R_{f_1}}(h_{B_1}(a))) = f_{R_{f_2}}(\mathcal{X}(l))^{-1}h_{B_1}(a)$$

Proof: For any $a \in B_1$,

$$\begin{array}{ll} (\mathcal{X}(l))^{-1}(f_{R_{f_1}}(h_{B_1}(a))) & \\ = (\mathcal{X}(l))^{-1}(h_{B_1}(f_1(a))) & \text{by Theorem 5} \\ = h_{B_2}(l(f_1(a))) & \text{by Theorem 1} \\ = h_{B_2}(f_2(l(a))) & \text{since } l \text{ is a homomorphism} \\ = f_{R_{f_2}}(h_{B_2}(l(a))) & \text{by Theorem 1} \\ = f_{R_{f_2}}(\mathcal{X}(l))^{-1}h_{B_1}(a) & \text{by Theorem 5.} \end{array}$$

□

Theorem 7. *Let n, k_{X_1}, k_{X_2} be as in Theorem 5. Then, for any $x, y \in X_1$,*

(a) $k_{X_1}(x)R_{f_{R_1}}k_{X_1}(y) \quad \Rightarrow \quad \mathcal{X}(n^{-1})k_{X_1}(x)R_{f_{R_2}}\mathcal{X}(n^{-1})k_{X_1}(y)$

(b) *If $\mathcal{X}(n^{-1})k_{X_1}(x)R_{f_{R_2}}k_{X_2}(y_2)$ then for some $y_1 \in X_1$,*
$k_{X_1}(x)R_{f_{R_1}}k_{X_1}(y)$ and $\mathcal{X}(n^{-1})(k_{X_1}(y_1)) = k_{X_2}(y_2)$.

Proof: For any $x, y \in X_1$,

$$\begin{array}{ll} k_{X_1}(x)R_{f_{R_1}}k_{X_1}(y) & \\ \Leftrightarrow xR_1y & \text{by Theorem 2} \\ \Rightarrow n(x)R_2n(y) & \text{since } n \text{ is a bounded morphism} \\ \Leftrightarrow k_{X_2}(n(x))R_{f_{R_2}}k_{X_2}(n(y)) & \text{by Theorem 2} \\ \Leftrightarrow \mathcal{X}(n^{-1})k_{X_1}(x)R_{f_{R_2}}\mathcal{X}(n^{-1})k_{X_1}(y) & \text{by Theorem 5.} \end{array}$$

$$\begin{array}{ll} \mathcal{X}(n^{-1})k_{X_1}(x)R_{f_{R_2}}k_{X_2}(y_2) & \\ \Leftrightarrow k_{X_2}(n(x))R_{f_{R_2}}k_{X_2}(y_2) & \text{by Theorem 5} \\ \Leftrightarrow n(x)R_2y_2 & \text{by Theorem 2} \\ \Rightarrow \exists y_1,\ xR_1y_1 \ \wedge\ n(y_1) = y_2 & \text{since } n \text{ is a bounded morphism} \\ \Leftrightarrow \exists y_1,\ k_{X_1}(x)R_{f_{R_1}}k_{X_1}(y) \ \wedge & \\ \qquad \mathcal{X}(n^{-1})(k_{X_1}(y_1)) = k_{X_2}(y_2) & \text{by Theorems 2 and 5.} \end{array}$$

□

All of the preceding results can be cast into a categorical framework as an equivalence between the categories of modal algebras and frames. For example, Theorem 5 is then simply a statement of the definition of the natural transformations involved in an equivalence (or more generally, adjunction).

In order to extend this to a Duality via Truth, we need a logical language. Let Lan_M be a modal language whose formulas are built from propositional variables taken from an infinite denumerable set Var, with the classical propositional operations of negation ($\neg$), disjunction ($\vee$), conjunction ($\wedge$), and with a modal operator ($\Diamond$). We slightly abuse the language by denoting the operations in modal algebras and the classical propositional operations of Lan_M with the same symbols.

The class Alg_M of modal algebras provides an algebraic semantics for Lan_M. Let (B, f) be a modal algebra. A valuation on B is a function $v : Var \to B$ which assigns elements of B to propositional variables and extends homomorphically to all the formulas of Lan_M, that is

$$v(\neg\alpha) = -\alpha,\ v(\alpha \vee \beta) = v(\alpha) \vee v(\beta),\quad v(\Diamond\alpha) = f(v(\alpha)).$$

The notion of truth determined by this semantics is as follows. A formula α in Lan_M is true in an algebra (B, f) whenever $v(\alpha) = 1$ for every v in B. A formula $\alpha \in \mathsf{Lan}_M$ is true in the class Alg_M iff it is true in every algebra $B \in \mathsf{Alg}_M$.

The class Frm of frames provides a well known frame semantics for Lan_M. A model based on a frame (X, R) is a system $M = (X, R, m)$, where $m : Var \to 2^X$ is a meaning function. The satisfaction relation $\models$ is defined as usual. We say that in a model M state $x \in X$ satisfies a formula whenever the following conditions are satisfied:

$$\begin{aligned} M, x \models p &\text{ iff } x \in m(p), \text{ for every } p \in Var \\ M, x \models \alpha \vee \beta &\text{ iff } M, x \models \alpha \text{ or } M, x \models \beta, \\ M, x \models \neg\alpha &\text{ iff } \text{ not } M, x \models \alpha, \\ M, x \models \Diamond\alpha &\text{ iff } \exists y \text{ such that } M, y \models \alpha \text{ and } xRy. \end{aligned}$$

A notion of truth of formulas based on this semantics is defined as usual. A formula $\alpha \in \mathsf{Lan}_M$ is true in a model M whenever for every $x \in X$ we have $M, x \models \alpha$. A formula $\alpha \in \mathsf{Lan}_M$ is true in a frame (X, R) iff α is true in every model based on this frame. And finally a formula $\alpha \in \mathsf{Lan}_M$ is true in the class Frm of frames iff it is true in every frame $X \in \mathsf{Frm}$.

It is easy to see that the complex algebra theorem holds:

Theorem 8. *A formula $\alpha \in \mathsf{Lan}_M$ is true in every model based on a frame (X, R) iff α is true in the modal complex algebra $(2^X, f_R)$ of that frame.*

Proof: Let (X, R) be any frame. The result is established by taking the meaning function m on any model (X, R, m) based on (X, R) to coincide with the valuation function on the modal complex algebra $(2^X, f_R)$ of (X, R). □

Finally, we prove the Duality via Truth theorem between modal algebras and frames.

Theorem 9. *A formula $\alpha \in \mathsf{Lan}_M$ is true in every algebra of Alg_M iff α is true in every frame of* Frm.

Proof: Let (B, f) be any modal algebra. Then any valuation v on B can be extended to a valuation $h \circ v$ on $2^{\mathcal{X}(B)}$ and thus

$$\begin{array}{lll} \alpha \text{ is true in } (B, f) & \text{iff} & \alpha \text{ is true in } (2^{\mathcal{X}(B)}, f_{R_f}) \\ & \text{iff} & \alpha \text{ is true in every model based on } (\mathcal{X}(B), R_f) \\ & \text{iff} & \alpha \text{ is true in } (\mathcal{X}(B), R_f). \end{array}$$

By the duality, every frame in Frm is of the form $(\mathcal{X}(B), R_f)$ for some modal algebra (B, f) in Alg_M. On the other hand, let (X, R) be any frame. Then

$$\begin{array}{lll} \alpha \text{ is true in } (X, R) & \text{iff} & \alpha \text{ is true in every model based on } (X, R) \\ & \text{iff} & \alpha \text{ is true in } (2^X, R_f). \end{array}$$

By the duality, every modal algebra in Alg_M is of the form $(2^X, f_R)$ for some frame (X, R) in Frm. □

The final part of the Duality via Truth involves establishing a correspondence between preservation of truth with respect to Alg_M and with respect to Frm_M. As a consequence of Theorem 5, we have

$$(\mathcal{X}(l))^{-1} \circ h_{B_1} \circ v_1 = h_{B_2} \circ v_2 \qquad \text{and} \qquad \mathcal{X}(n^{-1}) \circ \mathcal{X}(m_1) = \mathcal{X}(m_2),$$

where $v_2 = l \circ v_1$ and $m_2 = n^{-1} \circ m_1$. That is, the following diagrams commute.

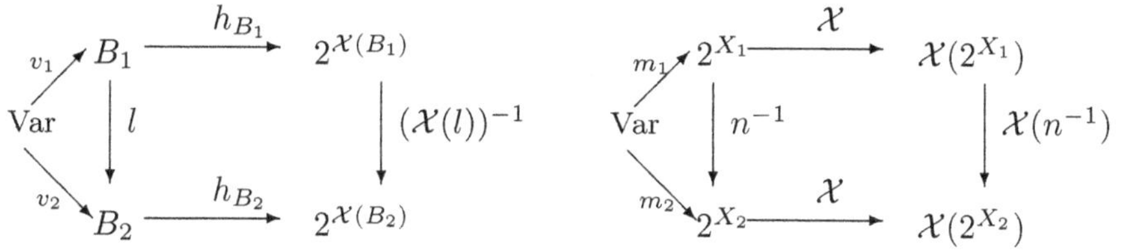

Hence, we have the following equivalence of preservation of truth.

Theorem 10. *Any homomorphism between algebras in Alg_M preserves truth with respect to Alg_M iff any bounded morphism between frames in Frm preserves truth with respect to* Frm.

Proof: Let $l : B_1 \to B_2$ be a homomorphism between modal algebras (B_1, f_1) and (B_2, f_2). Then $\mathcal{X}(l) : \mathcal{X}(B_2) \to \mathcal{X}(B_1)$ is a bounded morphism between the frames $(\mathcal{X}(B_1), R_{f_1})$ and $(\mathcal{X}(B_2), R_{f_2})$. Suppose l preserves truth in Alg_M, that is, for any formula α and any valuation v_1 on B_1,

$$v_1(\alpha) = 1 \qquad \text{iff} \qquad l \circ v_1(\alpha) = 1.$$

The valuation v_1 can be extended to a meaning function $h_{B_1} \circ v_1$ on $(\mathcal{X}(B_1), R_{f_1})$, and the valuation $l \circ v_1$ can be extended to a meaning function $h_{B_2} \circ l \circ v_1$

on $(\mathcal{X}(B_2), R_{f_2})$. By Theorem 5, it follows that for any formula α and any $F \in \mathcal{X}(B_1)$,

$$F \in h_{B_2} \circ l \circ v_1(\alpha) \quad \text{iff} \quad F \in (\mathcal{X}(l))^{-1} \circ h_{B_1} \circ v_1(\alpha) \quad \text{iff} \quad \mathcal{X}(l)(F) \in h_{B_1} \circ v_1(\alpha).$$

That is, the bounded morphism $\mathcal{X}(l)$ preserves truth with respect to Frm.

On the other hand, let $n : X_1 \to X_2$ be a bounded morphism between frames (X_1, R_1) and (X_2, R_2). Then $n^{-1} : 2^{X_2} \to 2^{X_1}$ is a homomorphism between the modal algebras $(2^{X_1}, f_{R_1})$ and $(2^{X_2}, f_{R_2})$. Suppose n preserves truth with respect to Frm, that is, for any formula α and any meaning function m_1 on X_1,

$$x \in m_1(\alpha) \qquad \text{iff} \qquad n(x) \in n \circ m_1(\alpha).$$

As a valuation function on 2^{X_1} take meaning function m_1 and as a valuation function on 2^{X_2} take the meaning function $n \circ m_1$. Then,

$$n \circ m_1(\alpha) = X_2 \qquad \text{iff} \qquad n^{-1}(n \circ m_1)(\alpha) = n^{-1}(X_2) \qquad \text{iff} \qquad m_1(\alpha) = X_1$$

That is, the homomorphism n^{-1} preserves truth with respect to Alg_M. □

Another representation of a modal algebra (B, f) is provided in [JoT51] by a canonical extension B^σ of B algebra. The canonical extension of the operator f is a map $f^\sigma : B^\sigma \to B^\sigma$ defined by $f^\sigma(\{y\}) = \bigcap\{h(f(a)) \mid a \in y\}$, for $y \in \mathcal{X}(B)$. It follows that

$$x \in f^\sigma(\{y\}) \;\text{ iff }\; \forall a,\; a \in y \;\Rightarrow\; f(a) \in x \;\text{ iff }\; y \subseteq f^{-1}(x).$$

Observe that here in fact we have a definition of a relation on a set of prime filters of B. This is precisely a relation of the canonical frame of the modal algebra. Next, for $Z \in \mathcal{X}(B)$ we define $f^\sigma(Z) = \bigcup\{f^\sigma(\{y\}) : y \in Z\}$. It follows that

$$x \in f^\sigma(h(a)) \;\text{ iff }\; \exists y,\; a \in y \,\wedge\, x \in f^\sigma(\{y\}) \;\text{ iff }\; \exists y,\; a \in y \,\wedge\, y \subseteq f^{-1}(x).$$

That is, $f^\sigma(h(a))$ provides a definition of the modal operator in the complex algebra of the canonical frame of (B, f). The canonical extension of the modal algebra (B, f) is then the algebra (B^σ, f^σ). It is known that f^σ is a complete modal operator on B^σ.

3 Information Algebras and Information Frames for Reasoning About Similarity

In this section we extend the results of Section 2 to a class of information algebras which are extensions of modal algebras with an indexed family of unary operators satisfying certain properties inspired by information systems.

Typically, in an information system objects are described in terms of some attributes and their values. The queries to an information system often have the form of a request for finding a set of objects whose sets of attribute values satisfy some conditions. This leads to the notion of information relation determined by a set of attributes. Let $a(x)$ and $a(y)$ be sets of values of an attribute a of the objects x and y. We may want to know a set of those objects from an information system whose sets of values of all (or some) of the attributes from a subset A of attributes are equal (or disjoint, or overlap etc.). To represent such queries we define, first, information relations on the set of objects and, second, information operators determined by those relations. For example, a relation of similarity of objects is defined as:

$$(x, y) \in \mathrm{sim}(a) \quad \text{iff} \quad a(x) \cap a(y) \neq \emptyset.$$

Next, we can extend this relation to any subset A of attributes so that a quantification over A is added:

$$(x, y) \in \mathrm{sim}(A) \quad \text{iff} \quad a(x) \cap a(y) \neq \emptyset \qquad \text{for all (some) } a \in A.$$

Relations defined with the universal (existential) quantifier are referred to as strong (weak) relations.

In an abstract setting as an index set we take a set of sets 2^{Par}, where each set $P \subseteq \mathrm{Par}$ is intuitively viewed as a set of attributes of objects in an information system. Then strong or weak relations are defined axiomatically.

An *information frame of weak similarity* (denoted FW-SIM in [DeO02]) is a binary relational structure $(X, \{R_P \mid P \subseteq \mathrm{Par}\})$ where the binary relations $R_P \subseteq X \times X$ (for each $P \subseteq \mathrm{Par}$) satisfy the following properties:

MF1 $R_{P \cup Q} = R_P \cup R_Q$
MF2 $R_\emptyset = \emptyset$
MF3 R_P is weakly reflexive (i.e., $\forall x, \forall y,\ xRy \Rightarrow xRx$)
MF4 R_P is symmetric (i.e., $\forall x, \forall y,\ xRy \Rightarrow yRx$).

Properties MF1 and MF2 reflect the intuition of weak relations; properties MF3 and MF4 are the abstract characterisation of similarity relations derived from an information system. By $\mathsf{Frm}_{\mathsf{WSIM}}$ we denote the class of weak similarity frames.

An *information algebra of weak similarity* (denoted AW-SIM in [DeO02]) is a Boolean algebra B with a family $\{f_P \mid P \subseteq \mathrm{Par}\}$ of additive normal unary operators satisfying the following additional properties:

MA1 $f_{P \cup Q}(x) = f_P(x) \vee f_Q(x)$
MA2 $f_\emptyset(x) = 0$
MA3 $x \wedge f_P(1) \leq f_P(x)$
MA4 $x \leq f_P^d f_P(x)$.

Properties MA1-MA4 will be shown below to be the algebraic counterparts of the properties MF1-MF4 on information relations.

Lemma 2. *Let $(B, \{f_P \mid P \subseteq \mathrm{Par}\})$ be an information algebra of weak similarity. For each operator f_P ($P \subseteq \mathrm{Par}$), the corresponding binary relation R_{f_P} over $\mathcal{X}(B)$ satisfies properties* MF1 - MF4.

Proof: We prove MF1-MF3; MF4 is well known from modal correspondence theory [vaB84].

MF1 For any $F, G \in \mathcal{X}(B)$, $FR_{f_{P\cup Q}}G$ iff $G \subseteq f_{P\cup Q}^{-1}(F)$ iff $(f_{P\cup Q}^d)^{-1}(F) \subseteq G$. But $(f_{P\cup Q}^d)^{-1}(F) = (f_P^d)^{-1}(F) \cap (f_Q^d)^{-1}(F)$. The propositional logic formulae $\alpha \wedge \beta \rightarrow \gamma$ and $\alpha \rightarrow \gamma \vee \beta \rightarrow \gamma$ are equivalent. Thus, $(f_P^d)^{-1}(F) \subseteq G$ or $(f_Q^d)^{-1}(F) \subseteq G$, that is, $G \subseteq (f_P)^{-1}(F)$ or $G \subseteq (f_Q)^{-1}(F)$. Thus $FR_{f_P}G$ or $FR_{f_Q}G$.

MF2 $FR_{f_\emptyset}G$ iff $G \subseteq f_\emptyset^{-1}(F)$ iff $F \subseteq \emptyset$. The latter is always false since F is a prime filter. Thus $R_{f_\emptyset} = \emptyset$.

MF3 Take any $F, G \in \mathcal{X}(B)$ with $FR_{f_P}G$ and $F(-R_{f_P})F$. Then $G \subseteq f_P^{-1}(F)$ and $F \not\subseteq f_P^{-1}(F)$. So, for some $w \in F$, $f_P(w) \notin F$ hence by MA3 $f_P(1) \notin F$. Also, since G is non-empty and up-closed, $1 \in G \subseteq f_P^{-1}(F)$ and hence $f_P(1) \in F$, which provides the required contradiction. □

Lemma 3. *Let $(X, \{R_P \mid P \subseteq \mathrm{Par}\})$ be an information frame. For each binary relation R_P ($P \subseteq \mathrm{Par}$), the corresponding unary operator f_{R_P} over 2^X satisfies properties* MA1 - MA4.

Proof: We prove MA1-MA3; MA4 is well known from modal correspondence theory [vaB84].

MA1 For $A \subseteq X$, $x \in f_{R_{P\cup Q}}(A)$ iff $R_{P\cup Q}(x) \cap A \neq \emptyset$ iff $R_{P\cup Q}(x) \not\subseteq -A$. But $R_{P\cup Q}(x) = R_P(x) \cup R_Q(x)$, so $R_P(x) \not\subseteq -A$ or $R_Q(x) \not\subseteq -A$, that is, $R_P(x) \cap A \neq \emptyset$ or $R_Q(x) \cap A \neq \emptyset$. Thus $x \in f_{R_P}(A)$ or $x \in f_{R_Q}(A)$.

MA2 $x \in f_{R_\emptyset}(A)$ iff $A \subseteq R_\emptyset(x)$ iff $A \subseteq X$. The latter is always true so $f_{R_\emptyset}(A) = X$.

MA3 Take any $x \in X$ such that $x \in A \cap f_{R_P}(X)$. Then $x \in A$ and $R_P(x) \cap X \neq \emptyset$. So for some $y \in X$, xR_Py. Hence, by property MF3 of R_P, xR_Px. Since $x \in A$ it follows that $x \in f_{R_P}(A)$, as required. □

A representation theorem analogous to Theorem 3 holds for weak similarity algebras and weak similarity frames. Also Theorems 5, 6, 7 are applicable to weak similarity algebras and weak similarity frames, these being special modal algebras and special frames, respectively.

The language $\mathsf{Lan}_{\mathsf{WSIM}}$ relevant for algebras and frames of weak similarity is an extension of the modal language Lan_M with a family of $\{\langle R_P \rangle \mid P \subseteq \mathrm{Par}\}$ of modal operators. Algebraic semantics of the language is provided by the class $\mathsf{Alg}_{\mathsf{WSIM}}$ and the frame semantics by the class $\mathsf{Frm}_{\mathsf{WSIM}}$. The notion of a model based on a frame of $\mathsf{Frm}_{\mathsf{WSIM}}$, satisfaction relation, and the notions of truth in a model, in a frame and in a class of frames are analogous to the respective notions in Section 2. In view of Lemma 3 the complex algebra theorem (**CA**) holds for $\mathsf{Frm}_{\mathsf{WSIM}}$. From the representation theorem and (**CA**) we obtain a Duality via Truth theorem, and also the equivalence of preservation of truth in Theorem 10.

4 Sufficiency Algebras and Frames

As second case study for the Duality via Truth approach we consider, in this section, a duality between sufficiency algebras and frames. These algebras were introduced in [DuO01] for reasoning about incomplete information and expressing algebraically certain properties of binary relations, such as irreflexivity or co-reflexivity, defined in terms of the complement of the relation.

A *sufficiency algebra* (B, g) is a Boolean algebra B endowed with an unary operator g over B that is co-additive (i.e. $g(a \vee b) = g(a) \wedge g(b)$) and co-normal (i.e. $g(0) = 1$). Let Alg_S denote the class of sufficiency algebras. The class Frm of frames adequate for providing Duality via Truth for sufficiency algebras is the same as in the case of modal algebras.

Given any frame (X, R), the binary relation R over X induces antitone operators, including $g_R : 2^X \to 2^X$ defined by

$$g_R(A) = \{x \in X \mid R(x) \cup A \neq X\} \qquad \text{for } A \subseteq X,$$

and its dual, namely, $g_R^d : 2^X \to 2^X$ defined by

$$g_R^d(A) = \{x \in X \mid A \subseteq R(x)\} \qquad \text{for } A \subseteq X.$$

Observing that these operators may be defined in terms of the monotone operators in Section 2 by $g_R(A) = f_{-R}(-A)$ and $g_R^d(A) = -f_{-R}(A)$ it follows that g_R is co-normal and co-additive, and g_R^d is co-full (i.e. $g^d(1) = 0$) and co-multiplicative (i.e. $g^d(a \wedge b) = g^d(a) \vee g^d(b)$). Hence, from a frame (X, R) we may define a sufficiency algebra $(2^X, g_R)$.

Next we show that any sufficiency algebra in turn gives rise to a frame. In the case of a co-normal and completely co-additive operator g over a powerset Boolean algebra 2^X, a relation r_g over X may be defined, as in [DuO01], by

$$x r_g y \quad \text{iff} \quad x \in g(\{y\}), \qquad \text{for } x, y \in X.$$

In general, as in Section 2 we invoke Stone's representation theorem and then define a binary relation over $\mathcal{X}(B)$. Let (B, g) be a sufficiency algebra. From the operator g define a binary relation R_g over $\mathcal{X}(B)$ by

$$F R_g G \quad \text{iff} \quad g(G) \cap F \neq \emptyset, \quad \text{for } F, G \in \mathcal{X}(B).$$

It is an easy exercise to show that the definition of r_g corresponds to that of R_g in the general case.

We now show how a sufficiency algebra can be recovered from the frame it gave rise to. That is, if we start with a sufficiency algebra (B, g), form its canonical frame $(\mathcal{X}(B), R_g)$ and form the complex algebra $(2^{\mathcal{X}(B)}, g_{R_g})$ of that, then this last sufficiency algebra contains an isomorphic copy of the original sufficiency algebra. For this it suffices to show that the Stone mapping $h : B \to 2^{\mathcal{X}(B)}$ preserves operators g over B. That is,

Theorem 11. *For any sufficiency algebra (B, g) and $a \in B$,*

$$h(g(a)) = g_{R_g}(h(a)).$$

Proof: For any $a \in B$,

$$\begin{aligned} g_{R_g}(h(a)) &= \{F \in \mathcal{X}(B) \mid h(a) \subseteq R_g(F)]\} \\ &= \{F \in \mathcal{X}(B) \mid (\forall G \in \mathcal{X}(B))[a \in G \text{ and } g(G) \cap F \neq \emptyset]\}. \end{aligned}$$

To show that this is equal to $h(g(a)) = \{F \in \mathcal{X}(B) \mid g(a) \in F\}$ we have to show that $g(a) \notin F$ iff $(\exists G \in \mathcal{X}(B))[a \in G$ and $g(G) \cap F = \emptyset]$.

The right-to-left direction is easy, because if $a \in G$ and $g(G) \cap F = \emptyset$ then $g(a) \in g(G)$ so $g(a) \notin F$. For the left-to-right direction, assume $g(a) \in F$. Consider the set $Z_g = \{b \in B \mid g^d(b) \notin F\}$. Let F' be the filter generated by $Z_g \cup \{a\}$, that is, $F' = \{b \in B \mid \exists a_1, \ldots, a_n \in Z_g, \ \ a_1 \wedge \ldots \wedge a_n \wedge a \leq b\}$. Then F' is proper. Suppose otherwise. Then for some $a_1, \ldots, a_n \in Z_g$, $a_1 \wedge \ldots \wedge a_n \wedge a = 0$, i.e., $a \leq -(a_1 \wedge \ldots \wedge a_n) = -a_1 \vee \ldots \vee -a_n$. Since g is antitone, $g(-a_1 \vee \ldots \vee -a_n) \leq g(a)$. Thus $g(-a_1) \wedge \ldots \wedge g(-a_n) \leq g(a)$, that is, $-g^d(a_1) \wedge \ldots \wedge -g^d(a_n) \leq g(a)$. By definition of Z_g we have $g^d(a_1), \ldots, g^d(a_n) \notin F$ so $-g^d(a_1), \ldots, -g^d(a_n) \in F$. Since F is a filter, $-g^d(a_1) \wedge \ldots \wedge -g^d(a_n) \in F$ and hence $g(a) \in F$ which contradicts the original assumption. So, by ([DaP90], p188), there is a prime filter G containing F'. Since $a \in F'$, $a \in G$ and hence $g \in h(a)$. Also $g(G) \cap F = \emptyset$ since if there is some $b \in B$ with $b \in g(G)$ and $b \in F$, then $b = g(c)$ for some $c \in G$ and thus $g(c) \in F$, so $g^d(-c) \notin F$ hence $-c \in Z_g \subseteq F' \subseteq G$ and thus $c \notin G$, which is a contradiction. □

On the other hand a frame can be recovered from the sufficiency algebra it gave rise to. That is, if we start with a frame (X, R), form its complex algebra $(2^X, g_R)$ and form the canonical frame $(\mathcal{X}(2^X), R_{g_R})$ of that, then this last frame contains an isomorphic copy of the original frame. For this we show the mapping $k : X \to \mathcal{X}(2^X)$ preserves structure. That is,

Theorem 12. *For any frame (X, R) and any $x, y \in X$,*

$$xRy \ \text{ iff } \ k(x) R_{g_R} k(y).$$

Proof: Note, for any $x, y \in X$,

$$k(x)R_{g_R}k(y) \quad \text{iff} \quad g_R(k(y)) \cap k(x) \neq \emptyset \quad \text{iff} \quad \{g_R(Y) \mid y \in Y\} \cap \{Z \mid x \in Z\} \neq \emptyset.$$

We now prove the desired double implication. For the left-to-right direction, suppose xRy. Then $\{y\} \subseteq R(x)$, so $x \in g_R(\{y\})$. Hence $g_R(\{y\}) \in g_R(k(y)) \cap k(x)$. Thus $k(x)R_{g_R}k(y)$. For the right-to-left direction, suppose $k(x)R_{g_R}k(y)$. Since $y \in \{y\}$, by the above, $g_R(\{y\}) \in \{Z \mid x \in Z\}$. Thus $x \in g_R(\{y\})$, that is, $\{y\} \subseteq R(x)$, that is, xRy. □

Therefore, we have a Jónsson/Tarski duality between sufficiency algebras and frames.

Theorem 13.

(a) *Any sufficiency algebra (B, g) is isomorphic to a subalgebra of the complex algebra of its canonical frame $(2^{\mathcal{X}(B)}, g_{R_g})$.*

(b) *Any frame (X, R) is isomorphic to a substructure of the canonical frame of its complex algebra $(\mathcal{X}(2^X), R_{g_R})$.*

Analogous results to Theorems 5, 6, 7 can be proved for sufficiency algebras and frames by invoking Theorems 11 and 12.

The language adequate for discussing a duality between sufficiency algebras Alg_S and frames of Frm is a propositional language Lan_S whose formulas are built with classical propositional connectives and the sufficiency operator $[\![\ ]\!]$. The frame semantics for Lan_S is defined as for the modal language. Let (X, R) be a frame and let $M = (X, R, m)$ be a model based on that frame. The satisfaction relation extends to the formulas with the sufficiency operator as follows:

$$M, x \models [\![R]\!]\alpha \quad \text{iff} \quad \forall y \text{ if } M, y \models \alpha \text{ then } xRy.$$

The notions of truth of a formula in a model, in a frame, and in a class of frames are defined as in the case of the modal logic. Using analogous reasoning to that for modal logic, we can prove the complex algebra theorem and Duality via Truth theorem, and also the equivalence of preservation of truth.

Theorem 14. *A formula $\alpha \in \mathsf{Lan}_S$ is true in every model based on a frame (X, R) iff α is true in the sufficiency complex algebra $(2^X, g_R)$ of that frame.*

Theorem 15. *A formula $\alpha \in \mathsf{Lan}_S$ is true in every algebra of Alg_S iff α is true in every frame of Frm.*

Theorem 16. *Any homomorphism between algebras in Alg_S preserves truth with respect to Alg_S iff any bounded morphism between frames in Frm preserves truth with respect to Frm.*

The canonical extension of a sufficiency algebra (B, g) is defined as follows. Let B^σ be the canonical extension of the Boolean algebra B and let h be the Stone embedding. Then the canonical extension of the operator g is a map $g^\sigma : B^\sigma \to B^\sigma$ defined by $g^\sigma(\{y\}) = \bigcup\{h(g(a)) \mid a \in y\}$, for $y \in \mathcal{X}(B)$. We have that

$$x \in g^\sigma(\{y\}) \quad \text{iff} \quad \exists a,\ a \in y \ \wedge\ g(a) \in x \quad \text{iff} \quad y \cap g(x) \neq \emptyset.$$

As in the case of modal algebras, this provides a definition of a relation on $\mathcal{X}(B)$. Next, for $Z \in \mathcal{X}(B)$ we define $g^\sigma(Z) = \bigcap\{g^\sigma(\{y\}) : y \in Z\}$. It follows that

$$x \in g^\sigma(h(a)) \quad \text{iff} \quad \forall y,\ a \in y \ \Rightarrow\ x \in g^\sigma(\{y\}) \quad \text{iff} \quad \forall y,\ a \in y \ \Rightarrow\ y \cap g(x) \neq \emptyset.$$

That is, $g^\sigma(h(a))$ provides a definition of the sufficiency operator in the complex algebra of the canonical frame of (B, g). The canonical extension of the sufficiency algebra (B, g) is then the algebra (B^σ, g^σ). It is known that g^σ is a completely co-additive sufficiency operator on B^σ.

5 Information Algebras and Information Frames of Strong Right Orthogonality

As before the representation results of Theorem 13 can be extended to some information algebras based on sufficiency algebras. Here the relations derived from an information systems are strong relations of right orthogonality defined as follows. For objects x and y of an information system and an attribute a,

$$(x, y) \in \mathrm{rort}(a) \quad \text{iff} \quad a(x) \subseteq -a(y).$$

For a subset A of attributes we may define strong (weak) relations by

$$(x, y) \in \mathrm{rort}(A) \quad \text{iff} \quad a(x) \subseteq -a(y) \qquad \text{for all (some) } a \in A.$$

An abstract characterisation of strong relations of right orthogonality derived from an information system may be defined as follows. An *information frame of strong right orthogonality* (denoted FS-RORT in [DeO02]) is a binary relational structure $(X, \{R_P \mid P \subseteq \mathrm{Par}\})$ where the binary relations $R_P \subseteq X \times X$ (for each $P \subseteq \mathrm{Par}$) satisfy the following properties:

SF1 $R_{P \cup Q} = R_P \cap R_Q$
SF2 $R_\emptyset = X \times X$
SF3 R_P is co-weakly reflexive (i.e., $\forall x, \forall y,\ x(-R)y \Rightarrow x(-R)x$)
SF4 R_P is symmetric (i.e., $\forall x, \forall y,\ xRy \Rightarrow yRx$).

Let $\mathsf{Frm}_{\mathsf{AS-RORT}}$ denote the class of all information frames of strong right orthogonality. On the other hand an *information algebra of strong right orthogonality* (denoted AS-RORT in [DeO02]) is a Boolean algebra B with a family $\{g_P \mid P \subseteq \mathrm{Par}\}$ of sufficiency operators satisfying the following additional properties:

SA1 $g_{P \cup Q}(x) = g_P(x) \wedge g_Q(x)$
SA2 $g_\emptyset(x) = 1$
SA3 $x \wedge g_P(x) \leq g_P(1)$
SA4 $x \leq g_P g_P(x)$.

Let $\mathsf{Alg}_{\mathsf{AS-RORT}}$ denote the class of all information algebras of strong right orthogonality.

Lemma 4. *Let $(B, \{g_P \mid P \subseteq \mathrm{Par}\})$ be an information algebra of strong right orthogonality. For each g_P ($P \subseteq \mathrm{Par}$), the corresponding binary relation R_{g_P} over $\mathcal{X}(B)$ satisfies properties* SF1 - SF4.

Proof: We prove SF1-SF3; SF4 is shown in [DeO02].

SF1. $FR_{g_{P \cup Q}}G$ iff $g_{P \cup Q}(G) \cap F \neq \emptyset$ iff $g_P(G) \cap g_Q(G) \cap F \neq \emptyset$ iff $g_P(G) \cap F \neq \emptyset$ and $g_Q(G) \cap F \neq \emptyset$ iff $FR_{g_P}G$ and $F_{g_Q}G$.
SF2. $FR_{g_\emptyset}G$ iff $g_\emptyset(G) \cap F \neq \emptyset$ iff $2^X \cap F \neq \emptyset$ iff $F \neq \emptyset$. The latter is always true since F is a prime filter. Thus $R_{g_\emptyset} = \mathcal{X}(B) \times \mathcal{X}(B)$.

SF3. Take any $F, G \in \mathcal{X}(B)$ with $F(-R_{g_P})G$ and $FR_{g_P}F$. Then $g_P(G) \cap F = \emptyset$ and $g_P(F) \cap F \neq \emptyset$. So, for some $w, z \in F$, $g_P(z) = w$. Thus $z \wedge g_P(z) \in F$ and hence, by SA3, $g_P(1) \in F$. Also, since $1 \in G$, $g_P(1) \in g_P(G)$ and hence, $g_P(1) \notin F$, which provides the required contradiction. □

Lemma 5. *Let* $(X, \{R_P \mid P \subseteq \text{Par}\})$ *be an information frame of strong right orthogonality. For each binary relation* R_P $(P \subseteq \text{Par})$, *the corresponding sufficiency operator* g_{R_P} *over* 2^X *satisfies properties* SA1 - SA4.

Proof: We prove SA1-SA3; SA4 is shown in [DeO02].

SA1. $x \in g_{R_{P\cup Q}}(A)$ iff $A \subseteq R_{P\cup Q}(x)$ iff $A \subseteq R_P(x) \cap R_Q(x)$ iff $A \subseteq R_P(x)$ and $A \subseteq R_Q(x)$ iff $x \in g_{R_P}(A) \cap g_{R_Q}(A)$, where the third double implication holds by definition of intersection and greatest lower bound.

SA2. $x \in g_{R_\emptyset}(A)$ iff $A \subseteq R_\emptyset(x)$ iff $A \subseteq X$. The latter is always true so $g_{R_\emptyset}(A) = X$.

SA3. Take any $x \in X$ such that $x \in A \cap g_{R_P}(A)$ and $x \notin g_{R_P}(X)$. Then $x \in A$ and $A \subseteq R_P(x)$ and $X \not\subseteq R_P(x)$. So xR_Px and for some $y \in X$ $x(-R)_Py$. Thus, xR_Px and, by property SF3 of R_P, $x(-R)x$, which provides the required contradiction. □

A representation theorem analogous to Theorem 13 holds for algebras and frames of strong right orthogonality. Also Theorems 5, 6, 7 are applicable to algebras and frames of strong right orthogonality, these being special sufficiency algebras and special frames, respectively.

The language $\mathsf{Lan}_{\mathsf{AS-RORT}}$ relevant for algebras and frames of strong right orthogonality is an extension of the modal language Lan_M with a family of $\{[\![R_P]\!] \mid P \subseteq \text{Par}\}$ of sufficiency operators. Algebraic semantics of the language is provided by the class $\mathsf{Alg}_{\mathsf{AS-RORT}}$ and the frame semantics by the class $\mathsf{Frm}_{\mathsf{AS-RORT}}$. The notion of a model based on a frame of $\mathsf{Frm}_{\mathsf{AS-RORT}}$, satisfaction relation, and the notions of truth in a model, in a frame and in a class of frames are analogous to the respective notions in Section 2. In view of Lemma 5 the complex algebra theorem (**CA**) holds for $\mathsf{Frm}_{\mathsf{AS-RORT}}$. From the representation theorem and (**CA**) we obtain a Duality via Truth theorem, and also the equivalence of preservation of truth.

6 Conclusion

We presented a Duality via Truth results for modal algebras, sufficiency algebras and for two classes of information algebras based on modal or sufficiency algebras, respectively. The main idea of Duality via Truth is to 'lift' the concepts of complex algebra and canonical frame so that they are assigned to an abstract frame and a general algebra, not only to a canonical frame and complex algebra. Once a Duality via Truth is established for a formal language with algebraic and frame semantics, a natural question arises regarding a suitable deduction mechanism for the language. Duality via Truth theorem guarantees that once we prove a completeness theorem with respect to one of the semantics, then we get it with respect to the other semantics too. Often, once the algebraic semantics of the language is given, a Hilbert-style axiomatisation can be derived from it. However, in the paper we do not consider any deduction methods for the presented languages.

Other Duality via Truth results can be found in [OrV03] (for lattice based languages with modal, sufficiency, necessity and dual sufficiency operators). The complete proof systems for these languages are also presented there. Duality via Truth results for a language of lattice-based relation algebras and for languages of substructural logics can be easily developed based on the representation results presented in [OrR05a, DOR03, Rew03, DORV05]

References

[BrR01] Brink, C. and I. Rewitzky. [2001]. *A Paradigm for Program Semantics: Power Sructures and Duality.* Stanford: CSLI Publications.

[DaP90] Davey, B.A. and H.A. Priestley. [1990]. *Introduction to Lattices and Order.* Cambridge: Cambridge University Press.

[DeO02] Demri, S.P. and E.S. Orłowska. [2002.] Incomplete Information: Structure, Inference, Complexity. EATCS Monographs in Theoretical Computer Science, Springer, Heidelberg.

[DOR03] Düntsch, I., E. Orłowska and A. Radzikowska. [2003]. Lattice-based relation algebras and their representability. In: Swart, H., Orlowska, E., Roubens, M., and Schmidt, G. (eds) Theory and Applications of Relational Structures as Knowledge Instruments. Lecture Notes in Computer Science 2929, 2003, 231-255.

[DORV05] Düntsch, I., E. Orłowska, A. Radzikowska and D. Vakarelov. [2005]. Relational representation theorems for some lattice-based structures. *Journal of Relational Methods in Computer Science*, vol. 1.

[DuO01] Düntsch, I. and E. Orłowska. [2001]. Beyond modalities: sufficiency and mixed algebras. In: E. Orlowska and A. Szalas (eds) Relational Methods for Computer Science Applications, Physica Verlag, Heidelberg, 263-285.

[JoT51] Jónsson, B. and A. Tarski. [1951]. Boolean algebras with operators I. *American Journal of Mathematics* **73**. p 891-939.

[OrR05a] Orłowska, E. and A. Radzikowska. [2005]. Relational representability for algebras of substructural logics. Proceedings of the 8th International Seminar RelMiCS, St.Catharines, Canada, February 22-26. p 189-195.

[OrR05b] Orłowska, E. and I. Rewitzky. [2005]. Duality via Truth: Semantic frameworks for lattice-based logics. To appear in Logic Journal of the IGPL. (28 pages).

[OrV03] Orłowska, E. and D. Vakarelov. [2003]. Lattice-based modal algebras and modal logics. In: D. Westerstahl, L.M. Valds-Villanueva and P. Hajek (eds) Proceedings of the 12th International Congress of Logic, Methodology and Philosophy of Science, Oviedo, Spain, 2003, Elsevier, to appear. Abstract in the Volume of Abstracts, 22-23.

[Rew03] Rewitzky, I. [2003]. Binary multirelations. In *Theory and Application of Relational Structures as Knowledge Instruments.* (eds: H de Swart, E Orlowska, G Schmidt, M Roubens). *Lecture Notes in Computer Science 2929.* p 259-274.

[vaB84] van Benthem, J. [1984]. Correspondence Theory. In *Handbook of Philosophical Logic. Volume II.*. ed. D Gabbay and F Guenthner. p 167-247. Amsterdam: Kluwer Academic Publishers.

Duality Theory for Projective Algebras

Alasdair Urquhart*

University of Toronto
urquhart@cs.toronto.edu

1 Introduction

Projective algebras were introduced by Everett and Ulam [4] as an algebraic formulation of the operations of projection and product on a two-dimensional algebra of relations. Although they were among the first structures to be investigated in the modern revival of the algebraic logic tradition, they have been somewhat overshadowed by their close kin, cylindric algebras and relation algebras. Chin and Tarski [2] showed that they can be viewed as two-dimensional cylindric algebras with special properties. Nevertheless, projective algebras are attractive as a natural axiomatic version of projection and product, and have a charm of their own. Ulam and Bednarek's report of 1977 [11] has some interesting suggestions on the use of these algebras in the theory of parallel computation.

In the present paper, we show that there is a duality theory for these algebras that throws light on their structure. As an application of the theory, we give simple solutions for some of Ulam's problems about projective algebras [12]. Solutions to these problems were given earlier by Faber, Erdős and Larson [3, 8, 9] using some more complicated constructions.

2 Projective Algebras

Everett and Ulam [4] define a projective algebra as an algebraic structure $\mathfrak{A} = \langle A, \vee, \wedge, \neg, 0, 1, p_0, p_1, \Box, \mathbf{a}\rangle$ that satisfies the following postulates, for $a, b, c \in A$ and $\epsilon \in \{0, 1\}$:

1. $\mathfrak{B} = \langle A, \vee, \wedge, \neg, 0, 1\rangle$ is a non-trivial Boolean algebra with least element 0 and greatest element 1.
2. p_0 and p_1 are one-place operations defined on A, $\Box$ is a partial two-place operation on A and $\mathbf{a}$ is an atom in $\mathfrak{B}$.
3. $p_\epsilon(a \vee b) = p_\epsilon(a) \vee p_\epsilon(b)$.
4. $p_0p_1 1 = p_1p_0 1 = \mathbf{a}$
5. $p_\epsilon a = 0 \Leftrightarrow a = 0$.
6. $p_\epsilon p_\epsilon a = p_\epsilon a$.
7. For $a \leq p_0 1$ and $b \leq p_1 1$, $a\Box b$ is defined.
8. For $a, b \neq 0$, $a \leq p_0 1$, $b \leq p_1 1$, $p_0(a\Box b) = a$ and $p_1(a\Box b) = b$.

* The author gratefully acknowledges the support of the National Sciences and Engineering Research Council of Canada.

W. MacCaull et al. (Eds.): RelMiCS 2005, LNCS 3929, pp. 33–47, 2006.

9. For $a \leq p_0 1$ and $b \leq p_1 1$, $a \Box 0 = 0 = 0 \Box b$.
10. $a \leq p_0 a \Box p_1 a$.
11. $(p_0 1) \Box \mathbf{a} = p_0 1$, and $\mathbf{a} \Box (p_1 1) = p_1 1$.
12. For $a, b \leq p_0 1$, $(a \vee b) \Box (p_1 1) = (a \Box (p_1 1)) \vee (b \Box (p_1 1))$.
13. For $a, b \leq p_1 1$, $(p_0 1) \Box (a \vee b) = ((p_0 1) \Box a) \vee ((p_0 1) \Box b)$.

The algebraic content of these postulates can be understood by defining a *concrete projective algebra* as follows. Let X and Y be non-empty sets, and $\mathfrak{B}$ a family of subsets of the cross-product $X \times Y$, closed under finite Boolean operations, and containing $0 = \emptyset$ and $1 = X \times Y$. Let $x_0 \in X$, $y_0 \in Y$, and define $\mathbf{a} = \{\langle x_0, y_0 \rangle\}$. For $A \in \mathfrak{B}$, define $p_0 A = Dom(A) \times \{y_0\}$ and $p_1 A = \{x_0\} \times Ran(A)$. Finally, if $A = U \times \{y_0\}$, for $U \subseteq X$ and $B = \{x_0\} \times V$, for $V \subseteq Y$, then we define $A \Box B$ as $U \times V$.

Theorem 1. *A concrete projective algebra satisfies the postulates of Everett and Ulam.*

Proof. The postulates are easily verified; for details, see [4, §1]. □

We have defined a projective algebra here using the original postulates of Everett and Ulam, in which the algebraic counterpart of the product operation is only partial, being defined only on those elements of the structure that (in a concrete projective algebra) are the images of the sets X and Y under the embedding maps $U \longmapsto U \times \{y_0\}$ and $V \longmapsto \{x_0\} \times V$. We have chosen to adhere to the original formulation as it seems to bring out the algebraic content of the theory more clearly; however,the postulates are easily modified to make □ a totally defined operation, since $p_0 a \Box p_1 b$ is a total operation extending □. A formulation using this operation as a primitive is given in Chapter 5 of [7]; the same chapter also contains a proof of equivalence of this postulate set to the cylindric algebraic formulation of Chin and Tarski [7, §5.2].

We now list without proof some of the immediate consequences of the postulates. For proofs, see Everett and Ulam [4, pp. 79-82]. We abbreviate $p_0 1$ and $p_1 1$ as 1_0 and 1_1.

Theorem 2. *In any projective algebra, the following hold:*

1. $a \leq b \Rightarrow p_\epsilon a \leq p_\epsilon b$.
2. $p_\epsilon a \leq 1_\epsilon$.
3. $1_\epsilon \wedge \neg p_\epsilon a \leq p_\epsilon \neg a$.
4. $p_\epsilon (a \wedge b) \leq p_\epsilon a \wedge p_\epsilon b$.
5. *If* $b \neq 0$, *then* $p_0 p_1 b = \mathbf{a} = p_1 p_0 b$.
6. $p_\epsilon \mathbf{a} = \mathbf{a} \leq 1_0 \wedge 1_1$.
7. $a \Box b = 0$ *if and only if* $a = 0$ *or* $b = 0$.
8. $a \leq 1_\epsilon \Leftrightarrow \exists b (p_\epsilon b = a) \Leftrightarrow p_\epsilon a = a$.
9. $0 < a \leq b \leq 1_0$ *and* $0 < c \leq d \leq 1_1$ *if and only if* $0 < a \Box c \leq b \Box d$.
10. *The operation* $a^* = p_0 a \Box p_1 a$ *is a closure operator, that is to say, it satisfies the conditions:*

$$0^* = 0;\ a \leq a^*;\ a^{**} = a^*;\ a \leq b \Rightarrow a^* \leq b^*;\ 1^* = 1.$$

11. $a, b \leq 1_0$ *and* $c, d \leq 1_1$ *imply* $(a \wedge b)\Box(c \wedge d) = (a\Box c) \wedge (b\Box d)$.
12. $a \leq 1_0$, $b \leq 1_1$ *implies* $a\Box b = (a\Box 1_1) \wedge (1_0\Box b)$.
13. $c \leq a\Box b$ *implies* $p_0 c \leq a$, $p_1 c \leq b$.
14. *For* $a, b, c, d > 0$, $((a\Box b) \vee (c\Box d))^* = (a \vee c)\Box(b \vee d)$.
15. $a \leq 1_0$ *implies* $(a\Box 1_1) \wedge 1_0 = a$.
16. *If* $0 < a \leq 1_0$ *and* $0 < b \leq 1_1$ *then* $a^* = a\Box\mathbf{a} = a$ *and* $b^* = \mathbf{a}\Box b = b$.
17. $\mathbf{a}^* = \mathbf{a} = \mathbf{a}\Box\mathbf{a} = (\mathbf{a}\Box 1_1) \wedge (1_0\Box\mathbf{a}) = 1_0 \wedge 1_1$.
18. *If* $p_0 c = \mathbf{a}$, *then* $c \leq 1_1$, *and if* $p_1 c = \mathbf{a}$, *then* $c \leq 1_0$.
19. *If* $a, b \leq 1_0$ *and* $b \leq 1_1$, *then* $(a \vee b)\Box c = (a\Box c) \vee (b\Box c)$, *and similarly for* $a, b \leq 1_1$, $c \leq 1_0$.
20. *If* $a, b \leq 1_0$ *and* $c, d \leq 1_1$, *then*

$$(a \vee b)\Box(c \vee d) = (a\Box c) \vee (a\Box d) \vee (b\Box c) \vee (b\Box d).$$

21. $\neg(a\Box 1_1) = (\neg a \wedge 1_0)\Box 1_1$; $\neg(1_0\Box b) = 1_0\Box(\neg b \wedge 1_1)$.

3 Duality Theory

In this section, we describe a family of structured topological spaces that are dual to the family of projective algebras. The spaces in question are equipped with operators that generalize the pairing and projection operators of a concrete projective algebra. The principal difference between the conventional pairing operation and the operation in the spaces defined below is that pairing is a multi-operation (that is to say, the result of a pairing operation (x, y) is a set of points, rather than a single ordered pair).

Definition 1. *A* projective space *is a structure* $\mathcal{S} = \langle S, \mathcal{T}, \pi_0, \pi_1, (\cdot,\cdot), \mathbf{p}\rangle$ *satisfying the conditions:*

1. $\langle S, \mathcal{T}\rangle$ *is a Stone space (that is, a compact totally disconnected Hausdorff space).*
2. *For* $\epsilon = 0, 1$, π_ϵ *is a continuous map on the space* $\langle S, \mathcal{T}\rangle$.
3. *For* A *a clopen subset of* S, $P_\epsilon(A) = \{\pi_\epsilon(x) : x \in A\}$ *is clopen; we write* S_ϵ *for* $P_\epsilon(S)$.
4. $\mathbf{p}$ *is an isolated point in the space* $\langle S, \mathcal{T}\rangle$.
5. $\pi_0\pi_1(x) = \mathbf{p} = \pi_1\pi_0(x)$.
6. $\pi_\epsilon\pi_\epsilon(x) = \pi_\epsilon(x)$.
7. *For* $x \in S_0, y \in S_1$, $(x, y) \neq \emptyset$.
8. *If* $x \in S_0, y \in S_1$, *and* $z \in (x, y)$, *then* $\pi_0(z) = x$ *and* $\pi_1(z) = y$.
9. *For all* $x \in S$, $x \in (\pi_0(x), \pi_1(x))$.
10. *If* $x \in S_0$, *then* $(x, \mathbf{p}) = \{x\}$, *and if* $x \in S_1$, *then* $(\mathbf{p}, x) = \{x\}$.

Definition 2. *If* $\mathcal{S} = \langle S, \mathcal{T}, \pi_0, \pi_1, (\cdot,\cdot), \mathbf{p}\rangle$ *is a projective space, then we define the* dual algebra $\mathfrak{A}(\mathcal{S})$ *as follows:*

1. $\langle A, \vee, \wedge, \neg, 0, 1\rangle$ *is the algebra of all clopen subsets of* S.
2. *The operations* P_ϵ *are defined by* $P_\epsilon(A) = \{\pi_\epsilon(x) : x \in A\}$.

3. *For* $A \subseteq S_0$ *and* $B \subseteq S_1$*, we define* $A \Box B = \bigcup\{(x, y) : x \in A, y \in B\}$.
4. *The atom* $\mathbf{a}$ *is the singleton* $\{\mathbf{p}\}$.

Theorem 3. *If* $\mathcal{S}$ *is a projective space, then the dual algebra* $\mathfrak{A}(\mathcal{S})$ *is a projective algebra.*

Proof. We need to show that $\mathfrak{A}(\mathcal{S})$ is closed under the operations. If $A \subseteq S$ is clopen, then $P_\epsilon(A)$ is also clopen by Condition 3 of Definition 1. Now assume that $A \subseteq S_0$ and $B \subseteq S_1$ are clopen. It follows from Conditions 8 and 9 of Definition 1 that $\pi_0^{-1}(A) = A \Box S_1$, $\pi_1^{-1}(B) = S_0 \Box B$, and $A \Box B = (A \Box S_1) \cap (S_0 \Box B)$, showing that $A \Box B$ is open, since π_0 and π_1 are continuous. To show that $A \Box B$ is closed, we have $\neg(A \Box S_1) = (\neg A \cap S_0) \Box S_1 = \pi_0^{-1}(\neg A \cap S_0)$, which is an open set. Similarly, $\neg(S_0 \Box B) = \pi_1^{-1}(\neg B \cap S_1)$, showing that $\neg(A \Box B)$ is open, since $\neg(A \Box B) = \neg(A \Box S_1) \cup \neg(S_0 \Box B) = \pi_0^{-1}(\neg A \cap S_0) \cup \pi_1^{-1}(\neg B \cap S_1)$.

The remaining postulates of a projective algebra are easily verified using the conditions of Definition 1. □

Definition 3. *If* $\mathfrak{A} = \langle A, \vee, \wedge, \neg, 0, 1, p_0, p_1, \Box, \mathbf{a}\rangle$ *is a projective algebra, then the* dual space *of* $\mathfrak{A}$, $\mathcal{S}(\mathfrak{A}) = \langle S, \mathcal{T}, \pi_0, \pi_1, (\cdot, \cdot), \mathbf{p}\rangle$ *is defined as follows.*

1. S *is the family of all prime filters* ∇ *in* $\mathfrak{A}$.
2. *The topology* $\mathcal{T}$ *on* S *is the Stone space of* $\mathfrak{A}$*, that is, the topology for which the family of all sets* $\phi(a) = \{\nabla \in S : a \in \nabla\}$*, where* $a \in A$*, is a base.*
3. *For* $\nabla \in S$, $\pi_\epsilon(\nabla) = \{b \in A : \exists a \in \nabla,\ p_\epsilon(a) \leq b\}$.
4. *Let* $\nabla_0 \in S_0 = \pi_0(S)$*, and* $\nabla_1 \in S_1 = \pi_1(S)$*. Then we define* (∇_0, ∇_1) *to be the set of all prime filters* ∇ *such that*

$$\{(a \wedge 1_0) \Box (b \wedge 1_1) : a \in \nabla_0, b \in \nabla_1\} \subseteq \nabla.$$

5. *The point* $\mathbf{p}$ *is the prime filter generated by the atom* $\mathbf{a}$.

Theorem 4. *The dual space of a projective algebra is a projective space.*

Proof. We successively verify the conditions in Definition 1. Most of the proofs are already to be found in McKinsey's 1947 paper [10]. Conditions 1 and 4 are immediate by construction, so we start by verifying Condition 2.

First, we observe that $\pi_\epsilon(\nabla)$ is in fact a prime filter; this is proved as Lemmas 3 and 4 of McKinsey [10]. To show that the map π_ϵ is continuous, it is sufficient to show that for $a \in A$, the set $\pi_\epsilon^{-1}(\phi(a))$ is open. This follows from the equation

$$\pi_\epsilon^{-1}(\phi(a)) = \bigcup_{b \in A} \{\phi(b) : p_\epsilon(b) \leq a\}.$$

To verify Condition 3 in Definition 1, we observe that since $\langle S, \mathcal{T}\rangle$ is a Boolean space, every clopen subset of S is of the form $\phi(a)$, for a in A, and that $P_\epsilon(\phi(a)) = \phi(p_\epsilon(a))$, by Lemmas 27 and 28 of [10].

Condition 5 of the Definition is Lemma 12, and Condition 6 is Lemma 10 of [10]. Condition 7 is Lemma 15 and Condition 8 is Lemma 21 of [10]. Condition 9 is Lemma 24, and Condition 10 is Lemmas 18 and 19 of [10]. □

Theorem 5. *If $\mathfrak{A}$ is a projective algebra, then $\phi(a) = \{\nabla \in S : a \in \nabla\}$ is an isomorphism between $\mathfrak{A}$ and $\mathfrak{A}(\mathcal{S}(\mathfrak{A}))$.*

Proof. The map ϕ is a Boolean isomorphism between the Boolean algebra $\mathfrak{B} = \langle A, \vee, \wedge, \neg, 0, 1\rangle$ and the algebra of all clopen subsets of the dual space $\mathcal{S}(\mathfrak{A})$ [5, p. 78]. Hence, it is sufficient to show that ϕ is an isomorphism with respect to the remaining operations.

Lemmas 27 and 28 of McKinsey [10] show that $\phi(p_\epsilon(a)) = P_\epsilon(\phi(a))$, while his Lemma 29 shows that for $a \leq 1_0$ and $b \leq 1_1$, $\phi(a\Box b) = \phi(a)\Box\phi(b)$. Finally, $\phi(\mathbf{a}) = \{\mathbf{p}\}$, by the definition of $\mathbf{p}$. □

Theorem 6. *If $\mathcal{S} = \langle S, \mathcal{T}, \pi_0, \pi_1, (\cdot,\cdot), \mathbf{p}\rangle$ is a projective space, then $\psi(x) = \{A \in \mathfrak{A}(\mathcal{S}) : x \in A\}$ is an isomorphism between $\mathcal{S}$ and $\mathcal{S}(\mathfrak{A}(\mathcal{S}))$.*

Proof. The map ψ is a homeomorphism between the Stone space $\langle S, \mathcal{T}\rangle$ and its second dual [5, p. 79]. Hence, we only need to verify that ψ is an isomorphism with respect to the operations and the constant $\mathbf{p}$.

To prove that ψ is an isomorphism with respect to the projection operator π_ϵ, we first establish the equation

$$\{\pi_\epsilon(x)\} = \bigcap\{P_\epsilon(B) : x \in B, B \in \mathfrak{A}(\mathcal{S})\},$$

for $x \in S$. The left-to-right inclusion holds by definition. Now assume that $y \neq \pi_\epsilon(x)$. Since π_ϵ is continuous, the pre-image of y, $\pi_\epsilon^{-1}(y)$ is closed. Hence, we can separate x and $\pi_\epsilon^{-1}(y)$ by a set $B \in \mathfrak{A}(\mathcal{S})$, that is to say, $x \in B$, and $\pi_\epsilon^{-1}(y) \cap B = \emptyset$. It follows that $y \notin \bigcap\{P_\epsilon(B) : x \in B, B \in \mathfrak{A}(\mathcal{S})\}$, completing the proof of the equation.

Now if $A \in \mathfrak{A}(\mathcal{S})$, and $\pi_\epsilon(x) \in A$, then $\{\pi_\epsilon(x)\}$ is a closed subset of A. Using the preceding equation and compactness, we find that there are $B_1, \ldots, B_k \in \mathfrak{A}(\mathcal{S})$, so that $x \in B = B_1 \cap \cdots \cap B_k$, and $P_\epsilon(B) \subseteq P_\epsilon(B_1) \cap \cdots \cap P_\epsilon(B_k) \subseteq A$. Finally, using this last result, we can calculate for $A \in \mathfrak{A}(\mathcal{S})$,

$$\begin{aligned} A \in \psi(\pi_\epsilon(x)) &\Leftrightarrow \pi_\epsilon(x) \in A \\ &\Leftrightarrow \exists B \in \mathfrak{A}(\mathcal{S})[x \in B, P_\epsilon(B) \subseteq A] \\ &\Leftrightarrow \exists B \in \mathfrak{A}(\mathcal{S})[B \in \psi(x), P_\epsilon(B) \subseteq A] \\ &\Leftrightarrow A \in \pi_\epsilon(\psi(x)), \end{aligned}$$

showing that $\psi(\pi_\epsilon(x)) = \pi_\epsilon(\psi(x))$.

To show that ψ is an isomorphism with respect to the pairing operation, let us assume that $x \in S_0$ and $y \in S_1$. We claim that $\psi(x, y) = (\psi(x), \psi(y))$. First, let us assume that ∇ is a prime filter in $\mathfrak{A}(\mathcal{S})$ so that $\nabla \in \psi(x, y)$, that is, for some $z \in (x, y)$, $\psi(z) = \nabla$. For $A, B \in \mathfrak{A}(\mathcal{S})$, if $x \in A$ and $y \in B$, then $z \in (x, y) \subseteq (A \cap S_0)\Box(B \cap S_1)$, showing that $(A \cap S_0)\Box(B \cap S_1) \in \nabla$. Thus, we have shown

$$\{(A \cap S_0)\Box(B \cap S_1) : A \in \psi(x), B \in \psi(y)\} \subseteq \nabla,$$

that it to say, $\nabla \in (\psi(x), \psi(y))$. For the converse, assume that $\nabla \in (\psi(x), \psi(y))$. Choose $z \in S$ so that $\psi(z) = \nabla$. We claim that $\pi_0(z) = x$. To show this, assume that $x \in A$, where $A \in \mathfrak{A}(\mathcal{S})$. Then by assumption, $z \in (A \cap S_0)\Box S_1$, so that $\pi_0(z) \in A$. It follows that $\pi_0(z) = x$, since distinct elements of S are separated by a set in $\mathfrak{A}(\mathcal{S})$. We can prove that $\pi_1(z) = y$ by an exactly similar proof, and hence $z \in (\pi_0(z), \pi_1(z)) = (x, y)$, showing that $\nabla \in \psi(x, y)$.

Finally, the point $\mathbf{p}$ is mapped into the principal filter generated by $\mathbf{a} = \{\mathbf{p}\}$, by construction. □

Definition 4. *Let* $\mathcal{S}_1 = \langle S_1, \mathcal{T}_1, \pi_0, \pi_1, (\cdot,\cdot), \mathbf{p}_1\rangle$ *and* $\mathcal{S}_2 = \langle S_2, \mathcal{T}_2, \rho_0, \rho_1, [\cdot,\cdot], \mathbf{p}_2\rangle$ *be projective spaces. Then a function* ψ *is a* projective map *from* S_1 *to* S_2 *if it satisfies the following conditions:*

1. ψ *is a continuous map from* S_1 *to* S_2.
2. *For* $x \in S_1$, $\psi(x) = \mathbf{p}_2 \Leftrightarrow x = \mathbf{p}_1$.
3. $\psi(\pi_\epsilon(x)) = \rho_\epsilon(\psi(x))$.
4. $\psi(x) = \rho_\epsilon(y) \Rightarrow \exists z(\psi(z) = y \wedge \pi_\epsilon(z) = x)$.
5. $x \in (u, v) \Rightarrow \psi(x) \in (\psi(u), \psi(v))$.
6. $\psi(x) \in (y, z) \Rightarrow \exists uv[\psi(u) = y \wedge \psi(v) = z \wedge x \in (u, v)]$.

If ψ is a projective map between projective spaces $\mathcal{S}_1$ and $\mathcal{S}_2$, then we define the *dual map of* ψ to be the mapping between $\mathfrak{A}(\mathcal{S}_2)$ and $\mathfrak{A}(\mathcal{S}_1)$ given by the definition $\phi(A) = \psi^{-1}(A)$, for $A \subseteq S_2$.

Theorem 7. *If* ψ *is a projective map between projective spaces* $\mathcal{S}_1$ *and* $\mathcal{S}_2$, *then the mapping* ϕ *dual to* ψ *is a morphism in the category of projective algebras.*

Proof. Since ψ is a continuous map, it is immediate that $\phi(A)$ is a clopen subset of S_1, for A a clopen subset of S_2. It remains to show that ϕ preserves the projective algebraic operations.

To show that ϕ preserves the projection operation P_ϵ, assume that x is in $P_\epsilon(\phi(A))$, so that for some y, $\psi(y) \in A$ and $\pi_\epsilon(y) = x$. Then $\psi(x) = \psi(\pi_\epsilon(y)) = \rho_\epsilon(\psi(y))$ by Definition 4, Condition 3, so that $x \in \phi(P_\epsilon(A))$. Conversely, assume that $x \in \phi(P_\epsilon(A))$, so that for some $y \in A$, $\rho_\epsilon(y) = \psi(x)$. By Condition 4 of Definition 4, there is an element $z \in S_1$ so that $\psi(z) = y$ and $\pi_\epsilon(z) = x$, showing $x \in P_\epsilon(\phi(A))$.

To show that ϕ preserves the operation $\Box$, assume that $A, B \subseteq S_2$, where $A \subseteq \rho_0(S_2)$ and $B \subseteq \rho_1(S_2)$. Assume, first, that $x \in \phi(A\Box B)$, so that for some $y \in A$ and $z \in B$, $\psi(x) \in (y, z)$. By Condition 6 of Definition 4, there are $u, v \in S_1$ so that $\psi(u) = y$, $\psi(v) = z$ and $x \in (u, v)$. Hence, $u \in \phi(A)$ and $v \in \phi(B)$, so that $x \in \phi(A)\Box\phi(B)$. Conversely, assume that $x \in \phi(A)\Box\phi(B)$, so that there are $u, v \in S_1$ with $x \in (u, v)$, $\psi(u) \in A$ and $\psi(v) \in B$. Then by Definition 4, Condition 5, $\psi(x) \in (\psi(u), \psi(v))$, so that $\psi(x) \in A\Box B$, that is to say, $x \in \phi(A\Box B)$.

Finally, by Condition 2 of Definition 4,

$$\phi(\mathbf{a}_2) = \phi(\{\mathbf{p}_2\}) = \psi^{-1}(\{\mathbf{p}_2\}) = \{\mathbf{p}_1\} = \mathbf{a}_1. \qquad \Box$$

If $\mathfrak{A}_1$ and $\mathfrak{A}_2$ are projective algebras, and ϕ a projective morphism from $\mathfrak{A}_1$ to $\mathfrak{A}_2$, then we define the map dual to ϕ to be the map ψ from $\mathcal{S}(\mathfrak{A}_2)$ to $\mathcal{S}(\mathfrak{A}_1)$ given by $\psi(\nabla) = \phi^{-1}(\nabla)$, where ∇ is a prime filter in $\mathfrak{A}_2$.

Theorem 8. *If $\mathfrak{A}_1$ and $\mathfrak{A}_2$ are projective algebras, and ϕ a projective morphism from $\mathfrak{A}_1$ to $\mathfrak{A}_2$, then the map dual to ϕ is a projective map from $\mathcal{S}(\mathfrak{A}_2)$ to $\mathcal{S}(\mathfrak{A}_1)$.*

Proof. The map ψ is a map in the category of Stone spaces, so continuity follows immediately. The second condition of Definition 4 is also easily verified. It remains to verify Conditions 3 to 6.

For Condition 3, we need to show that for a prime filter ∇ in $\mathfrak{A}_2$, $\psi(\pi_\epsilon(\nabla)) = \rho_\epsilon(\psi(\nabla))$. If $a \in \rho_\epsilon(\psi(\nabla))$, then for some b, $\phi(b) \in \nabla$ and $p_\epsilon(b) \leq a$. Hence, $p_\epsilon(\phi(b)) = \phi(p_\epsilon(b)) \leq \phi(a)$, showing that $a \in \psi(\pi_\epsilon(\nabla))$. Thus $\rho_\epsilon(\psi(\nabla)) \subseteq \psi(\pi_\epsilon(\nabla))$, from which the condition follows.

For Condition 4, assume that $\psi(\nabla_1) = \rho_\epsilon(\nabla_2)$, for ∇_1 a prime filter in $\mathfrak{A}_2$, and ∇_2 a prime filter in $\mathfrak{A}_1$. We define a subset of A_2 by

$$\Delta = \{\phi(a) : a \in \nabla_2\} \cup \{\neg c : \exists b(p_\epsilon(c) \leq b \wedge b \notin \nabla_1)\}.$$

We need to show that Δ generates a proper filter in $\mathfrak{A}_2$. If this condition fails, then there are $a_1, \dots, a_m \in \nabla_2$ and $c_1, \dots, c_n \in \mathfrak{A}_2$ so that for any c_j, there exists a b_j so that $p_\epsilon(c_j) \leq b_j$ and $b_j \notin \nabla_1$. Setting $a = a_1 \wedge \cdots \wedge a_m$, $c = c_1 \vee \cdots \vee c_n$ and $b = b_1 \vee \cdots \vee b_n$, we have $\phi(a) \leq c$, where $a \in \nabla_2$, $p_\epsilon(c) \leq b$, for $b \notin \nabla_1$. Then

$$\phi(p_\epsilon(a)) = p_\epsilon(\phi(a)) \leq p_\epsilon(c) \leq b,$$

and since $p_\epsilon(a) \in \psi(\nabla_2)$, it follows that $\phi(p_\epsilon(a)) \in \nabla_1$, showing that $b \in \nabla_1$, a contradiction. Hence, there is a prime filter ∇_3 extending Δ. By construction, $\nabla_2 \subseteq \psi(\nabla_3)$, so that $\psi(\nabla_3) = \nabla_2$. Last, if $b \in \pi_\epsilon(\nabla_3)$, then for some $c \in \nabla_3$, $p_\epsilon(c) \leq b$, so $b \in \nabla_1$, by the definition of Δ, showing that $\pi_\epsilon(\nabla_3) \subseteq \nabla_1$, hence $\pi_\epsilon(\nabla_3) = \nabla_1$; this completes the proof of the fourth condition.

With a view to proving Condition 5 of Definition 4, let us assume that $\nabla_1 \in (\nabla_2, \nabla_3)$, where $\nabla_1, \nabla_2, \nabla_3$ are prime filters in $\mathfrak{A}_2$. By Definition 3, this means that

$$\{(c \wedge 1_0)\Box(d \wedge 1_1) : c \in \nabla_2, d \in \nabla_3\} \subseteq \nabla_1.$$

Now if $a \in \phi^{-1}(\nabla_2)$ and $b \in \phi^{-1}(\nabla_3)$ then

$$\phi[(a \wedge 1_0)\Box(b \wedge 1_1)] = [(\phi(a) \wedge 1_0)\Box(\phi(b) \wedge 1_1)] \in \nabla_1,$$

so that $(a \wedge 1_0)\Box(b \wedge 1_1) \in \phi^{-1}(\nabla_1)$, that is, $\psi(\nabla_1) \in (\psi(\nabla_2), \psi(\nabla_3))$.

The last condition to be verified is Condition 6, which is the most complicated to prove. To this end, let us assume that $\psi(\nabla) \in (\Gamma, \Delta)$, where ∇ is a prime filter in $\mathfrak{A}_2$, and Γ, Δ are prime filters in $\mathfrak{A}_1$. We have to show that there are prime filters Ω and Π in $\mathfrak{A}_2$ so that $\psi(\Omega) = \Gamma$, $\psi(\Pi) = \Delta$, and $\nabla \in (\Omega, \Pi)$.

Define $\mathcal{F}$ to be the family of all pairs of sets $\langle \Sigma, \Theta \rangle$, where $\Sigma, \Theta \subseteq \mathfrak{A}_2$ satisfying the conditions:

1. $0 \notin \Sigma \cup \Theta$.
2. $\{\phi(a) : a \in \Gamma\} \subseteq \Sigma$.
3. $\{\phi(b) : b \in \Delta\} \subseteq \Theta$.
4. For $a \in \Sigma$, $b \in \Theta$, $(a \wedge 1_0)\Box(b \wedge 1_1) \in \nabla$.

We need to argue first that $\mathcal{F}$ is non-empty. Set $\Sigma_0 = \{\phi(a) : a \in \Gamma\}$ and $\Theta_0 = \{\phi(b) : b \in \Delta\}$. We claim that $\langle \Sigma_0, \Theta_0 \rangle \in \mathcal{F}$. If $0 \in \Sigma_0$, then $\phi(a) = 0$, for $a \in \Gamma$ so $0 \in \psi(\nabla)$, a contradiction; similarly $0 \notin \Theta_0$. If $a \in \Gamma$ and $b \in \Delta$, then $\phi[(a \wedge 1_0)\Box(b \wedge 1_1)] \in \nabla$, since $\psi(\nabla) \in (\Gamma, \Delta)$, showing that $[(\phi(a) \wedge 1_0)\Box(\phi(b) \wedge 1_1)] \in \nabla$, completing the proof that $\langle \Sigma_0, \Theta_0 \rangle \in \mathcal{F}$.

Order $\mathcal{F}$ by containment; that is to say, $\langle \Sigma_1, \Theta_2 \rangle \leq \langle \Sigma_2, \Theta_2 \rangle$ if and only if $\Sigma_1 \subseteq \Sigma_2$ and $\Theta_1 \subseteq \Theta_2$. Every chain in $\mathcal{F}$ has an upper bound, so by Zorn's Lemma, there is a maximal element $\langle \Omega, \Pi \rangle$ in $\mathcal{F}$. We now have to prove that Ω and Π are prime filters in $\mathfrak{A}_2$. We carry out the proof in detail for Ω; the proof for Π is exactly similar.

We need to show that Ω is a filter. Assume that $a \in \Omega$, and $a \leq b$. Then it is easy to show using Theorem 2, Part 9, that $\langle \Omega \cup \{b\}, \Pi \rangle$ is in $\mathcal{F}$, from which it follows by maximality that $b \in \Omega$. To show closure under $\wedge$, assume that $a, b \in \Omega$. Then for any $c \in \Pi$, $(a \wedge 1_0)\Box(c \wedge 1_1)$ and $(b \wedge 1_0)\Box(c \wedge 1_1)$ are in ∇, so that by Theorem 2, Part 11, $(a \wedge b \wedge 1_0)\Box(c \wedge 1_1) \in \nabla$. It follows that $\langle \Omega \cup \{a \wedge b\}, \Pi \rangle$ is in $\mathcal{F}$, so by maximality, $a \wedge b \in \Omega$.

To show that Ω is prime, assume that $a \vee b \in \Omega$, but $a \notin \Omega$ and $b \notin \Omega$. If $a = 0$ or $b = 0$, then it follows immediately that either $a \in \Omega$ or $b \in \Omega$. Consequently, we can assume that $a \neq 0$ and $b \neq 0$. Since by maximality, $\langle \Omega \cup \{a\}, \Pi \rangle \notin \mathcal{F}$ and $\langle \Omega \cup \{b\}, \Pi \rangle \notin \mathcal{F}$, it follows that $\langle \Omega \cup \{a\}, \Pi \rangle$ and $\langle \Omega \cup \{b\}, \Pi \rangle$ fail to satisfy the third condition defining the members of $\mathcal{F}$. Thus there are $c, d \in \Pi$ so that $(a \wedge 1_0)\Box(c \wedge 1_1) \notin \nabla$ and $(b \wedge 1_0)\Box(d \wedge 1_1) \notin \nabla$. It follows by Theorem 2, Part 9, that $[(a \wedge 1_0)\Box(c \wedge d \wedge 1_1)] \vee [(b \wedge 1_0)\Box(c \wedge d \wedge 1_1)] \notin \nabla$, so by Theorem 2, Part 19, $[((a \vee b) \wedge 1_0)\Box(c \wedge d \wedge 1_1)] \notin \nabla$, contrary to assumption.

By construction, if $a \in \Gamma$, then $\phi(a) \in \Omega$, so that $a \in \psi(\Omega)$, showing that $\psi(\Omega) = \Gamma$; similarly, $\psi(\Pi) = \Delta$. Finally, since by the definition of $\mathcal{F}$, $\Delta \in (\Omega, \Pi)$, so the proof of Condition 6 is complete. □

Theorem 9. *The category of projective spaces and projective maps is dual to the category of projective algebras and projective homomorphisms.*

Proof. The duality theory is defined as a special case of the Stone duality for Boolean algebras. The dual of a projective homomorphism and the dual of a projective map are defined exactly as in the Boolean case. Hence, we can prove that the second dual of a homomorphism is isomorphic to the homomorphism, and the second dual of a projective map is isomorphic to the map itself, by the same argument as in the Boolean case [5, §20]. □

4 Completion and Embedding

Borrowing some terminology from the literature of modal logic, let us define a *projective frame* to be a structure satisfying the conditions defining a projective space, but with all references to topology removed.

Definition 5. *A* projective frame *is a structure* $\mathcal{F} = \langle S, \pi_0, \pi_1, (\cdot,\cdot), \mathbf{p} \rangle$ *satisfying the conditions:*

1. *For* $\epsilon = 0, 1$, π_ϵ *is a function on the space* S*; we write* S_ϵ *for* $P_\epsilon(S)$.
2. $\mathbf{p}$ *is a point in* S.
3. $\pi_0\pi_1(x) = \mathbf{p} = \pi_1\pi_0(x)$.
4. $\pi_\epsilon\pi_\epsilon(x) = \pi_\epsilon(x)$.
5. *For* $x \in S_0, y \in S_1$, $(x, y) \neq \emptyset$.
6. *If* $x \in S_0, y \in S_1$, *and* $z \in (x, y)$, *then* $\pi_0(z) = x$ *and* $\pi_1(z) = y$.
7. *For all* $x \in S$, $x \in (\pi_0(x), \pi_1(x))$.
8. *If* $x \in S_0$, *then* $(x, \mathbf{p}) = \{x\}$, *and if* $x \in S_1$, *then* $(\mathbf{p}, x) = \{x\}$.

If $\mathfrak{A}$ is a projective algebra, then we define the *frame of* $\mathfrak{A}$ as the frame defined by Definition 3; that is to say, the frame of an algebra is defined in the same way as its dual space, but with the topology omitted.

Figure 1 shows a typical example of a finite projective frame. The sets (x, y) are indicated by boxes, while the projection maps are sketched as arrows. The subset S_0 is the bottom row of boxes, while the S_1 is the lefthand column. Thus in this picture, S_0 serves as the x-axis, S_1 as the y-axis, and the atom $\mathbf{a}$ as the origin.

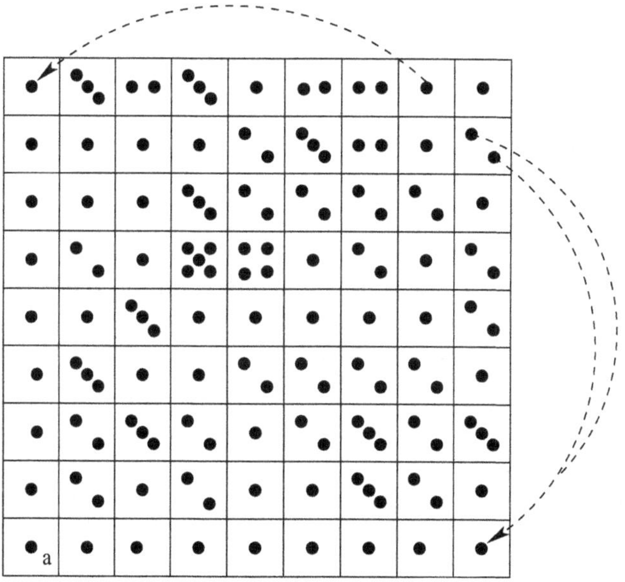

Fig. 1. An example of a finite projective frame

Let us say that a projective frame is *non-degenerate* if the sets $S_0 \setminus \{\mathbf{a}\}$ and $S_1 \setminus \{\mathbf{a}\}$ are both non-empty. Reflection on the diagrams of finite projective frames easily leads to the conclusion that there is a one-to-one correspondence between finite projective frames and equivalence classes of finite matrices with positive integer entries, where we count two matrices as equivalent if one can be derived from the other by permutation of rows and columns.

If $\mathcal{F}$ is a projective frame, then we can define a complete atomic projective algebra $\mathcal{CA}(\mathcal{F})$ by taking as the underlying Boolean algebra the family of all subsets of the frame $\mathcal{F}$, while the operations on $\mathcal{CA}(\mathcal{F})$ are given as in Definition 2.

Theorem 10. *Every projective algebra can be embedded in a complete atomic projective algebra.*

Proof. This is the main theorem of McKinsey's 1947 paper [10], and we prove it in exactly the same way as McKinsey. If $\mathfrak{A}$ is a projective algebra, then by Theorem 5, the map ϕ is an embedding of $\mathfrak{A}$ into the complete atomic algebra $\mathcal{CA}(\mathcal{F})$, where $\mathcal{F}$ is the frame of $\mathfrak{A}$. □

McKinsey's embedding theorem can be extended to a categorical duality between the category of projective frames and the category of complete atomic projective algebras. If $\mathfrak{A}$ is a complete atomic projective algebra, then the *frame of* $\mathfrak{A}$ can be defined as the projective frame defined on the atoms of $\mathfrak{A}$ by using definitions adapted from those of Definition 3. The morphisms in the category of frames are defined as in Definition 4 (omitting the topological conditions), while the morphisms in the category of complete atomic projective algebras must preserve infinite meets and joins.

Theorem 11. *The category of projective frames is the dual of the category of complete atomic projective algebras.*

Proof. We omit the proof of the theorem, as it largely parallels the proof of the duality theorem of the previous section. □

It is not hard to see that the frames of concrete projective algebras are exactly those in which (x, y) is a unit set, for all elements x, y in the frame; we call these *concrete frames*. We can employ this characterization together with the duality theory for frames to give a simple proof of the main theorem [4, §4] of Everett and Ulam, as extended by McKinsey [10].

Theorem 12. *Every projective algebra can be embedded in a concrete projective algebra.*

Proof. Let $\mathcal{S} = \langle S, \rho_0, \rho_1, [\cdot,\cdot], \mathbf{p}\rangle$ be a projective frame. We shall construct a concrete projective frame $\mathcal{S}^* = \langle S^*, \pi_0, \pi_1, (\cdot,\cdot), \mathbf{p}^*\rangle$ and a surjective frame morphism from $\mathcal{S}_1$ onto $\mathcal{S}_2$. The construction is a simplified version of the original method employed by Everett and Ulam [4, §4].

To define the universe of the frame $\mathcal{S}^*$, we define two sets

$$X = \{\langle s, q\rangle : s \in S_0, q \in S, (s = \mathbf{p} \Rightarrow q = \mathbf{p})\},$$

$$Y = \{\langle t, q\rangle : t \in S_1, q \in S, (t = \mathbf{p} \Rightarrow q = \mathbf{p})\},$$

and then define the universe of the concrete frame to be $S^* = X \times Y$, and $\mathbf{p}^* = \langle \mathbf{p}, \mathbf{p}\rangle$. If $x = \langle\langle s, q\rangle, \langle t, r\rangle\rangle$ is an element of S^*, then $\pi_0(x) = \langle\langle s, q\rangle, \langle \mathbf{p}, \mathbf{p}\rangle\rangle$ and $\pi_1(x) = \langle\langle \mathbf{p}, \mathbf{p}\rangle, \langle t, q\rangle\rangle$. Finally, if $x = \langle\langle s, q\rangle, \langle \mathbf{p}, \mathbf{p}\rangle\rangle$ is an element of S_0^*

and $y = \langle\langle \mathbf{p}, \mathbf{p}\rangle, \langle t, r\rangle\rangle$ an element of S_1^*, we define the pairing operation on the concrete frame by $[x, y] = \{\langle\langle s, q\rangle, \langle t, r\rangle\rangle\}$. With these definitions, it is easy to verify that S^* is a concrete projective frame.

The family $\{(x, y) : x, y \in S\}$ forms a partition of the set S. Let $C(x, y)$ be a choice function for this family, that is to say, $C(x, y) \in (x, y)$ for all $x, y \in S$. In addition, let θ be a function from $S \times S$ to S satisfying the condition

$$\forall x \in S\, \forall y \in S \exists z \in S(\theta(y, z) = x).$$

A function satisfying this condition is easily constructed by employing a commutative group associated with the elements of S; details are to be found in Everett and Ulam [4, §4].

Now define the map ψ from S^* to S as follows. For $\langle\langle s, q\rangle, \langle t, r\rangle\rangle$ an element of S^*,

$$\psi(\langle\langle s, q\rangle, \langle t, r\rangle\rangle) = \begin{cases} \theta(q, r) & \text{if } \theta(q, r) \in (s, t) \\ C(q, r) & \text{otherwise.} \end{cases}$$

We must now show that ψ is a morphism in the category of frames; that is, we have to verify Conditions 2 to 6 of Definition 4.

First, we observe that for $x = \langle\langle s, q\rangle, \langle t, r\rangle\rangle$ an element of S^*, $\psi(x) \in (s, t)$ by definition. Now let $x = \langle\langle s, q\rangle, \langle \mathbf{p}, \mathbf{p}\rangle\rangle$ be an element of S_0^* and $y = \langle\langle \mathbf{p}, \mathbf{p}\rangle, \langle t, r\rangle\rangle$ be an element of S_1^*. Then $\psi(x) \in (s, p) = \{s\}$, so $\psi(x) = s$; similarly, $\psi(y) = t$. It follows that if $x = \langle\langle s, q\rangle, \langle t, r\rangle\rangle$ is in $[y, z]$, where $y \in S_0^*$ and $z \in S_1^*$, then $\pi_0(x) = y$, $\pi_1(x) = z$, $\psi(y) = s$ and $\psi(z) = t$, showing that $\psi(x) \in (\psi(y), \psi(z))$, and verifying Condition 5.

For Condition 3, assume that $x = \langle\langle s, q\rangle, \langle t, r\rangle\rangle$, so that $\pi_0(x) = \langle\langle s, q\rangle, \langle \mathbf{p}, \mathbf{p}\rangle\rangle$, from which it follows that $\psi(\pi_0(x)) = s$. Secondly, we have $\psi(x) \in (s, t)$, so that $\rho_0(\psi(x)) \in (s, \mathbf{p}) = \{z\}$, so $\psi(\pi_0(x)) = \rho_0(\psi(x))$. Similarly, $\psi(\pi_1(x)) = \rho_1(\psi(x))$, completing the proof of Condition 3.

For Condition 2, let $x = \langle\langle s, q\rangle, \langle t, r\rangle\rangle$, and $\psi(x) = \mathbf{p}$. Then $s = \psi(\pi_0(x)) = \rho_0(\psi(x)) = \mathbf{p}$, so by the definition of X, $q = \mathbf{p}$. Similarly, $t = r = \mathbf{p}$, so that $x = \mathbf{p}^*$, concluding the proof.

For Condition 4, let $x = \langle\langle s, q\rangle, \langle t, r\rangle\rangle$, and assume $\psi(x) = \rho_0(y)$. Then $s = \psi(\pi_0(x)) = \rho_0(\psi(x)) = \rho_0\rho_0(y) = \rho_0(y)$. Now $s = \psi(x) \in (s, t)$, hence $\rho_1(s) = t$, showing that $\mathbf{p} = \rho_1(\rho_0(s)) = t$, from which it follows that $r = \mathbf{p}$. We have $y \in (s, u)$, for some u. Choose v so that $\theta(s, v) = y$, and set $z = \langle\langle s, q\rangle, \langle u, v\rangle\rangle$. Then $\psi(z) = \theta(s, v) = y$, and $\pi_0(z) = \langle\langle s, q\rangle, \langle \mathbf{p}, \mathbf{p}\rangle\rangle = x$, completing the proof of the Condition.

Finally, for Condition 6, assume that $\psi(x) \in (y, z)$, where $x = \langle\langle s, q\rangle, \langle t, r\rangle\rangle$. Then $s = \psi(\pi_0(x)) = \rho_0(\psi(x)) = y$, and similarly $t = z$. Set $u = \langle\langle s, q\rangle, \langle \mathbf{p}, \mathbf{p}\rangle\rangle$ and $v = \langle\langle \mathbf{p}, \mathbf{p}\rangle, \langle t, r\rangle\rangle$. Then $\psi(u) = s = y$, and $\psi(v) = t = z$, and $x \in [u, v]$, completing the proof of the Conditions of Definition 4.

Since the map ϕ dual to the ψ just constructed is an embedding in the category of complete atomic projective algebras, and Theorem 10 embeds any projective algebra in a complete atomic projective algebra, it follows that any projective algebra can be embedded in a concrete projective algebra. □

5 Finitely Generated Algebras

In this section, we show how the duality theory developed above can be employed to solve some of Ulam's problems, namely those relating to finitely generated projective algebras. The matrices corresponding to the frames of the algebras are easy to visualize and manipulate. Earlier solutions to these problems are to be found in the work of Faber, Erdős and Larson [3, 8, 9], based on some more complicated constructions.

Theorem 13. *There are $2^{\aleph_0}$ non-isomorphic projective algebras generated by one element.*

Proof. The proof is an adaptation of a construction described by Henkin, Monk and Tarski [6, p. 253] showing that the two-dimensional cylindric set algebra of all relations on a finite set is generated by a single element. Bednarek and Ulam in [1] use a very similar idea to prove the slightly weaker result that the concrete algebra of all subsets of $S \times S$, where S is a non-empty finite set, can be generated by two elements.

Let X be a set of positive natural numbers. We show how to construct a frame $\mathcal{S}(X)$ corresponding to X in such a way that distinct sets give rise to non-isomorphic frames. We start from the concrete frame defined over the set $\omega \times \omega$, with $\mathbf{a} = \{\langle 0,0\rangle\}$, and then selectively add to certain sets (x, y) an extra element, so that the frame is composed of a collection of one-element and two-element sets; in the visual imagery employed earlier, each box in the frame contains either one or two elements. If $X \subseteq \omega \setminus \{0\}$ is a set of positive integers, then we add the extra elements in such a way that there is a column in the frame containing k 2-element boxes if and only if $k \in X$. More explicitly, let the boxes to which we add the extra elements be the first k boxes in column $2k$, $(2k, 1), (2k, 2), \ldots (2k, k-1), (2k, k)$, omitting $\langle 2k, 0\rangle$. Let a_{xy} be the original element in a box, and b_{xy} the added element, so that if $(x, y) = \{a_{xy}\}$ was the original box, then the box with the added element is $\{a_{xy}, b_{xy}\}$. We use the notation $[x, y]$ to denote the pairing operation on the frame $\mathcal{S}(X)$; that is, $[x, y]$ is the contents of the box with coordinates x, y.

Since the number of 2-element boxes in a column is preserved under isomorphism of frames, it follows that if $X \neq Y$, then $\mathcal{S}(X)$ and $\mathcal{S}(Y)$ are non-isomorphic. Figure 2 shows the construction for $X = \omega \setminus \{0\}$.

It is easy to see that a subalgebra of $\mathcal{CA}(\mathcal{S}(X))$ containing all of the unit sets uniquely determines $\mathcal{S}(X)$, and hence X, so that if $\mathfrak{A}_1$ and $\mathfrak{A}_2$ are subalgebras of $\mathcal{CA}(\mathcal{S}(X))$ and $\mathcal{CA}(\mathcal{S}(Y))$ satisfying this condition, then they cannot be isomorphic. Hence, it only remains to prove that all of the singletons in $\mathcal{S}(X)$ can be generated from a single element in $\mathcal{CA}(\mathcal{S}(X))$.

Let $A = \{a_{xy} : (x > y > 0) \vee (x \text{ even}, y = 0) \vee (x = 0, y \text{ even})\}$. The unfilled circles in Figure 2 represent the elements of A. We now need to argue that all singletons in the frame are in the subalgebra of $\mathcal{CA}(\mathcal{S}(X))$ generated by A. Let S be the universe of the frame (that is, $\omega \times \omega$ together with the added elements b_{xy}), and let $S_0 = P_0(S)$, $S_1 = P_1(S)$. Cylindrification operations C_0 and C_1 can be defined by setting $C_0(G) = P_0(G)\Box S_1$, and $C_1(G) = S_0 \Box P_1(G)$,

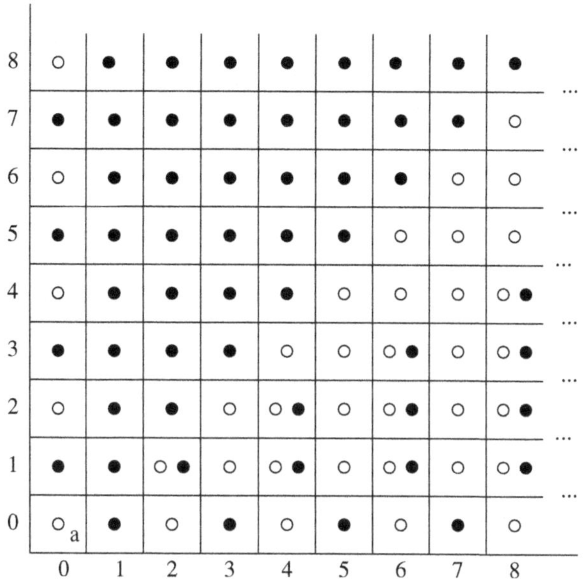

Fig. 2. The frame $\mathcal{S}(X)$ for $X = \omega \setminus \{0\}$

for $G \subseteq S$. It is easy to see that $C_0(G) = \bigcup\{[a,b] : \exists y(G \cap [a,y] \neq \emptyset)\}$, and $C_1(G) = \bigcup\{[a,b] : \exists x(G \cap [x,b] \neq \emptyset)\}$.

It is sufficient to show that all singletons corresponding to elements in S_0 and S_1 can be generated (that is to say, it is sufficient to generate all of the coordinates of points), for all other singletons can be generated by appropriate use of $\Box$ and intersection. We define the auxiliary sets $B = C_0(S_0 \cap A)$ and $C = C_1(S_1 \cap A)$; B is the set of all a_{xy} and b_{xy} with x even, while C is the set of all a_{xy} and b_{xy} with y even. Now define $X_1 = S \setminus S_1$, $Y_1 = S \setminus S_0$, $X_{k+1} = C_0(Y_k \cap A)$, $Y_{k+1} = C_1[((X_{k+1} \cap Y_k) \setminus (A \cup B)) \cup Z_k]$, where $Z_k = (Y_k \cap C)$ for k odd, and $Z_k = \emptyset$ for k even. Then we can prove by induction on k that $X_k = C_0\{a_{x0} : x \geq k\}$, and $Y_k = C_1\{a_{0y} : y \geq k\}$; the construction is easily completed using this sequence of sets. $\Box$

Theorem 13 provides an answer to the fourth of Ulam's questions about projective algebras, namely, "How many nonisomorphic projective algebras exist with k generators?" [12, p. 13] for $k > 0$. It also provides an answer to Ulam's second question: "Does there exist a universal countable projective algebra, i.e., a countable projective algebra such that every countable projective algebra is isomorphic to some subalgebra of it?" [12, p. 13].

Corollary 1. *There is no universal countable projective algebra.*

Proof. Suppose that such an algebra $\mathfrak{A}$ exists. Then there would be at most countably many non-isomorphic projective algebras with one generator, since every such algebra would be a finitely generated subalgebra of $\mathfrak{A}$. However, this contradicts Theorem 13. $\Box$

The duality theory can be employed in addition to provide an easy solution to Ulam's first problem, namely: "Given a countable class of sets in the plane, does there exist a finite number of sets which generate a projective algebra containing all sets of this countable class?" [12, p. 13].

To answer this question, we can carry out a construction similar to the one employed in Theorem 13. In the previous construction, each box in the frame contains either one or two elements. In the next construction, we again start with the concrete frame defined over the set $\omega \times \omega$, with $\mathbf{a} = \{\langle 0, 0 \rangle\}$, but in this case, although the number of elements in each box is finite, there is no upper bound on such numbers. To be more specific, let us add elements to the boxes (a, a) on the diagonals in such a way that for $a > 0$, the box $[a, a]$ in the new frame $\mathcal{S}$ contains a elements. Let $\mathcal{S}_\omega$ be the resulting frame. Further, let $\mathfrak{B}_\omega$ be the Boolean algebra of finite and co-finite subsets of $\mathcal{S}_\omega$, and $\mathfrak{A}_\omega$ the projective algebra generated by $\mathfrak{B}_\omega$

Theorem 14. *The projective algebra $\mathfrak{A}_\omega$ is not a subalgebra of any finitely generated algebra of subsets of the frame $\mathcal{S}_\omega$.*

Proof. Let $A_1, \ldots, A_k$ be a finite set of subsets of $\mathcal{S}_\omega$, and let $B = [a, b]$ be a given box in the frame $\mathcal{S}_\omega$. Define $\mathcal{F}_{ab}$ to be the family of all subsets of B of the form $B \cap C$, where C is in the Boolean algebra generated by $A_1, \ldots, A_k$. We claim that if D is any element of the projective algebra generated by $A_1, \ldots, A_k$, that $D \cap B$ is already in $\mathcal{F}_{ab}$.

The claim is proved by induction on the complexity of the set D. If D is in the Boolean algebra generated by $A_1, \ldots, A_k$, then it is true by definition. If D is of the form $P_\epsilon(E)$, then $B \cap D$ is either empty or equal to B, while if D is of the form $E \Box F$, then $B \cap D$ is again equal to B or the empty set.

Now since the Boolean algebra generated by $A_1, \ldots, A_k$ has at most 2^{2^k} elements, it follows that if there are more than 2^{2^k} elements in a box $[a, a]$, that there is at least one subset of $[a, a]$ in $\mathfrak{A}_\omega$ that is not in the projective algebra generated by $A_1, \ldots, A_k$. □

The algebra in Theorem 14 is not in fact a concrete algebra of subsets of the plane, as required by the statement of Ulam's problem. However, the representation provided by Theorem 5 is an embedding of $\mathfrak{A}_\omega$ into the concrete algebra of all subsets of the plane, so this easy construction answers Ulam's question.

References

1. A.R. Bednarek and S.M. Ulam. Generators for algebras of relations. *Bulletin of the American Mathematical Society*, 82:781–782, 1976.
2. Louise H. Chin and Alfred Tarski. Remarks on projective algebras. *Bulletin of the American Mathematical Society*, pages 80–81, 1948. Abstract.
3. Paul Erdős, Vance Faber, and Jean Larson. Sets of natural numbers of positive density and cylindric set algebras of dimension 2. *Algebra Universalis*, 12:81–92, 1981.

4. C.J. Everett and S. Ulam. Projective algebra I. *American Journal of Mathematics*, 68:77–88, 1946.
5. P.R. Halmos. *Lectures on Boolean Algebras.* Van Nostrand, Princeton, NJ, 1963. Van Nostrand Mathematical Studies #1.
6. L. Henkin, J.D. Monk, and A. Tarski. *Cylindric Algebras Part I.* North-Holland, Amsterdam, 1971. Studies in Logic and Foundations of Mathematics Volume 64.
7. L. Henkin, J.D. Monk, and A. Tarski. *Cylindric Algebras Part II.* North-Holland, Amsterdam, 1985. Studies in Logic and Foundations of Mathematics Volume 115.
8. Jean A. Larson. The number of one-generated diagonal-free cylindric set algebras of finite dimension greater than two. *Algebra Universalis*, 16:1–16, 1983.
9. Jean A. Larson. The number of one-generated cylindric set algebras of dimension greater than two. *Journal of Symbolic Logic*, 50:59–71, 1985.
10. J.C.C. McKinsey. On the representation of projective algebras. *American Journal of Mathematics*, 70:375–384, 1947.
11. S.M. Ulam and A.R. Bednarek. On the theory of relational structures and schemata for parallel computation. In *Analogies between Analogies: The mathematical reports of S.M. Ulam and his Los Alamos collaborators*, pages 477–508. University of California Press, 1990. Los Alamos Report LA-6734-MS, May 1977.
12. Stanislaw Ulam. *A Collection of Mathematical Problems.* John Wiley and Sons, 1960. Paperback reprint 1964 Wiley Science Editions (with an added preface) entitled *Problems in Modern Mathematics.*

Relational Approach to Boolean Logic Problems

Rudolf Berghammer and Ulf Milanese

Institut für Informatik und Praktische Mathematik,
Universität Kiel, Olshausenstraße 40, D-24098 Kiel

Abstract. We present a method for specifying and implementing algorithms for Boolean logic problems. It is formally grounded in relational algebra. Specifications are written in first-order set theory and then transformed systematically into relation-algebraic forms which can be executed directly in RelView, a computer system for the manipulation of relations and relational programming. Our method yields programs that are correct by construction. It is illustrated by some examples.

1 Introduction

For many years, relational algebra (see [15, 13]) has been used successfully for formal problem specification, prototyping, and program development. This is mainly due to the fact that many fundamental structures and datatypes of discrete mathematics and computer science can be modeled naturally by relations and, hence, many computations on them can be reduced to relation-algebraic computations. Boolean logic is also a fundamental system of discrete mathematics and computer science. Of course, here relations are used, for instance, the satisfiability relation $\models$ between assignments and formulae. But when having a look through relevant literature, *relational algebra* in the sense of [15, 13] does not seem to play a role. Especially in the case of algorithmic problems, binary decision diagrams (BDDs; see [3]) and so-called BDD-packages (like BuDDy [10] and CUDD [14]) are currently the standard tools.

We present a relational approach to Boolean logic problems and illustrate its usefulness by solving some algorithmic problems. The design of the algorithms starts from mathematical problem specifications, which are based on a relational model of Boolean formulae in conjunctive normal form (CNF). With the aid of rigorous calculations the specifications are transformed gradually into relation-algebraic descriptions which, essentially, are given by expressions of relational algebra. Finally, the latter are translated into the language of the computer system RelView (see [2]) and then can be executed directly. In this way programs are built up very quickly and their correctness is guaranteed by construction.

RelView also uses (reduced ordered) BDDs to implement relations in a very efficient way. Details are presented in [9, 11]. Hence, our programs can be seen as BDD-algorithms at a very high level. We found their development relatively easy. Based on the experience made during the development of the RelView system, however, we think that programming at the level of BDD-packages is too

W. MacCaull et al. (Eds.): RelMiCS 2005, LNCS 3929, pp. 48–59, 2006.

low-level to be effective. It is error prone and the rather complex BDD implementation, in particular the use of shared pointer structures with dynamically allocated and freed cells, makes these errors difficult to track down. Errors are often observable only after a long system operation time or under high load and so refrain from being detected by standard test scenarios.

The remainder of the paper is organized as follows. Section 2 collects some relational preliminaries. In Section 3 we introduce a relational model of CNF-formulae and develop algorithms for computing all satisfying assignments, solving the MAX-SAT problem, and testing the independence of CNF-formulae from variables. In Section 4 we explain how these algorithms can be implemented in the RELVIEW tool, present a small example, and report on the results of our practical experiments. Section 5 contains some concluding remarks.

2 Relational Preliminaries

We write $R : X \leftrightarrow Y$ if R is a relation with domain X and range Y, i.e., a subset of $X \times Y$. If the sets of R's *type* $X \leftrightarrow Y$ are finite, we may consider R as a Boolean matrix. Since this interpretation is well suited for many purposes, in the following we often use matrix terminology and matrix notation. Especially, we write R_{xy} instead of $(x, y) \in R$. We assume the reader to be familiar with the basic operations R^{T} (*transposition*), $\overline{R}$ (*complement*), $R \cup S$ (*union*), $R \cap S$ (*intersection*), RS (*composition*), $R \subseteq S$ (*inclusion*), and the special relations O (*empty relation*), L (*universal relation*), and I (*identity relation*).

There are some relation-algebraic possibilities to model sets. Our first modeling uses *vectors*, which are relations v with $v = v\mathsf{L}$. Since for a vector the range is irrelevant we consider mostly vectors $v : X \leftrightarrow \mathbf{1}$ with a specific singleton set $\mathbf{1} = \{\perp\}$ as range and omit in such cases the second subscript, i.e., write v_x instead of $v_{x\perp}$. Such a vector can be considered as a Boolean matrix with exactly one column, i.e., as a Boolean column vector, and *represents* the subset $\{x \in X \mid v_x\}$ of X. In [13] it is shown that if $R : X \leftrightarrow X$ is a quasi-order and $v : X \leftrightarrow \mathbf{1}$ represents a subset Y of X, then the set of *greatest elements* of Y with respect to R is represented by the vector $\mathit{GreEl}(R, v) := v \cap \overline{\overline{R}^{\mathsf{T}} v} : X \leftrightarrow \mathbf{1}$.

A vector is a *point* if it is non-empty and injective. This means that it represents a singleton subset of its domain or an element from it if we identify a singleton set with the only element it contains. In the matrix model a point $v : X \leftrightarrow \mathbf{1}$ is a Boolean column vector in which exactly one component is true.

We will also use injective mappings for modeling subsets. Given an injective mapping as relation $\imath : Y \leftrightarrow X$, we may consider Y as a subset of X by identifying it with its image under $\imath$. If Y is actually a subset of X and $\imath$ is the identity mapping from Y to X, then the vector $\imath^{\mathsf{T}}\mathsf{L} : X \leftrightarrow \mathbf{1}$ represents Y as subset of X in the sense above. Clearly, the transition in the other direction is also possible, i.e., the generation of an injective mapping $\mathit{inj}(v) : Y \leftrightarrow X$ fulfilling $\mathit{inj}(v)_{yx}$ iff $y = x$ from a given vector $v : X \leftrightarrow \mathbf{1}$ representing the subset Y of X.

As a third possibility to model subsets of a given set X we will use the set-theoretic *membership relation* $\varepsilon : X \leftrightarrow 2^X$, defined by ε_{xY} iff $x \in Y$. Such relations lead to a column-wise representation of sets of subsets. If $v : 2^X \leftrightarrow \mathbf{1}$ represents a subset $\mathfrak{S}$ of 2^X in the sense above, then for all $x \in X$ and $Y \in \mathfrak{S}$ we get the equivalence of $(\varepsilon\, inj(v)^{\mathsf{T}})_{xY}$ and $x \in Y$. Hence, the elements of $\mathfrak{S}$ are represented precisely by the columns of $\varepsilon\, inj(v)^{\mathsf{T}} : X \leftrightarrow \mathfrak{S}$. Generally, we say that a subset $\mathfrak{S}$ of 2^X is *represented by the columns* of $R : X \leftrightarrow \mathfrak{S}$ if R_{xY} is equivalent to $x \in Y$ for all $x \in X$ and $Y \in \mathfrak{S}$.

The relation $\varepsilon\, inj(v)^{\mathsf{T}}$ is a special case of *range restriction*. The general case is given by $R : X \leftrightarrow Y$ and a vector v with domain Y describing a subset Z of Y. Then the range restriction $R\, inj(v)^{\mathsf{T}} : X \leftrightarrow Z$ of R with respect to v restricts the range Y of R to the subset Z. Using component-wise notation this means that for all $x \in X$ and $z \in Z$ we have R_{xz} iff $(R\, inj(v)^{\mathsf{T}})_{xz}$.

Reduced ordered BDDs allow a very compact implementation of membership relations. In [9] an implementation of $\varepsilon : X \leftrightarrow 2^X$ is developed with the number of BDD-nodes in $\mathcal{O}(|X|)$. The same holds for the so-called *size-comparison relation* $S : 2^X \leftrightarrow 2^X$, which relates two sets $A, B \in 2^X$ iff $|A| \leq |B|$. Here $\mathcal{O}(|X|^2)$ BDD-nodes suffice for an implementation; see [11] for details.

3 Satisfiability and Related Problems

In this section we show how to model Boolean formulae in conjunctive normal form using relations. Based on this approach, we solve the satisfiability problem SAT, the MAX-SAT problem, and the independence problem for variables.

3.1 A Relational Model for CNF-Formulae

Let $\mathcal{V}$ be a set of Boolean variables. A *literal* over $\mathcal{V}$ is either a variable $x \in \mathcal{V}$ or the negation $\neg x$ of a variable $x \in \mathcal{V}$, and a *clause* over $\mathcal{V}$ is a disjunction $\lambda_1 \vee \ldots \vee \lambda_m$ of $m \geq 1$ literals. In the following we only consider Boolean formulae φ over $\mathcal{V}$ in CNF. This means that φ is of the form $\gamma_1 \wedge \ldots \wedge \gamma_n$ for some $n \geq 1$, where each γ_i, $1 \leq i \leq n$, is a clause over $\mathcal{V}$.

Now, assume $\mathcal{K}^\varphi$ to be the clause set of the CNF-formula φ. Then φ can be modeled by a pair $P^\varphi : \mathcal{V} \leftrightarrow \mathcal{K}^\varphi$ and $N^\varphi : \mathcal{V} \leftrightarrow \mathcal{K}^\varphi$ of relations such that

$$P^\varphi_{x\gamma} \;:\Longleftrightarrow\; x \text{ is a literal of } \gamma \qquad\qquad N^\varphi_{x\gamma} \;:\Longleftrightarrow\; \neg x \text{ is a literal of } \gamma\,.$$

To enhance readability, in the following we omit the superscript φ and write $\mathcal{K}$, P, and N instead of $\mathcal{K}^\varphi$, P^φ, and N^φ.

For illustrative purposes we consider a simple example, taken from [7]. Suppose φ to be the following CNF-formula over the set $\mathcal{V} := \{a, b, c, d\}$.

$$\begin{gathered}(a \vee \neg b) \wedge (a \vee c \vee \neg d) \wedge (b \vee \neg a) \wedge (a \vee d \vee \neg c) \\ \wedge\ (b \vee c \vee \neg d) \wedge (a \vee \neg c) \wedge c \wedge (a \vee d) \wedge (\neg a \vee \neg c) \wedge a\end{gathered} \tag{1}$$

In RELVIEW relations can be depicted as Boolean matrices, where it is additionally possible to label rows and columns for explanatory purposes. Using this feature, RELVIEW displays the relations P and N of φ as subsequently shown.

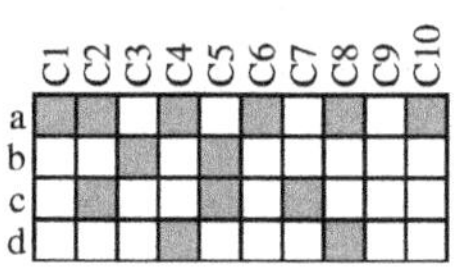

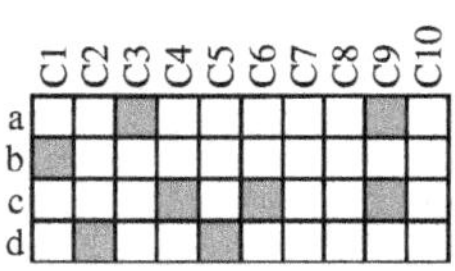

To simplify presentation, in the sequel we always suppose that a clause neither contains a variable and its negation as literals nor contains duplicates of literals. These restrictions are harmless. Under this point of view our modeling consist of two steps. First, we go to a so-called *clause normal form* by neglecting associativity, commutativity, and idempotency of conjunction and disjunction. It represents a CNF-formula as set of clauses, where each clause is divided into the set of positive literals (variables) and negative literals (negated variables). After that, we describe this splitting of the clauses by the relations P and N.

3.2 Computing Satisfying Assignments

Usually, an assignment is defined as a function $a : \mathcal{V} \to \mathbb{B}$ that assigns a Boolean value *tt* of *ff* to every variable. Since there is a 1-1-correspondence between a and the set $\{x \in \mathcal{V} \mid a(x) = tt\}$, in the following we define assignments to be subsets of $\mathcal{V}$. Then an assignment $A \in 2^{\mathcal{V}}$ *satisfies* a variable $x \in \mathcal{V}$, denoted by $A \models x$, iff $x \in A$. The other cases of the relation $\models$ are defined as usual: $A \models \neg\varphi$ iff the relationship $A \models \varphi$ does not hold, $A \models \varphi \vee \psi$ iff $A \models \varphi$ or $A \models \psi$, and $A \models \varphi \wedge \psi$ iff $A \models \varphi$ and $A \models \psi$.

Testing satisfiability of CNF-formulae is one of the paradigmatic NP-complete problems and many algorithms have been developed for it (see e.g., [5, 12] for an overview). For clauses there exists a simple syntactic criterion, viz.:

Theorem 3.1. *Let $A \in 2^{\mathcal{V}}$ be an assignment and γ be a clause over $\mathcal{V}$. Then A satisfies γ iff a literal of γ is a variable from A or the negation of a variable from $\mathcal{V} \setminus A$.*

Proof. Assume the clause γ to be of the form $\lambda_1 \vee \ldots \vee \lambda_m$ for some $m \geq 1$.

For proving "$\Rightarrow$", suppose $A \models \gamma$. Then there exists i, $1 \leq i \leq m$, such that $A \models \lambda_i$. The literal λ_i is either a variable x or a negation $\neg x$ of a variable x. If λ_i equals x, then $x \in A$ follows from $A \models x$, and if λ_i equals $\neg x$, then $x \notin A$ follows from $A \models \neg x$, since the latter is equivalent to $A \not\models x$.

To prove "$\Leftarrow$", assume a variable x with $x \in A$. If there exists i, $1 \leq i \leq m$, such that λ_i equals x, then $x \in A$ shows $A \models \lambda_i$, which in turn implies $A \models \gamma$. Similarly one treats the case of λ_i being of the form $\neg x$ with $x \notin A$. □

Assume the CNF-formula φ over $\mathcal{V}$ to be modeled by the relations $P : \mathcal{V} \leftrightarrow \mathcal{K}$ and $N : \mathcal{V} \leftrightarrow \mathcal{K}$ as introduced in Section 3.1. We want to compute a vector of type $2^{\mathcal{V}} \leftrightarrow \mathbf{1}$ representing the satisfying assignments $\mathfrak{S} := \{A \in 2^{\mathcal{V}} \mid A \models \varphi\}$ as subset of $2^{\mathcal{V}}$ and, after that, a column-wise representation of $\mathfrak{S}$ as relation of type $\mathcal{V} \leftrightarrow \mathfrak{S}$. To obtain these goals, we take an assignment $A \in 2^{\mathcal{V}}$ and calculate as follows, where γ and x range over $\mathcal{K}$ and $\mathcal{V}$, respectively, the membership relation ε is of type $\mathcal{V} \leftrightarrow 2^{\mathcal{V}}$, and the universal vector L is of type $\mathcal{K} \leftrightarrow \mathbf{1}$. In the first step the calculation uses the definition of satisfiability, in the second step Theorem 3.1 in combination with the definition of the two relations P and N, and in the third step the definition of the membership relation; the remaining steps apply predicate logic and some well-known correspondences between certain kinds of logical and relation-algebraic constructions.

$$\begin{aligned}
A \models \varphi &\iff \forall \gamma : A \models \gamma \\
&\iff \forall \gamma : (\exists x : x \in A \wedge P_{x\gamma}) \vee (\exists x : x \notin A \wedge N_{x\gamma}) \\
&\iff \forall \gamma : (\exists x : \varepsilon_{xA} \wedge P_{x\gamma}) \vee (\exists x : \overline{\varepsilon}_{xA} \wedge N_{x\gamma}) \\
&\iff \forall \gamma : (P^{\mathsf{T}}\varepsilon)_{\gamma A} \vee (N^{\mathsf{T}}\overline{\varepsilon})_{\gamma A} \\
&\iff \forall \gamma : (P^{\mathsf{T}}\varepsilon \cup N^{\mathsf{T}}\overline{\varepsilon})_{\gamma A} \\
&\iff \neg\exists \gamma : \neg(P^{\mathsf{T}}\varepsilon \cup N^{\mathsf{T}}\overline{\varepsilon})_{\gamma A} \\
&\iff \neg\exists \gamma : \overline{P^{\mathsf{T}}\varepsilon \cup N^{\mathsf{T}}\overline{\varepsilon}}^{\mathsf{T}}_{A\gamma} \wedge \mathsf{L}_{\gamma} \\
&\iff \overline{\overline{P^{\mathsf{T}}\varepsilon \cup N^{\mathsf{T}}\overline{\varepsilon}}^{\mathsf{T}}\mathsf{L}}_{A}
\end{aligned}$$

If we remove the subscript A from the last expression of this calculation following the vector representation of sets introduced in Section 2 and apply after that some well-known relation-algebraic rules to transpose[1] only a "row vector" instead of a relation of type $\mathcal{K} \leftrightarrow 2^{\mathcal{V}}$, we get the relation-algebraic description

$$SatVect(P, N) := \overline{\mathsf{L}^{\mathsf{T}}\,\overline{P^{\mathsf{T}}\varepsilon \cup N^{\mathsf{T}}\overline{\varepsilon}}}^{\mathsf{T}} : 2^{\mathcal{V}} \leftrightarrow \mathbf{1} \tag{2}$$

of the vector representing the subset $\mathfrak{S}$ of $2^{\mathcal{V}}$. Hence, the CNF-formula φ is satisfiable iff $SatVect(P, N) \neq \mathsf{O}$. In this case the column-wise representation

$$SatSet(P, N) := \varepsilon\, inj(SatVect(P, N))^{\mathsf{T}} : \mathcal{V} \leftrightarrow \mathfrak{S} \tag{3}$$

of $\mathfrak{S}$ is an immediate consequence of the technique shown in Section 2.

3.3 Solving the Max-Sat Problem

Given again a CNF-formula φ over the variables $\mathcal{V}$, the objective of Max-Sat is to find an assignment that satisfies the maximal number of clauses of φ. A classical approach to solve this NP-hard problem is branch and bound; see e.g., [7]. In the following we show how to solve Max-Sat using relational algebra. Again we assume φ to be modeled by the relations $P : \mathcal{V} \leftrightarrow \mathcal{K}$ and $N : \mathcal{V} \leftrightarrow \mathcal{K}$.

Suppose $\varepsilon : \mathcal{V} \leftrightarrow 2^{\mathcal{V}}$ to be the membership relation with respect to variables. In Section 3.2 we have already shown that the relationships $(P^{\mathsf{T}}\varepsilon \cup N^{\mathsf{T}}\overline{\varepsilon})_{\gamma A}$ and $A \models \gamma$ are equivalent. Now, let in addition $\epsilon : \mathcal{K} \leftrightarrow 2^{\mathcal{K}}$ be the membership relation with respect to the clauses of φ. Then we are able to calculate for each subset C of the set $\mathcal{K}$ as follows, where A ranges over $2^{\mathcal{V}}$, γ ranges over $\mathcal{K}$, and L is the universal vector of type $2^{\mathcal{V}} \leftrightarrow \mathbf{1}$. The second step of the calculation uses the definition of ϵ and the equivalence just mentioned and the remaining steps

[1] Using a BDD-implementation, transposition of a relation with domain or range $\mathbf{1}$ only means to exchange domain and range, the BDD remains unchanged; see [11].

again base on predicate logic and certain correspondences between logical and relation-algebraic constructions.

$$
\begin{aligned}
C \text{ is satisfiable} &\iff \exists A : \forall \gamma : \gamma \in C \rightarrow A \models \gamma \\
&\iff \exists A : \forall \gamma : \epsilon_{\gamma C} \rightarrow (P^{\mathsf{T}}\varepsilon \cup N^{\mathsf{T}}\,\overline{\varepsilon})_{\gamma A} \\
&\iff \exists A : \neg \exists \gamma : \epsilon_{\gamma C} \wedge \neg (P^{\mathsf{T}}\varepsilon \cup N^{\mathsf{T}}\,\overline{\varepsilon})_{\gamma A} \\
&\iff \exists A : \neg \exists \gamma : \epsilon^{\mathsf{T}}_{C\gamma} \wedge \overline{P^{\mathsf{T}}\varepsilon \cup N^{\mathsf{T}}\,\overline{\varepsilon}}_{\gamma A} \\
&\iff \exists A : \overline{\epsilon^{\mathsf{T}}\,\overline{P^{\mathsf{T}}\varepsilon \cup N^{\mathsf{T}}\,\overline{\varepsilon}}}_{CA} \wedge \mathsf{L}_A \\
&\iff (\overline{\epsilon^{\mathsf{T}}\,\overline{P^{\mathsf{T}}\varepsilon \cup N^{\mathsf{T}}\,\overline{\varepsilon}}\;\mathsf{L}})_C
\end{aligned}
$$

If we remove the subscript C from the last expression of this calculation and apply after that, as in Section 3.2, some simple transposition rules to improve efficiency in view of a BDD-implementation of relations, this leads to $(\mathsf{L}^{\mathsf{T}}\,\overline{\overline{\varepsilon^{\mathsf{T}}P \cup \overline{\varepsilon}^{\mathsf{T}}N\,\epsilon}})^{\mathsf{T}} : 2^{\mathcal{K}} \leftrightarrow \mathbf{1}$ as a relation-algebraic description of the vector representing the set of those clause sets which are satisfiable. Next, we use the size-comparison relation $S : 2^{\mathcal{K}} \leftrightarrow 2^{\mathcal{K}}$, which relates two clause sets C_1 and C_2 iff $|C_1| \leq |C_2|$. It is obvious how to combine S with the relation-algebraic description of the satisfiable clause sets and the relation-algebraic description of the greatest elements in order to compute the vector representation of the subset $\mathfrak{C}$ of $2^{\mathcal{K}}$ containing the largest satisfiable clause sets. Here is the result.

$$
\mathit{MaxSatVectCl}(P,N) := \mathit{GreEl}(S, (\mathsf{L}^{\mathsf{T}}\,\overline{\overline{\varepsilon^{\mathsf{T}}P \cup \overline{\varepsilon}^{\mathsf{T}}N\,\epsilon}})^{\mathsf{T}}) : 2^{\mathcal{K}} \leftrightarrow \mathbf{1} \qquad (4)
$$

Now, we are almost done. To find an assignment that satisfies the maximal number of clauses of the CNF-formula φ we select a point $p : 2^{\mathcal{K}} \leftrightarrow \mathbf{1}$ contained in the vector $\mathit{MaxSatVectCl}(P,N)$. It represents a set C from $\mathfrak{C}$ as an element of $2^{\mathcal{K}}$. The little calculation

$$
(\epsilon p)_\gamma \iff \exists K : \epsilon_{\gamma K} \wedge p_K \iff \exists K : \gamma \in K \wedge K = C \iff \gamma \in C
$$

for all clauses $\gamma \in \mathcal{K}$ shows that the same set C is represented as a subset of the clause set $\mathcal{K}$ by the vector $\epsilon p : \mathcal{K} \leftrightarrow \mathbf{1}$. Hence, the restrictions $P\,\mathit{inj}(\epsilon p)^{\mathsf{T}} : \mathcal{V} \leftrightarrow C$ and $N\,\mathit{inj}(\epsilon p)^{\mathsf{T}} : \mathcal{V} \leftrightarrow C$ of $P : \mathcal{V} \leftrightarrow \mathcal{K}$ and $N : \mathcal{V} \leftrightarrow \mathcal{K}$, respectively, to the range $C \subseteq \mathcal{K}$ model the sub-formula ψ of φ which consists of the conjunction of the clauses of C. Consequently, the following vector represents the assignments satisfying ψ, i.e., solutions of MAX-SAT.

$$
\mathit{MaxSatVect}(P,N) := \mathit{SatVect}(P\,\mathit{inj}(\epsilon p)^{\mathsf{T}}, N\,\mathit{inj}(\epsilon p)^{\mathsf{T}}) : 2^{\mathcal{V}} \leftrightarrow \mathbf{1} \qquad (5)
$$

For the column-wise enumeration of the assignments satisfying ψ, in (5) only the relational function *SatVect* has to be replaced by *SatSet*.

3.4 Recognizing Independent Variables

Let φ be a Boolean formula over $\mathcal{V}$ and $x \in \mathcal{V}$ be a variable. Then φ is said to be *independent* of x if for all assignments $A \in 2^{\mathcal{V}}$ the relationships $A \cup \{x\} \models \varphi$ and

$A \setminus \{x\} \models \varphi$ are equivalent. Testing this property (or enumerating all variables with this property) is an important problem of Boolean logic and the contents of this section is its relation-algebraic solution. As in the previous sections, we assume φ to be given in CNF $\gamma_1 \wedge \ldots \wedge \gamma_n$.

We want to reduce the independence problem of the CNF-formula φ with respect to the variable x to satisfiability. For that reason we consider the formula $\varphi^x := \gamma_1^x \wedge \ldots \wedge \gamma_n^x$, which results from φ by removing from its clauses all literals x and $\neg x$. If a clause γ_i of φ is of the form x or $\neg x$, i.e., it disappears during this process, then the corresponding formula γ_i^x is defined as Boolean constant *false*. A formal way to obtain this is to represent *false* as $y \wedge \neg y$, where y is the next fresh variable. Then also φ^x is a formula over $\mathcal{V}$ in CNF, but the number of clauses of φ and φ^x may be different. The constant *false* is only introduced to obtain a 1-1 correspondence between the clauses of φ and the conjuncts of φ^x, which simplifies presentation. After these preparations, we are in a position to prove the following decisive fact.

Theorem 3.2. *The CNF-formula φ over $\mathcal{V}$ is independent of the variable $x \in \mathcal{V}$ iff the sets $\{A \in 2^{\mathcal{V}} \mid A \models \varphi^x\}$ and $\{A \in 2^{\mathcal{V}} \mid A \models \varphi\}$ are equal.*

Proof. In the first part of the proof of "$\Rightarrow$" we assume $A \in 2^{\mathcal{V}}$ to be an assignment with $A \models \varphi^x$. Then φ does not contain clauses of the form x or $\neg x$ and, as a consequence, each of its clauses γ_i, $1 \leq i \leq n$, is obtained from the corresponding clause γ_i^x of φ^x by adding x or $\neg x$ via a disjunction. This implies $A \models \varphi$ and, summing up, we have shown $\{A \in 2^{\mathcal{V}} \mid A \models \varphi^x\} \subseteq \{A \in 2^{\mathcal{V}} \mid A \models \varphi\}$.

In the second part, we assume that this is a proper inclusion and derive a contradiction. Therefore, let an assignment $A \in 2^{\mathcal{V}}$ be given such that $A \not\models \varphi^x$ and $A \models \varphi$. The first property implies $A \not\models \gamma_i^x$ for some i, $1 \leq i \leq n$, but $A \models \gamma_i$ holds due to the second property. Hence, γ_i^x and γ_i must be different.

Let γ_i be of the form $\lambda_1 \vee \ldots \vee \lambda_m$, where $m \geq 2$. Since clauses do not contain duplicates of literals and also the literals x and $\neg x$ cannot occur simultaneously in clauses, γ_i^x equals $\lambda_1 \vee \ldots \vee \lambda_{k-1} \vee \lambda_{k+1} \vee \ldots \vee \lambda_m$ for some k, $1 \leq k \leq m$. From $A \not\models \gamma_i^x$ we get $A \not\models \lambda_j$ for all $j \neq k$, which in turn implies $A \models \lambda_k$ due to $A \models \gamma_i$. Now, we distinguish two cases: If λ_k equals x, then $A \models \lambda_k$ is equivalent to $x \in A$ and the assumption $A \models \varphi$ becomes $A \cup \{x\} \models \varphi$. But for all $j \neq k$ the variable x does not occur in λ_j. Hence, for these literals $A \not\models \lambda_j$ is equivalent to $A \setminus \{x\} \not\models \lambda_j$ due to the coincidence lemma of logic. The property $A \setminus \{x\} \not\models \lambda_k$ is a consequence of $x \notin A \setminus \{x\}$. Altogether, we have $A \setminus \{x\} \not\models \lambda_j$ for all j, $1 \leq j \leq m$. As a consequence, $A \setminus \{x\} \not\models \gamma_i$ holds, which yields $A \setminus \{x\} \not\models \varphi$. This contradicts the assumption of φ being independent of x. If the literal λ_k equals $\neg x$, the proof is similar to the previous one.

It remains to consider the case of γ_i being a single literal. If γ_i is x, then $A \models \gamma_i$ yields $x \in A$ and the assumption $A \models \varphi$ becomes $A \cup \{x\} \models \varphi$. Now, the contradiction $A \setminus \{x\} \not\models \varphi$ follows from $x \notin A \setminus \{x\}$ and its consequence $A \setminus \{x\} \not\models \gamma_i$. In the same way one deals with the literal $\neg x$.

A proof of direction "$\Leftarrow$" is rather simple. Assume $A \in 2^{\mathcal{V}}$ to be an assignment. Then we have $A \cup \{x\} \models \varphi$ iff $A \setminus \{x\} \models \varphi$ due to $\{A \in 2^{\mathcal{V}} \mid A \models \varphi^x\} =$

$\{A \in 2^{\mathcal{V}} \mid A \models \varphi\}$ and the fact that x does not occur in φ^x (cf. coincidence lemma of logic). Hence, φ is independent of x. □

Now, let $P : \mathcal{V} \leftrightarrow \mathcal{K}$ and $N : \mathcal{V} \leftrightarrow \mathcal{K}$ be the relational model of φ. Furthermore, suppose the point $p : \mathcal{V} \leftrightarrow \mathbf{1}$ to represent the variable $x \in \mathcal{V}$. Guided by the definition of φ^x and Theorem 3.2, we first compute the vector representation of the set of those clauses which are of the form x or $\neg x$. So, assume $\gamma \in \mathcal{K}$. Then we can calculate as follows, where y ranges over $\mathcal{V}$ and the second step exploits that the relational complement $\overline{p}$ represents the set complement $\mathcal{V} \setminus \{x\}$.

$$\begin{aligned} \gamma \text{ is of the form } x \text{ or } \neg x &\iff \neg\exists y : (P_{y\gamma} \vee N_{y\gamma}) \wedge y \neq x \\ &\iff \neg\exists y : (P^{\mathsf{T}}_{\gamma y} \vee N^{\mathsf{T}}_{\gamma y}) \wedge \overline{p}_y \\ &\iff \overline{(P \cup N)^{\mathsf{T}}\,\overline{p}}_{\gamma} \end{aligned}$$

Hence, $\overline{(P \cup N)^{\mathsf{T}}\,\overline{p}} : \mathcal{K} \leftrightarrow \mathbf{1}$ is the vector representation of the set of all clauses which are equal to x or to $\neg x$. If this vector is not empty, then a conjunct of φ^x is *false* and Theorem 3.2 implies the following fact.

$$\varphi \text{ independent of } x \iff \mathit{SatVect}(P, N) = \mathsf{O} \tag{6}$$

In the case $\overline{(P \cup N)^{\mathsf{T}}\,\overline{p}} = \mathsf{O}$ the relational model $P^x : \mathcal{V} \leftrightarrow \mathcal{K}$ and $N^x : \mathcal{V} \leftrightarrow \mathcal{K}$ of the formula φ^x is obtained by deleting from P and N all pairs (x, γ), where $\gamma \in \mathcal{K}$. Relation-algebraically this means that P^x equals $P \cap \overline{p\mathsf{L}}$ and N^x equals $N \cap \overline{p\mathsf{L}}$, where $\mathsf{L} : \mathbf{1} \leftrightarrow \mathcal{K}$, and the characterization

$$\varphi \text{ independent of } x \iff \mathit{SatVect}(P, N) = \mathit{SatVect}(P \cap \overline{p\mathsf{L}}, N \cap \overline{p\mathsf{L}}) \tag{7}$$

is again a consequence of Theorem 3.2. Note that in the case $\overline{(P \cup N)^{\mathsf{T}}\,\overline{p}} = \mathsf{O}$ neither $P \cap \overline{p\mathsf{L}}$ nor $N \cap \overline{p\mathsf{L}}$ contains empty columns, because of the general assumption that no clause contains a variable and its negation as literals.

It should be remarked that, if there are clauses in φ which contain a variable as well as its negation as literals, their removal is very easy to model with relation-algebraic means, too. We only have to restrict the range of P and N with respect to the vector $(P \cap N)^{\mathsf{T}}\mathsf{L} : \mathcal{K} \leftrightarrow \mathbf{1}$ since the equivalence of $((P \cap N)^{\mathsf{T}}\mathsf{L})_\gamma$ and $\exists x : P_{x\gamma} \wedge N_{x\gamma}$ for all $\gamma \in \mathcal{K}$ proves that this vector represents the set of clauses which contain a variable and its negation as literals.

4 Implementation and Experimental Results

In the case of finite carrier sets the operations of relational algebra can be implemented very efficiently. At Kiel University we have developed a visual computer system for the manipulation of relations and relational programming, called RelView. It is written in C, uses – as already mentioned – BDDs for representing relations, and makes full use of the X-windows graphical user interface.

The main purpose of RelView is the evaluation of relation-algebraic expressions which are constructed from the relations of its workspace using pre-defined operations and tests, user-defined relational functions, and user-defined relational programs. Relational functions are of the form $F(X_1, \ldots, X_n) = t$, where F is the function name, the X_i, $1 \leq i \leq n$, are the formal parameters (standing for relations), and t is a relation-algebraic expression over the relations of the system's workspace that can additionally contain the formal parameters. A relational program is much like a function procedure in the programming languages Pascal or Modula 2, except that it only uses relations as data type.

For example, the relation-algebraic description (2) of the vector representing all satisfying assignments immediately leads to the following RELVIEW-program.

```
SatVect(P,N)
  DECL E
  BEG  E = epsi(O(P))
       RETURN -(L1n(P) * -(P^ * E | N^ * -E))^
  END.
```

A translation of the remaining relation-algebraic descriptions of Section 3 into RELVIEW-code is also straightforward.

Now, we return to the example of Section 3.1 and assume the two relations P and N for modeling the formula (1) to be stored in RELVIEW's workspace under the names `P` and `N`. Then the call `SatVect(P,N)` returns an empty Boolean vector of length 2^4. This means that the formula φ is not satisfiable.

To solve the MAX-SAT problem for the formula φ of (1), we can use again RELVIEW and compute the column-wise representation of the largest satisfiable clause sets and the assignments satisfying these clauses. The corresponding labeled Boolean matrices look as in the following pictures. From the Boolean matrix on the left we obtain that all clauses except c (the 7th one) or all clauses except $\neg a \vee \neg c$ (the 9th one) are satisfiable. The columns of the matrix in the middle represent the assignments $\{a, b\}$ and $\{a, b, d\}$ which satisfy all clauses except c, and the columns of the matrix on the right represent the assignments $\{a, b, c\}$ and $\{a, b, c, d\}$ which satisfy all clauses except $\neg a \vee \neg c$.

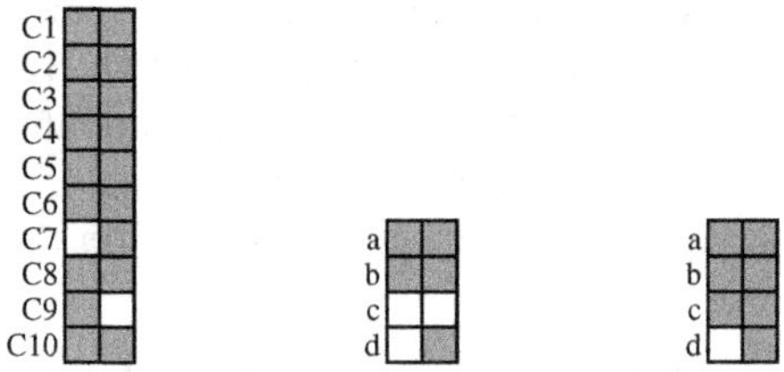

Besides this small example we have tested our approach with many further and larger examples. The tests have been carried out on a Sun Fire-880 workstation running Solaris 9 at 750 MHz and with 32 GByte main memory. From this memory, however, at most 4 GByte has actually been in use by RELVIEW.

Using RELVIEW's operations for random relations and random permutations (see [11]), we have generated a lot of uniformly distributed random instances of k-SAT. The following table shows some of the experimental results for 3-SAT with 50 variables. As the first column indicates, we only looked at formulae where the number of clauses to variable ratio equals 3.0 to 5.4. It is known (see e.g., [4]) that for random instances of k-SAT within a region around the so-called *critical value* α_k (where $\alpha_3 \approx 4.2$) there is a sharp transition from satisfiabilty to unsatisfiability which makes problems particularly hard for SAT-solvers.

	satisfying assignments			time (sec.)		
ratio	min	mean	max	min	mean	max
3.0	453615	12392518	53870974	99	280	593
3.6	1556	51766	233759	169	509	1451
4.2	0	5032	98009	375	720	1998
4.8	0	0	14	369	843	1501
5.4	0	0	0	280	817	1854

To be precise, 49 of the 50 instances with ratio 4.8 we have investigated proved to be unsatisfiable. In the remaining case we obtained 14 satisfying assignments.

Of course, in general RelView cannot compete in efficiency with special purpose tools for the problems we dealt with (although the complexities are usually the same). Nevertheless, our tests showed that in many cases our approach does not need too much time compared with other tools. Sometimes it is even superior. E.g., in the case of the 24 DIMACS aim-50 instances of www.satlib.org [6] (generated with the method of [1], 50 variables, between 80 and 300 clauses) the computation times of RelView are between 0.03 to 2.63 seconds, with 0.5 seconds as arithmetic mean. Comparing this with the corresponding times of the 25 SAT-solvers of the rank list of www.satlib.org for aim-50, RelView is between the numbers 9 (posit, 0.42 seconds) and 10 (asat, 0.60 seconds).

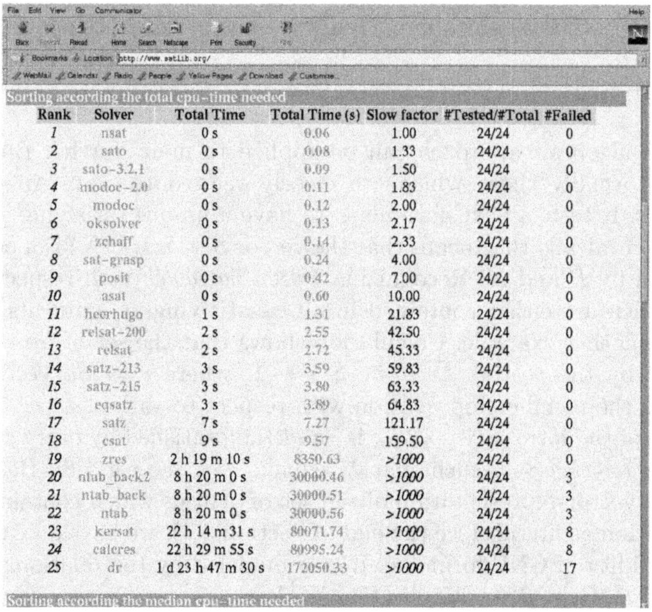
Sorting according the total cpu-time needed

Rank	Solver	Total Time	Total Time (s)	Slow factor	#Tested/#Total	#Failed
1	nsat	0 s	0.06	1.00	24/24	0
2	sato	0 s	0.08	1.33	24/24	0
3	sato-3.2.1	0 s	0.09	1.50	24/24	0
4	modoc-2.0	0 s	0.11	1.83	24/24	0
5	modoc	0 s	0.12	2.00	24/24	0
6	oksolver	0 s	0.13	2.17	24/24	0
7	zchaff	0 s	0.14	2.33	24/24	0
8	sat-grasp	0 s	0.24	4.00	24/24	0
9	posit	0 s	0.42	7.00	24/24	0
10	asat	0 s	0.60	10.00	24/24	0
11	heerhugo	1 s	1.31	21.83	24/24	0
12	relsat-200	2 s	2.55	42.50	24/24	0
13	relsat	2 s	2.72	45.33	24/24	0
14	satz-213	3 s	3.59	59.83	24/24	0
15	satz-215	3 s	3.80	63.33	24/24	0
16	eqsatz	3 s	3.89	64.83	24/24	0
17	satz	7 s	7.27	121.17	24/24	0
18	csat	9 s	9.57	159.50	24/24	0
19	zres	2 h 19 m 10 s	8350.63	>1000	24/24	0
20	ntab_back2	8 h 20 m 0 s	30000.46	>1000	24/24	3
21	ntab_back	8 h 20 m 0 s	30000.51	>1000	24/24	3
22	ntab	8 h 20 m 0 s	30000.56	>1000	24/24	3
23	kersat	22 h 14 m 31 s	80071.74	>1000	24/24	8
24	calcres	22 h 29 m 55 s	80995.24	>1000	24/24	8
25	dr	1 d 23 h 47 m 30 s	172050.33	>1000	24/24	17

Sorting according the median cpu-time needed

The aim-50 rank list has been established by means of a Pentium II PC running Linux at 400 MHz and with 512 MByte main memory. In practice, the computation times of RelView on a Pentium II PC and a Sun Fire-800 workstation are rather equal. But even if we assume that RelView runs on the PC half as fast than on the workstation, the tool is still faster than number 11 of the rank list (heerhugo, 1.31 seconds).

We also have applied the program `SatVect` to many other benchmarks presented at www.satlib.org. Here we used RelView in combination with a small C program for converting the DIMACS cnf-format into the ASCII file-format of RelView, but also Kure, a C library which has been developed in the course of the Ph.D. thesis [11] and consists of the functional core of RelView. For all 1000 uniform random 3-Sat instances with 20 variables and 91 clauses (critical value instances uf20-91) both systems computed the vectors describing the satisfying assignments in less than 0.06 seconds (arithmetic mean 0.037 seconds). The computation times of the 1000 uniform random 3-Sat instances with 50 variables and 218 clauses (critical value instances uf50-218 and uuf50-218) ranged from 77 to 3509 seconds; as arithmetic mean of all times we obtained 766 seconds.

We believe that the reason for the good behaviour of RelView when solving Sat is that during a run of `SatVect(P,N)` both input relations `P` and `N` are represented by relatively small BDDs, which have at most $|\mathcal{K}| * |\mathcal{V}|$ nodes. The membership relation is represented by another small BDD, complement and union of relations are not time consuming, and the same holds for the transposition of small relations. Therefore, we can only get large BDDs, which represent interim results, during the computation of the composition with the relation `L` in the `RETURN`-expression of `SatVect`. We found in experiments that our method is able to compete with Sat-solvers especially if the given formula is satisfiable. In the other case, Sat-solvers often stop the computation early with the result "unsatisfiable", whereas our approach always has to get to the end.

5 Conclusion

Our relation-algebraic approach can be applied to many further Boolean logic problems, especially those which are closely related to Sat. An example is Unique-Sat. It tests a CNF-formula φ to have a unique satisfying assignment. Relation-algebraically this means that the vector $SatVect(P, N)$ is a point, where φ is modeled by P and N. Recognizing *frozen variables* with respect to φ, i.e., variables which are either contained in all φ-satisfying assignments or in none of them, is another example. Calculation shows that the set of frozen variables is described by the vector $\overline{\overline{\varepsilon} v} \cap \overline{\varepsilon v} : \mathcal{V} \leftrightarrow \mathbf{1}$, where $v = SatVect(P, N)$ and $\varepsilon : \mathcal{V} \leftrightarrow 2^{\mathcal{V}}$ is the membership relation with respect to variables.

A clause of the form $\lambda_1 \vee \cdots \vee \lambda_n$ is *valid* (i.e., satisfied by every assignment) iff there are $1 \leq i, j \leq n$ such that λ_i equals $\neg\lambda_j$; see e.g., [8]. Based on this fact and the vector-representation of the set of clauses which contain a variable and its negation as literals (see the end of Section 3.4), we have an easy and fast check for validity of CNF-formulae. If φ is modeled by the relations $P : \mathcal{V} \leftrightarrow \mathcal{K}$ and $N : \mathcal{V} \leftrightarrow \mathcal{K}$, then φ is valid iff $\overline{(P \cup N)}^{\mathsf{T}} \mathsf{L} = \mathsf{L}$.

Presently, we investigate how to generalize our approach to arbitrary Boolean formulae. Using matrix terminology, this is based on the fact that all possible results $\{f(a_1,\ldots,a_n) \mid a_1,\ldots,a_n \in \mathbb{B}\}$ of an n-ary Boolean function $f : \mathbb{B}^n \to \mathbb{B}$ can be computed "in parallel" by applying a corresponding relational function to the n rows of the $n \times 2^n$ Boolean "membership matrix". Also, an interesting direction for future research is to find out how powerful our approach is in the case of practical applications, if these are reduced to Boolean logic problems as, for instance, the factorization of large numbers, asynchronous circuit synthesis, and SAT-based model checking. Besides this better insight into the weaknesses and strengths of our approach, we are also interested on further empirical comparisons with other SAT-solvers.

References

1. Asahiro Y, Iwama K, Miyano E: Random generation of test instances with controlled attributes. In: Johnson D.S., Trick M.A. (eds.): Cliques, Coloring, and Satisfiability: The Second DIMACS Implementation Challenge. DIMACS Series on Discr. Math. and Theoret. Comput. Sci. 26, 377-394 (1996)
2. Behnke R. et al.: RELVIEW — A system for calculation with relations and relational programming. In: Astesiano E. (ed.): Proc. 1st Conf. *Fundamental Approaches to Software Engineering*, LNCS 1382, Springer, 318-321 (1998)
3. Bryant R.E.: Symbolic Boolean manipulation with ordered binary decision diagrams. ACM Comp. Surveys 24, 293-318 (1992)
4. Crawford J.M., Auton L.D.: Experimental results on the crossover point in random 3SAT. Artificial Intelligence 81, 59-80 (1996)
5. Dantsin E., Hirsch E.A.: Algorithms for SAT and upper bounds of their complexity. Electr. Coll. Comp. Compl., Rep. 12, http://www.eccc.uni-trier.de/eccc (2001)
6. Hoos H.H., Stützle T.: SATLIB: An online resource for research on SAT. In: Gent I.P., v. Maaren H., Walsh T (eds.): SAT 2000, 283-292, IOS Press (2000)
7. Hromkovic J.: Algorithms for hard problems. Introduction to combinatorial optimization, randomization, approximation, and heuristics. EATCS Texts in Theoret. Comput. Sci., Springer (2001)
8. Huth M.R.A., Ryan M.D.: Logic in computer science. Cambr. Univ. Press (2000)
9. Leoniuk B.: ROBDD-based implementation of relational algebra with applications (in German). Ph.D. thesis, Inst. für Inf. und Prak. Math., Univ. Kiel (2001)
10. Lind-Nielson J.: BuDDy, a binary decision diagram package, version 2.2. Techn. Univ. of Denmark, http://www.itu.dk/research/buddy (2003)
11. Milanese U.: On the implementation of a ROBDD-based tool for the manipulation and visualization of relations (in German). Ph.D. thesis, Inst. für Inf. und Prak. Math., Univ. Kiel (2003)
12. Purdam P.: A survey of average time analysis of satisfiability algorithms. J. of Inf. Processing 13, 449-455 (1990)
13. Schmidt G., Ströhlein T.: Relations and graphs. Discrete Mathematics for Computer Scientists, EATCS Monographs on Theoret. Comput. Sci., Springer (1993)
14. Somenzi F.: CUDD: CU decision diagram package, release 2.3.1. Univ. of Colorado at Boulder, http://www.vlsi.colorado.edu/~fabio/CUDD (2001)
15. Tarski A.: On the calculus of relations. J. Symbolic Logic 6, 73-89 (1941)

Static Analysis of Programs Using Omega Algebra with Tests*

Claude Bolduc and Jules Desharnais

Département d'informatique et de génie logiciel,
Université Laval, Québec, QC, G1K 7P4, Canada
claude.bolduc.1@ulaval.ca, Jules.Desharnais@ift.ulaval.ca

Abstract. Recently, Kozen has proposed a framework based on Kleene algebra with tests for verifying that a program satisfies a security policy specified by a security automaton. A security automaton is used for the specification of linear safety properties on finite and infinite runs. This kind of property is very interesting for most common programs. However, it is not possible to specify liveness properties with security automata. In this paper, we use omega algebra with tests and automata on infinite words to extend the field of properties that can be handled by security automata in Kozen's framework.

1 Introduction

The static analysis of programs is recognized as a fundamental tool for the verification of programs. Static analysis regroups a lot of different methods such as abstract interpretation, model-checking and data flow analysis.

Recently, Kozen has proposed a framework based on Kleene algebra with tests (KAT) [7] for verifying that a program satisfies a security policy specified by a security automaton. A security automaton is used for the specification of linear safety properties on finite and infinite runs. This kind of property is very interesting for most common programs. However, it is not possible to specify liveness properties with security automata [11]. For the verification of reactive programs, this is problematic: a university research shows that at least 50% of the properties that a user wants to specify are liveness properties [5].

Omega algebra with tests (ωAT), an extension of KAT, was proposed by Cohen as a program specification and verification tool [3]. In this paper, we use ωAT and automata on infinite words to extend the field of properties that can be handled by security automata in Kozen's framework.

In Sect. 2, we present omega algebra and ωAT. In Sect. 3, we recall the definition of automata on infinite words. In Sect. 4, we describe our framework and state the soundness and the completeness of the method. We give an example in Sect. 5 and conclude in Sect. 6.

---

* This research was supported by NSERC (Natural Sciences and Engineering Research Council of Canada).

W. MacCaull et al. (Eds.): RelMiCS 2005, LNCS 3929, pp. 60–72, 2006.

2 Omega Algebra with Tests

We use Cohen's axiomatization of omega algebra [3]. An *omega algebra* is an algebraic structure $\langle W, +, \cdot, ^*, ^\omega, 0, 1\rangle$ satisfying the following axioms[1].

$$
\begin{array}{lll}
x + (y + z) = (x + y) + z & x(yz) = (xy)z & x + 0 = x \\
x(y + z) = xy + xz & x + y = y + x & x0 = 0 = 0x \\
(x + y)z = xz + yz & x + x = x & x1 = x = 1x \\
yx + z \leqslant x \rightarrow y^*z \leqslant x & 1 + xx^* \leqslant x^* & x \leqslant y \leftrightarrow x + y = y \\
xy + z \leqslant x \rightarrow zy^* \leqslant x & 1 + x^*x \leqslant x^* & \\
x \leqslant yx + z \rightarrow x \leqslant y^\omega + y^*z & x^\omega = xx^\omega &
\end{array}
$$

An *omega algebra with tests* is an algebraic structure $\langle W, B, +, \cdot, ^*, ^\omega, ^-, 0, 1\rangle$ such that $B \subseteq W$, $\langle W, +, \cdot, ^*, ^\omega, 0, 1\rangle$ is an omega algebra, and $\langle B, +, \cdot, ^-, 0, 1\rangle$ is a Boolean algebra. The elements of B are called *tests* and the elements of W are called *programs.* In this paper, the symbols $b, c, d, \dots$ range over tests and the symbols $p, q, r, \dots$ range over programs. The $^-$ operator is a complementation operator. It is defined only on tests by the two axioms $b\overline{b} = 0$ and $b + \overline{b} = 1$. It follows from the axioms that 1^ω is the largest element (top element) and that $b^\omega = b1^\omega$, for any $b \in B$. We write $\omega\mathsf{AT} \vdash \varphi$ if the formula φ is a theorem of $\omega\mathsf{AT}$ and $\omega\mathsf{AT} \nvdash \varphi$ if it is not.

We define P to be the set of all atomic (or primitive) programs and B to be the set of all atomic (or primitive) tests of an algebra. An $\omega\mathsf{AT}$ *expression over* P *and* B is a well-formed term over the sets P and B and the operators of $\omega\mathsf{AT}$. In the sequel, we denote the set of all $\omega\mathsf{AT}$ expressions over P and B by $\omega\text{-}\mathbf{Exp}_{\mathsf{P},\mathsf{B}}$.

A *Boolean expression over* B is a well-formed term over the set B and the operators of Boolean algebra. For a finite and nonempty B (say $\mathsf{B} \overset{\text{def}}{=} \{b_1, b_2, \dots, b_n\}$, with $n \in \mathbb{N}_+$), we define an *atom* over B to be a Boolean expression of the form $c_1c_2 \dots c_n$ where $(\forall i \mid 1 \leqslant i \leqslant n : c_i \in \{b_i, \overline{b_i}\})$. When B is empty, we define the constant 1 to be the only atom over $\emptyset$. In the sequel, we denote the set of all atoms over a finite B by $\mathsf{Atoms}_\mathsf{B}$.

A *regular expression* is an expression not containing the $^\omega$ operator. In the sequel, we denote the set of all regular expressions over P and B by $\mathbf{RExp}_{\mathsf{P},\mathsf{B}}$. An ω*-regular expression* is an expression such that no $^\omega$ appears in the argument of $^\omega$ or in the left argument of $\cdot$ [3]. Every ω-regular expression is equivalent (under the axioms of $\omega\mathsf{AT}$) to an expression of the form $p + (\sum i \mid i \in I : q_i r_i^\omega)$, where I is a finite index set and p and each q_i, r_i are regular expressions. When B is finite, we define a *standard* ω*-regular expression* to be an expression of the form $(\sum i \mid i \in I : q_i r_i^\omega)$, where I is a finite index set, each q_i, r_i are regular expressions and $(\forall i, \alpha \mid i \in I \wedge \alpha \in \mathsf{Atoms}_\mathsf{B} : \omega\mathsf{AT} \nvdash \alpha \leqslant r_i)$. The last condition in this definition may seem difficult to prove, but, in fact, this is very simple. Define the function $\mathsf{AP} : \mathbf{RExp}_{\mathsf{P},\mathsf{B}} \rightarrow 2^{\mathsf{Atoms}_\mathsf{B}}$ by induction on the structure of its argument as follows:

[1] In the sequel, we write xy instead of $x \cdot y$. The increasing precedence of the operators is $+, \cdot, ^*, ^\omega$.

$$
\begin{array}{ll}
\mathsf{AP}(0) \stackrel{\text{def}}{=} \emptyset, & \mathsf{AP}(b) \stackrel{\text{def}}{=} \{\alpha \mid \alpha \in \mathsf{Atoms_B} \wedge \alpha \leqslant b\} \quad \text{for } b \in \mathsf{B}, \\
\mathsf{AP}(1) \stackrel{\text{def}}{=} \mathsf{Atoms_B}, & \mathsf{AP}(\bar{b}) \stackrel{\text{def}}{=} \mathsf{Atoms_B} - \mathsf{AP}(b), \\
\mathsf{AP}(p) \stackrel{\text{def}}{=} \emptyset \quad \text{for } p \in \mathsf{P}, & \mathsf{AP}(p+q) \stackrel{\text{def}}{=} \mathsf{AP}(p) \cup \mathsf{AP}(q), \\
\mathsf{AP}(p^*) \stackrel{\text{def}}{=} \mathsf{Atoms_B}, & \mathsf{AP}(pq) \stackrel{\text{def}}{=} \mathsf{AP}(p) \cap \mathsf{AP}(q).
\end{array}
$$

It is easy to show that the condition $(\forall i, \alpha \mid i \in I \wedge \alpha \in \mathsf{Atoms_B} : \omega\mathsf{AT} \nvdash \alpha \leqslant r_i)$ is equivalent to the condition $(\forall i \mid i \in I : \mathsf{AP}(r_i) = \emptyset)$.

$\omega\mathsf{AT}$ is an algebraic formalism that can be used to represent simple programming constructs such as conditionals and loops; in fact, ω-regular expressions suffice for these programming constructs [4]. A program P is encoded by the $\omega\mathsf{AT}$ expression $H(P) + N(P)$, where H and N represent respectively the encoding of halting and non-halting executions of P as shown below:

$$
\begin{array}{rcl}
H(p) & \stackrel{\text{def}}{=} & p, \quad \text{for } p \in \mathsf{P} \cup \mathsf{B}, \\
H(p;q) & \stackrel{\text{def}}{=} & H(p) \cdot H(q), \\
H(\textbf{if } b \textbf{ then } p \textbf{ else } q) & \stackrel{\text{def}}{=} & H(b) \cdot H(p) + \overline{H(b)} \cdot H(q), \\
H(\textbf{while } b \textbf{ do } p) & \stackrel{\text{def}}{=} & (H(b) \cdot H(p))^* \cdot \overline{H(b)}, \\
N(p) & \stackrel{\text{def}}{=} & 0, \quad \text{for } p \in \mathsf{P} \cup \mathsf{B}, \\
N(p;q) & \stackrel{\text{def}}{=} & N(p) + H(p) \cdot N(q), \\
N(\textbf{if } b \textbf{ then } p \textbf{ else } q) & \stackrel{\text{def}}{=} & H(b) \cdot N(p) + \overline{H(b)} \cdot N(q), \\
N(\textbf{while } b \textbf{ do } p) & \stackrel{\text{def}}{=} & (H(b) \cdot H(p))^* \cdot H(b) \cdot N(p) + (H(b) \cdot H(p))^\omega.
\end{array}
$$

Hoare partial correctness assertions $\{b\}p\{c\}$ can be expressed by one of the four equivalent forms [4, 7]

$$bp = bpc, \qquad bp \leqslant pc, \qquad bp\bar{c} = 0, \qquad p \leqslant \bar{b}1^\omega + 1^\omega c. \tag{1}$$

Like KAT, $\omega\mathsf{AT}$ has many interesting models: language-theoretic, relational, trace-based, matrix, ... In this paper, we are particularly interested in trace-based models.

2.1 Trace Model

In the trace model, programs and tests are interpreted over sets of traces that form an $\omega\mathsf{AT}$. So, we extend the trace model for KAT [6] to allow infinite traces. As Kozen did for KAT, we define our trace model for $\omega\mathsf{AT}$ with *Kripke frames* [6]. A Kripke frame is defined over a set of atomic programs P and a set of atomic tests B. Formally, a Kripke frame is a structure $\langle K, \mathfrak{m}_K \rangle$ such that K is a set of states and $\mathfrak{m}_K$ is a map that binds atomic tests to sets of states and atomic programs to sets of state transitions. So, $\mathfrak{m}_K : \mathsf{B} \to 2^K$ and $\mathfrak{m}_K : \mathsf{P} \to 2^{K \times K}$.

Let $\mathcal{K} \stackrel{\text{def}}{=} \langle K, \mathfrak{m}_K \rangle$ be a Kripke frame. A *finite trace* in $\mathcal{K}$ is a sequence of the form $k_0 p_0 k_1 \dots k_{n-1} p_{n-1} k_n$ such that $n \geqslant 0$, $k_n \in K$ and $k_i \in K$, $p_i \in \mathsf{P}$, $(k_i, k_{i+1}) \in \mathfrak{m}_K(p_i)$ for $0 \leqslant i \leqslant n-1$. An *infinite trace* in $\mathcal{K}$ is a sequence of the form $k_0 p_0 k_1 p_1 k_2 \dots$ such that $k_i \in K$, $p_i \in \mathsf{P}$ and $(k_i, k_{i+1}) \in \mathfrak{m}_K(p_i)$ for $i \geqslant 0$. We denote by $\mathsf{Traces}_\mathcal{K}$ the set of all finite and infinite traces on $\mathcal{K}$. In the sequel, the symbols $\sigma, \tau, \dots$ range over traces. The first state of a trace σ is denoted by $\mathsf{first}(\sigma)$. The length of a trace is given by the number of atomic programs in it.

The *fusion product* $\sigma\tau$ of a finite trace σ of the form $k_0p_0k_1\ldots k_{n-1}p_{n-1}k_n$ with an arbitrary trace τ is defined as follows:

$$\sigma\tau \overset{\text{def}}{=} \begin{cases} k_0p_0k_1\ldots k_{n-1}p_{n-1}\tau & \text{if } \mathsf{first}(\tau) = k_n, \\ \text{undefined} & \text{otherwise.} \end{cases}$$

Using the fusion product, operations on sets of traces can be defined. We proceed in two steps, first imposing restrictions on the sets on which the operations are defined. Let S be a set of finite traces, T be a set of finite traces of positive length and V be an arbitrary set of traces. We define the following four operations (again, we omit the composition symbol $\cdot$ and use juxtaposition):

1. $SV \overset{\text{def}}{=} \{\sigma\tau \mid \sigma \in S \wedge \tau \in V \wedge \sigma\tau \text{ is defined}\}$;
2. $S^* \overset{\text{def}}{=} (\bigcup i \mid i \in \mathbb{N} : S^i)$, where $S^0 = K$ and $S^{i+1} = SS^i$;
3. $S^+ \overset{\text{def}}{=} SS^*$;
4. $T^\omega \overset{\text{def}}{=} \{\sigma_0\sigma_1\sigma_2\ldots \mid (\forall i \mid i \in \mathbb{N} : \sigma_i \in T \wedge \sigma_i\sigma_{i+1} \text{ is defined})\}$.

Now let

$$P \overset{\text{def}}{=} \{kpk' \mid p \in \mathsf{P} \wedge (k,k') \in \mathfrak{m}_K(p)\}. \tag{2}$$

Since P is a set of traces of length one, P^* and P^ω are defined. Note that $P^* \cup P^\omega = \mathsf{Traces}_{\mathcal{K}}$ is the set of all traces on $\mathcal{K}$. For any set S of traces, we define the following subsets of S:

$$S_K \overset{\text{def}}{=} S \cap K, \qquad S_+ \overset{\text{def}}{=} S \cap P^+, \qquad S_{\mathsf{f}} \overset{\text{def}}{=} S_K \cup S_+, \qquad S_\infty \overset{\text{def}}{=} S \cap P^\omega.$$

We can now define ST, S^* and S^ω for arbitrary sets of traces S and T:

1. $ST \overset{\text{def}}{=} \begin{cases} \emptyset & \text{if } T = \emptyset \\ S_{\mathsf{f}}T \cup S_\infty & \text{if } T \neq \emptyset, \end{cases}$
2. $S^* \overset{\text{def}}{=} S_{\mathsf{f}}^* \cup S_{\mathsf{f}}^* S_\infty$,
3. $S^\omega \overset{\text{def}}{=} S_+^\omega \cup S_+^* S_\infty \cup S_+^* S_K (P^* \cup P^\omega)$.

Note that subscripts have precedence over superscripts; for instance, $S_+^* = (S_+)^*$. With these operations, the structure $\langle 2^{\mathsf{Traces}_{\mathcal{K}}}, 2^K, \cup, \cdot, ^*, ^\omega, ^-, \emptyset, K\rangle$ forms an ωAT ($^-$ is complementation over 2^K). The tests are all the sets of traces of length 0. We call this algebra the *full trace algebra on* $\mathcal{K}$.

Like for the trace model in KAT [6], a canonical interpretation $[\![\ ]\!]_{\mathcal{K}}$ for ωAT expressions over P and B to the full trace algebra on $\mathcal{K}$ is defined by

$$[\![p]\!]_{\mathcal{K}} \overset{\text{def}}{=} \{kpk' \mid (k,k') \in \mathfrak{m}_K(p)\}, \quad \text{for } p \in \mathsf{P},$$
$$[\![b]\!]_{\mathcal{K}} \overset{\text{def}}{=} \mathfrak{m}_K(b), \quad \text{for } b \in \mathsf{B},$$

extended homomorphically. We denote the image of this homomorphism by $\mathsf{Tr}_{\mathcal{K}}$. In other words, a set S of traces belongs to $\mathsf{Tr}_{\mathcal{K}}$ if $(\exists p \mid p \in \omega\text{-}\mathbf{Exp}_{\mathsf{P},\mathsf{B}} : [\![p]\!]_{\mathcal{K}} = S)$. The image $\mathsf{Tr}_{\mathcal{K}}$ along with the operators defined for trace algebra forms an ωAT. By construction, $[\![\ ]\!]_{\mathcal{K}}$ is onto $\mathsf{Tr}_{\mathcal{K}}$. We denote the family of all $\mathsf{Tr}_{\mathcal{K}}$ by TR. We write $\mathsf{TR}, [\![\ ]\!] \models \varphi$ if, for all Kripke frames $\mathcal{K}$, the formula φ is true under the canonical interpretation $[\![\ ]\!]_{\mathcal{K}}$.

$\omega\mathsf{AT}$ is sound and complete for a restricted version of the Hoare theory of the trace model in the following sense [1]. Recall that the Hoare theory of a class of algebras is the set of Horn formulae of the form

$$r_1 = 0 \rightarrow r_2 = 0 \rightarrow \ldots \rightarrow r_n = 0 \rightarrow p = q$$

that are valid for every algebra of the class. In this paper, we present a soundness and completeness result for the Hoare theory of $\omega\mathsf{AT}$ for the trace model only for particular r_1, r_2, ..., r_n, p and q. Call an expression a *simple expression* if the expression does not contain the operators * and $^\omega$ and, for each occurrence of $\cdot$, at least one argument of $\cdot$ is a Boolean expression over B. For example, $bp + cqbd + b$ is a simple expression, but bpq is not. Let p and q be two standard ω-regular expressions on finite P and B and let R be a set of simple expressions on P and B. Let $\sum A$ stand for the sum of all elements of the set A. Then,

$$\begin{aligned} &\omega\mathsf{AT} \vdash (\textstyle\sum R) = 0 \rightarrow p = q \\ &\Leftrightarrow \mathsf{TR}, [\![\ ]\!] \models (\textstyle\sum R) = 0 \rightarrow p = q \\ &\Leftrightarrow \omega\mathsf{AT} \vdash p + (\textstyle\sum \mathsf{P})^*(\textstyle\sum R)(\textstyle\sum \mathsf{P})^\omega = q + (\textstyle\sum \mathsf{P})^*(\textstyle\sum R)(\textstyle\sum \mathsf{P})^\omega . \end{aligned} \tag{3}$$

Moreover, the problem to determine if $\omega\mathsf{AT} \vdash (\sum R) = 0 \rightarrow p = q$ is decidable [1]. These two results follow from the extension to $\omega\mathsf{AT}$ of the completeness and decidability results of Cohen for omega algebra [4] (similar to the completeness of KAT [8]), and properties of Kripke frames.

2.2 Constructions on Kripke Frames

The definition of $\omega\mathsf{AT}$ is very close to that of KAT. So, much of the results discovered for KAT extend naturally to $\omega\mathsf{AT}$ with some restrictions. For instance, almost all the results found by Kozen for the trace model in [6] extend easily to $\omega\mathsf{AT}$. In this section, we present two concepts coming from [6]: *induced subframes* and *coherent functions*. Prior to that, we present *canonical restricted homomorphisms* for the trace model.

Let $\mathcal{K} \stackrel{\text{def}}{=} \langle K, \mathfrak{m}_K \rangle$ and $\mathcal{L} \stackrel{\text{def}}{=} \langle L, \mathfrak{m}_L \rangle$ be two Kripke frames on P and B. For the trace model, we say that a function $h : 2^{\mathsf{Traces}_{\mathcal{K}}} \rightarrow 2^{\mathsf{Traces}_{\mathcal{L}}}$ is a *restricted homomorphism* iff h satisfies the following conditions for all $S, T \subseteq \mathsf{Traces}_{\mathcal{K}}$, $V \subseteq P^*$ (see (2)) and $W \subseteq K$:

$$\begin{array}{llll} h(S \cup T) = h(S) \cup h(T), & h(S^*) = h(S)^*, & h(\emptyset) = \emptyset, & h(\overline{W}) = \overline{h(W)}, \\ h(VT) = h(V)h(T), & h(V^\omega) = h(V)^\omega, & h(K) = L. & \end{array}$$

We say that a restricted homomorphism $h : 2^{\mathsf{Traces}_{\mathcal{K}}} \rightarrow 2^{\mathsf{Traces}_{\mathcal{L}}}$ is *canonical* iff h also satisfies the conditions

$$h([\![p]\!]_{\mathcal{K}}) = [\![p]\!]_{\mathcal{L}} \text{ for all } p \in \mathsf{P} \qquad \text{and} \qquad h([\![b]\!]_{\mathcal{K}}) = [\![b]\!]_{\mathcal{L}} \text{ for all } b \in \mathsf{B}.$$

The definition of a restricted homomorphism in the trace model is almost the same as the definition of a homomorphism, except that the function h commutes with the operators $\cdot$ and $^\omega$ only when the first argument is a set of finite traces. The reason is that a model of ωAT has to satisfy the law $x0 = 0$ for all x. In the case of the trace model, this yields the particular case $T = \emptyset$ in the definition of the operator $\cdot$. Unfortunately, this definition causes problems when looking for a homomorphism for this model (and other models of ωAT in general). To overcome this problem, we use restricted homomorphisms. Another solution (not used here) would be to define the trace model without the particular case $T = \emptyset$ and use lazy omega algebra or von Wright's refinement algebra instead of omega algebra [9, 12].

Canonical restricted homomorphisms have the following simple property.

Theorem 1. *Let* $\mathcal{K} \stackrel{\text{def}}{=} \langle K, \mathfrak{m}_K\rangle$ *and* $\mathcal{L} \stackrel{\text{def}}{=} \langle L, \mathfrak{m}_L\rangle$ *be two Kripke frames on* P *and* B. *Let* $h : 2^{\mathsf{Traces}_\mathcal{K}} \rightarrow 2^{\mathsf{Traces}_\mathcal{L}}$ *be a function. If* h *is a canonical restricted homomorphism, then, for all* ω*-regular terms* p *and* q, $h([\![p]\!]_\mathcal{K}) = [\![p]\!]_\mathcal{L}$.

We are now ready to present induced subframes and coherent functions.

Induced Subframes. Let $\mathcal{K} \stackrel{\text{def}}{=} \langle K, \mathfrak{m}_K\rangle$ be a Kripke frame on P and B. Let $L \subseteq K$ be a subset of K. We call the Kripke frame $\mathcal{L} \stackrel{\text{def}}{=} \langle L, \mathfrak{m}_L\rangle$ defined by

$$\begin{aligned} \mathfrak{m}_L(p) &\stackrel{\text{def}}{=} \mathfrak{m}_K(p) \cap (L \times L) \quad \text{for } p \in \mathsf{P}, \\ \mathfrak{m}_L(b) &\stackrel{\text{def}}{=} \mathfrak{m}_K(b) \cap L \qquad\qquad \text{for } b \in \mathsf{B} \end{aligned}$$

the *induced subframe on L*.

Kozen's result for KAT [6] can be extended to ωAT as follows.

Theorem 2. *Let* $\mathcal{L} \stackrel{\text{def}}{=} \langle L, \mathfrak{m}_L\rangle$ *be an induced subframe of* $\mathcal{K} \stackrel{\text{def}}{=} \langle K, \mathfrak{m}_K\rangle$ *on* P *and* B *such that for all atomic actions* $p \in \mathsf{P}$,

$$(\forall k, l \mid k \in K \wedge l \in L : (l, k) \in \mathfrak{m}_K(p) \ \Rightarrow\ k \in L).$$

Then, the function $h : 2^{\mathsf{Traces}_\mathcal{K}} \rightarrow 2^{\mathsf{Traces}_\mathcal{L}}$ *defined by* $h(S) \stackrel{\text{def}}{=} S \cap \mathsf{Traces}_\mathcal{L}$ *for* $S \subseteq \mathsf{Traces}_\mathcal{K}$ *is a canonical restricted homomorphism from* $\mathsf{Tr}_\mathcal{K}$ *to* $\mathsf{Tr}_\mathcal{L}$.

The proof of this result is similar to Kozen's proof [6], except that we must take care of infinite traces and that we must add an induction case for the restricted homomorphism proof [1].

Coherent Functions. Let $\mathcal{K} \stackrel{\text{def}}{=} \langle K, \mathfrak{m}_K\rangle$ and $\mathcal{L} \stackrel{\text{def}}{=} \langle L, \mathfrak{m}_L\rangle$ be two Kripke frames on P and B. We call a function $f : K \rightarrow L$ *coherent* if

1. $(k, k') \in \mathfrak{m}_K(p) \Rightarrow (f(k), f(k')) \in \mathfrak{m}_L(p)$, for $p \in \mathsf{P}$,
2. $k \in \mathfrak{m}_K(b) \Leftrightarrow f(k) \in \mathfrak{m}_L(b)$, for $b \in \mathsf{B}$.

A coherent function $f : K \rightarrow L$ can be extended to a function $f : \mathsf{Traces}_\mathcal{K} \rightarrow \mathsf{Traces}_\mathcal{L}$ by letting

$$\begin{aligned} f(k_0 p_0 k_1 \ldots k_{n-1} p_{n-1} k_n) &\stackrel{\text{def}}{=} f(k_0) p_0 f(k_1) \ldots f(k_{n-1}) p_{n-1} f(k_n), \\ f(k_0 p_0 k_1 p_1 k_2 \ldots) &\stackrel{\text{def}}{=} f(k_0) p_0 f(k_1) p_1 f(k_2) \ldots \end{aligned}$$

This extension of f is well defined, by Condition 1 of the definition of a coherent function.

For a coherent function $f : K \rightarrow L$ and $T \subseteq \mathsf{Traces}_{\mathcal{K}}$, $S \subseteq \mathsf{Traces}_{\mathcal{L}}$, define f^{-1} by the Galois connection $f(T) \subseteq S \Leftrightarrow T \subseteq f^{-1}(S)$, where f is extended to sets in the usual way. As a consequence, $f(f^{-1}(S)) \subseteq S$.

Kozen's result for KAT [6] can be extended to ωAT as follows.

Theorem 3. *If $f : K \rightarrow L$ is a coherent function, then the restriction $f^{-1} : \mathsf{Tr}_{\mathcal{L}} \rightarrow \mathsf{Tr}_{\mathcal{K}}$ is a canonical restricted homomorphism from $\mathsf{Tr}_{\mathcal{L}}$ to $\mathsf{Tr}_{\mathcal{K}}$.*

The proof of this result is similar to Kozen's proof [6]. We give the proof of the restricted homomorphism property for the $^{\omega}$ operator, assuming that the property holds for composition, i.e., $f^{-1}(V \cdot T) = f^{-1}(V) \cdot f^{-1}(T)$. We have to show $f^{-1}(V^{\omega}) = (f^{-1}(V))^{\omega}$.

1. Case $f^{-1}(V^{\omega}) \subseteq (f^{-1}(V))^{\omega}$: Using the ωAT law $x^{\omega} = xx^{\omega}$ and the restricted homomorphism property for composition, we have
$$f^{-1}(V^{\omega}) = f^{-1}(V \cdot V^{\omega}) = f^{-1}(V) \cdot f^{-1}(V^{\omega}).$$
The result then follows by ω-induction ($x \leqslant yx \rightarrow x \leqslant y^{\omega}$).
2. Case $(f^{-1}(V))^{\omega} \subseteq f^{-1}(V^{\omega})$:
$$\begin{array}{ll} & (f^{-1}(V))^{\omega} \subseteq f^{-1}(V^{\omega}) \\ \Leftrightarrow & \quad \langle \text{ Galois connection } \rangle \\ & f((f^{-1}(V))^{\omega}) \subseteq V^{\omega} \\ \Leftarrow & \quad \langle \text{ Omega induction } \rangle \\ & f((f^{-1}(V))^{\omega}) \subseteq V \cdot f((f^{-1}(V))^{\omega}) \\ \Leftarrow & \quad \langle\, x^{\omega} = xx^{\omega} \text{ and } f(f^{-1}(V)) \subseteq V \,\rangle \\ & f(f^{-1}(V) \cdot (f^{-1}(V))^{\omega}) = f(f^{-1}(V)) \cdot f((f^{-1}(V))^{\omega}) \end{array}$$
—this holds by extending the coherent function f to sets of traces

3 Automata on Infinite Words

Automata on infinite words are often used in program verification, in particular for the specification of linear properties. The interest in automata on infinite words in program verification has grown with the discovery of a strong link between linear temporal logics and Büchi automata (a special kind of automata on infinite words).

We denote by Σ^{ω} the set of all infinite words that can be generated from the alphabet Σ. Note that if the set Σ is finite, all infinite words of Σ^{ω} contain at least one element of Σ that appears infinitely often in the word.

An *automaton on infinite words* $\mathcal{I} \stackrel{\text{def}}{=} (Q, \Sigma, \delta, I, C)$ is an automaton such that (1) Q is a finite set of states; (2) Σ is the alphabet of the automaton; (3) $\delta : Q \times \Sigma \rightarrow 2^{Q}$ is the transition function; (4) $I \subseteq Q$ is the set of initial states; (5) C is a parameter used for determining when an infinite word is accepted.

An automaton on infinite words is *deterministic* iff it has a single initial state ($|I| = 1$) and $(\forall q, a \mid q \in Q \wedge a \in \Sigma : |\delta(q,a)| \leqslant 1)$. An automaton on infinite words is *complete* iff $(\forall q, a \mid q \in Q \wedge a \in \Sigma : \delta(q,a) \neq \emptyset)$. If an automaton is not complete, it can be completed by inserting a dummy state d and creating a transition labelled by $a \in \Sigma$ to this dummy state from each state $q \in Q \cup \{d\}$ such that $\delta(q,a) = \emptyset$.

We present two kinds of automata on infinite words that are interesting for program verification: Büchi automata and Rabin automata.

3.1 Büchi Automata

A *Büchi automaton* $\mathcal{B} \stackrel{\text{def}}{=} (Q, \Sigma, \delta, I, F)$ is an automaton on infinite words such that $F \subseteq Q$ is the set of accepting states. An infinite word $w \in \Sigma^\omega$ is accepted iff there is a run of the automaton on w that starts in one of the initial states and goes infinitely often through at least one state of F.

It has been shown that deterministic Büchi automata are less expressive than nondeterministic Büchi automata [10].

3.2 Rabin Automata

A *Rabin automaton* $\mathcal{R} \stackrel{\text{def}}{=} (Q, \Sigma, \delta, I, \mathcal{T})$ is an automaton on infinite words such that $\mathcal{T} \stackrel{\text{def}}{=} \{(G_1, R_1), (G_2, R_2), \ldots, (G_n, R_n)\}$, where $G_i \subseteq Q$ and $R_i \subseteq Q$ for $1 \leqslant i \leqslant n$. An infinite word $w \in \Sigma^\omega$ is accepted iff there is a run χ of the automaton on w and a pair $(G_i, R_i) \in \mathcal{T}$ such that χ starts in one of the initial states and goes infinitely often through at least one state of G_i and does not contain a state of R_i infinitely often.

Deterministic Rabin automata are as expressive as nondeterministic ones [10]. Also, there is an algorithm, namely the Safra construction, that constructs a deterministic Rabin automaton from a nondeterministic Büchi automaton with at most $2^{\mathcal{O}(n \log n)}$ states, where n is the number of states of the nondeterministic Büchi automaton [10].

4 Static Analysis of Programs Using Omega Algebra with Tests

In this section, we show how to verify that a program represented by a standard ω-regular expression r on *finite* sets of atomic tests B and atomic programs P satisfies a property defined by a *deterministic* automaton on infinite words specifying the acceptable runs of the program (which are the sequences accepted by the automaton). Note that our framework allows the use of nondeterministic Büchi automata since we just have to use the Safra construction to produce a deterministic Rabin automaton from a nondeterministic Büchi automaton. Unfortunately, this may yield an exponential blow-up of states.

Let $\mathcal{K} \stackrel{\text{def}}{=} \langle K, \mathfrak{m}_K \rangle$ be the Kripke frame over the atomic actions P and atomic tests B on which r is run. Let $\mathcal{I} \stackrel{\text{def}}{=} (Q, \Sigma, \delta, I, C)$ be a deterministic automaton on infinite words over alphabet P representing the property to verify. Let $U, V, \ldots$ range over elements of Q. The encoding of the automaton in ωAT is very similar to Kozen's encoding of security automata [7]. To ensure soundness of the method, the automaton $\mathcal{I}$ must be complete; if it is not, it can be completed by applying the method of Sect. 3.

First, create a new atomic test U for each $U \in Q$ (we assume that $Q \cap \mathsf{B} = \emptyset$). Let Q denote both the set of states of the automaton and the new set of atomic tests. Second, create the following hypotheses:

$$UV = 0, \qquad \text{for } U, V \in Q \text{ and } U \neq V, \tag{4}$$
$$Up \leqslant pV, \qquad \text{for } U \in Q,\, p \in \mathsf{P},\, V \in \delta(U, p). \tag{5}$$

Hypotheses (4) say that a program cannot be in two different states of the automaton at the same time. Hypotheses (5) describe transitions; note that $V \in \delta(U, p)$ is unique, since the automaton is deterministic and complete. We denote the conjunction of all hypotheses by $\mathcal{H}_{\mathcal{I}}$.

We can view $\mathcal{I}$ as a Kripke frame $\langle Q, \mathfrak{m}_I \rangle$ over atomic programs P and atomic tests Q by defining $\mathfrak{m}_I(p) \stackrel{\text{def}}{=} \{(U, V) \mid U \in Q \wedge V \in \delta(U, p)\}$ for all $p \in \mathsf{P}$ and $\mathfrak{m}_I(U) \stackrel{\text{def}}{=} \{U\}$ for all $U \in Q$. We can also construct the instrumented frame $\mathcal{K} \times \mathcal{I}$ over atomic actions P and atomic tests $\mathsf{B} \cup Q$ as in [7]. The states of $\mathcal{K} \times \mathcal{I}$ are $K \times Q$ and the function $\mathfrak{m}_{K \times I}$ is defined for all $p \in \mathsf{P}$, $b \in \mathsf{B}$ and $U \in Q$ by

$$\begin{aligned}
\mathfrak{m}_{K \times I}(p) &\stackrel{\text{def}}{=} \{((k, V), (k', V')) \mid (k, k') \in \mathfrak{m}_K(p) \wedge (V, V') \in \mathfrak{m}_I(p)\}, \\
\mathfrak{m}_{K \times I}(b) &\stackrel{\text{def}}{=} \{(k, V) \mid k \in \mathfrak{m}_K(b) \wedge V \in Q\}, \\
\mathfrak{m}_{K \times I}(U) &\stackrel{\text{def}}{=} \{(k, U) \mid k \in K\}.
\end{aligned}$$

The instrumented frame $\mathcal{K} \times \mathcal{I}$ simulates the following idea: when $\mathcal{K}$ changes its state by way of an atomic program $p \in \mathsf{P}$, $\mathcal{I}$ also changes its state by a transition labelled by p. Let $\pi_1 : \mathcal{K} \times \mathcal{I} \to \mathcal{K}$ and $\pi_2 : \mathcal{K} \times \mathcal{I} \to \mathcal{I}$ be the projections of the instrumented frame on its first and second component, respectively. The projection π_1 (resp. π_2) is coherent, ignoring tests in Q (resp. B) (see Sect. 2.2).

Let **secureTraces** be an ωAT expression representing, under the canonical interpretation of the trace model, the set of all traces in the instrumented frame that satisfy the property represented by $\mathcal{I}$. We say that a program r in this Kripke frame $\mathcal{K}$ respects the property specified by $\mathcal{I}$ iff

$$[\![Ir]\!]_{\mathcal{K} \times \mathcal{I}} \subseteq [\![\mathbf{secureTraces}]\!]_{\mathcal{K} \times \mathcal{I}}. \tag{6}$$

The expression **secureTraces** depends on the type of deterministic automaton that we have. For a Büchi automaton, a trace in the instrumented frame respects the property iff the projection π_2 of this trace in $\mathcal{I}$ starts in the initial state I and contains infinitely often at least one state of F. Thus,

$$\mathbf{secureTraces} \stackrel{\text{def}}{=} I((\textstyle\sum \mathsf{P})^*(\sum F)(\sum \mathsf{P}))^\omega. \tag{7}$$

For a Rabin automaton, a trace in the instrumented frame respects the property iff there exists a pair $(G_i, R_i) \in \mathcal{T}$ such that the projection π_2 of this trace in $\mathcal{I}$ starts in the initial state I and contains infinitely often at least one state of G_i and does not contain a state of R_i infinitely often. Let $\mathcal{T} = \{(G_1, R_1), (G_2, R_2), \ldots, (G_n, R_n)\}$ be such that $G_i, R_i \subseteq Q$, for $1 \leqslant i \leqslant n$. Let $\overline{R_i}$ stand for the complement of R_i with respect to Q, for $1 \leqslant i \leqslant n$. Then,

$$\begin{aligned}&\mathbf{secureTraces} \overset{\text{def}}{=} \\ &\qquad I(\textstyle\sum \mathsf{P})^*(\sum i \mid 1 \leqslant i \leqslant n : (((\sum \overline{R_i})(\sum \mathsf{P}))^*(\sum G_i)(\sum \overline{R_i})(\sum \mathsf{P}))^\omega).\end{aligned} \tag{8}$$

The following theorem states the soundness of the method.

Theorem 4. *Let $\mathcal{H}_{\text{prog}}$ be a set of hypotheses consisting of equations involving only ω-regular expressions on finite sets P and B, and let r be a standard ω-regular expression on these sets. If*

$$\omega\mathsf{AT} \vdash \mathcal{H}_{\mathcal{I}} \to \mathcal{H}_{\text{prog}} \to Ir \leqslant \mathbf{secureTraces}, \tag{9}$$

then program r respects the property specified by $\mathcal{I}$ for all Kripke frames $\mathcal{K}$ whose trace algebra $\mathsf{Tr}_{\mathcal{K}}$ satisfies $\mathcal{H}_{\text{prog}}$ under the canonical interpretation $[\![\ ]\!]_{\mathcal{K}}$.

The proof of this theorem is similar to Kozen's proof for security automata [7], since we have Theorem 3 and we know that π_1 (resp. π_2) is coherent, ignoring tests of Q (resp. B).

The following theorem states the completeness of the method.

Theorem 5. *Let P and B be finite. Let $\mathcal{H}_{\text{prog}}$ be a finite set of hypotheses on P and B containing only equations of the form $p = 0$ such that p is a simple expression. Let r be a standard ω-regular expression on P and B. If program r respects the property specified by $\mathcal{I}$ for all Kripke frames $\mathcal{K}$ whose trace algebra $\mathsf{Tr}_{\mathcal{K}}$ satisfies $\mathcal{H}_{\text{prog}}$ under the canonical interpretation $[\![\ ]\!]_{\mathcal{K}}$, then*

$$\omega\mathsf{AT} \vdash \mathcal{H}_{\mathcal{I}} \to \mathcal{H}_{\text{prog}} \to Ir \leqslant \mathbf{secureTraces}.$$

The proof of this theorem is similar to Kozen's proof for security automata [7], by (3) and Theorems 2 and 3.

To conclude this section, we remark that the restriction to finite sets of atomic tests, atomic programs and hypotheses in Theorem 5 is not a practical limitation, since modelling any real program requires only a finite number of tests, atomic programs and hypotheses.

5 Example

Let the atomic tests be $\mathsf{B} \overset{\text{def}}{=} \{a, b\}$ and the atomic programs be $\mathsf{P} \overset{\text{def}}{=} \{p, q\}$. Consider the program $r \overset{\text{def}}{=} a\bar{b}(p+q)^\omega$, with the hypotheses

$$\mathcal{H}_{\text{prog}} \overset{\text{def}}{=} \left\{ \begin{matrix} \bar{a}\bar{b} = 0, & \bar{a}bq = 0, & a\bar{b}q = 0, & \\ a\bar{b}p \leqslant p\bar{a}b, & \bar{a}bp \leqslant pab, & abp \leqslant p\bar{a}b, & abq \leqslant qa\bar{b} \end{matrix} \right\}.$$

As an aid to understanding it, this program may be represented by the labelled Kripke structure on the left below, where the atomic tests or their complement label the states.

We want to check that program r satisfies the following property: r *infinitely often executes the action* p *twice in a row*. The Büchi automaton on the right represents this property. Note that it is deterministic and complete.

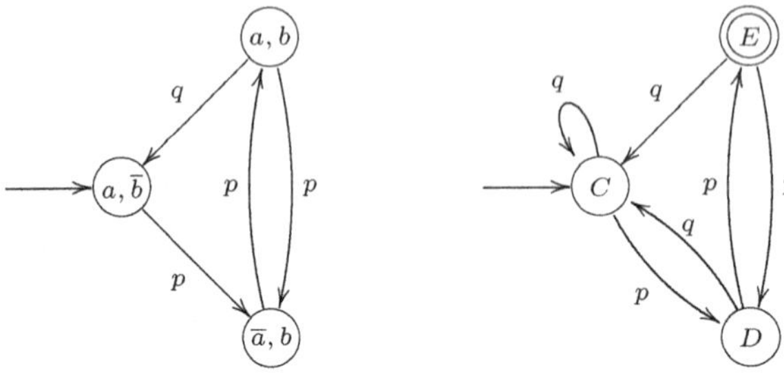

The automaton may be encoded in ωAT as explained in Sect. 4. This is done by introducing the atomic tests $\{C, D, E\}$ and the hypotheses

$$\mathcal{H}_\mathcal{I} \stackrel{\text{def}}{=} \left\{ \begin{matrix} CD = 0, & CE = 0, & DE = 0, \\ Cp \leqslant pD, & Cq \leqslant qC, & Dp \leqslant pE, & Dq \leqslant qC, & Ep \leqslant pD, & Eq \leqslant qC \end{matrix} \right\}.$$

After instantiating (9) and (7), the ωAT expression to prove is

$$\mathcal{H}_\mathcal{I} \to \mathcal{H}_{\text{prog}} \to Ca\bar{b}(p+q)^\omega \leqslant C((p+q)^*E(p+q))^\omega. \tag{10}$$

To prove it, first note that

$$\mathcal{H}_\mathcal{I} \to E(pp+qpp) \leqslant (pp+qpp)E \tag{11}$$

and

$$\mathcal{H}_{\text{prog}} \to ab(p+q)^\omega \leqslant (ab(pp+qpp))^\omega \tag{12}$$

are theorems of ωAT. Formula (11) is easy to prove using distributivity of $\cdot$ on $+$ and some hypotheses of $\mathcal{H}_\mathcal{I}$. By omega induction (i.e., $x \leqslant yx \to x \leqslant y^\omega$), (12) follows from $ab(p+q)^\omega \leqslant ab(pp+qpp)ab(p+q)^\omega$, which is proved as follows by using the law $x^\omega = xx^\omega$ a few times, some hypotheses of $\mathcal{H}_{\text{prog}}$ and basic laws of omega algebra:

$$\begin{aligned} ab(p+q)^\omega &\leqslant abab(p+q)(p+q)(p+q)^\omega \leqslant ab(ppab + qp\bar{a}b)(p+q)^\omega \\ &= ab(ppab(p+q)^\omega + qp\bar{a}b(p+q)(p+q)^\omega) \\ &\leqslant ab(ppab(p+q)^\omega + qppab(p+q)^\omega) = ab(pp+qpp)ab(p+q)^\omega. \end{aligned}$$

With these two results, the proof of (10) is short:

$$\begin{aligned} & Ca\bar{b}(p+q)^\omega \\ = \quad & \quad \langle \text{ Law of } \omega\mathsf{AT}\text{: } x^\omega = xx^\omega \ \rangle \\ & Ca\bar{b}(p+q)(p+q)(p+q)^\omega \\ = \quad & \quad \langle \text{ Distributivity of} \cdot \text{on} + \rangle \\ & (Ca\bar{b}pp + Ca\bar{b}pq + Ca\bar{b}q(p+q))(p+q)^\omega \\ \leqslant \quad & \quad \langle \text{ Hypotheses of } \mathcal{H}_{\text{prog}}\text{: } a\bar{b}q = 0,\ a\bar{b}p \leqslant p\bar{a}b,\ \bar{a}bq = 0 \text{ and } \bar{a}bp \leqslant pab \ \rangle \end{aligned}$$

$$
\begin{array}{ll}
 & Cppab(p+q)^\omega \\
\leqslant & \quad \langle\ (12)\ \rangle \\
 & Cpp(ab(pp+qpp))^\omega \\
\leqslant & \quad \langle\ \text{Idempotency of} \cdot \text{on tests, } ab \leqslant 1\text{, and Hypotheses of } \mathcal{H}_\mathcal{I}\text{: } Cp \leqslant pD \\
 & \quad \text{and } Dp \leqslant pE\ \rangle \\
 & CppE(pp+qpp)^\omega \\
\leqslant & \quad \langle\ (11) \text{ and law of } \omega\mathsf{AT}\text{: } cs \leqslant sc \rightarrow cs^\omega \leqslant (cs)^\omega\ \rangle \\
 & Cpp(E(pp+qpp))^\omega \\
\leqslant & \quad \langle\ pp \leqslant (p+q)^*,\ pp+qpp \leqslant (p+q)(p+q)^* \text{ and monotonicity}\ \rangle \\
 & C(p+q)^*(E(p+q)(p+q)^*)^\omega \\
= & \quad \langle\ \text{Law of } \omega\mathsf{AT}\text{: } (xy)^\omega = x(yx)^\omega\ \rangle \\
 & C((p+q)^*E(p+q))^\omega .
\end{array}
$$

This shows that program r satisfies the desired property.

A Rabin automaton representing the same property as the Büchi automaton can also be used. It has exactly the same structure as the previous Büchi automaton except that the parameter $\mathcal{T}$ is $\{(\{E\}, \emptyset)\}$. So, to prove that the program r satisfies the desired property, it suffices to show (see (8))

$$
\begin{aligned}
\mathcal{H}_\mathcal{I} \rightarrow \mathcal{H}_{\text{prog}} \rightarrow Ca\bar{b}(p+q)^\omega \leqslant \\
C(p+q)^*(((C+D+E)(p+q))^*E(C+D+E)(p+q))^\omega ,
\end{aligned}
$$

which is easy.

6 Conclusion

This paper extends Kozen's framework for the static analysis of programs by using $\omega\mathsf{AT}$ instead of KAT. We have shown how to use $\omega\mathsf{AT}$ to verify that a program represented by a standard ω-regular expression satisfies a property represented by a deterministic automaton on infinite words. This allows us to handle liveness properties. We have presented theorems stating that the method is sound and complete over trace models.

Our investigations show that Kozen's framework is robust and extendible. In this paper, we concentrate only on Büchi and Rabin automata, but we can also handle other types of automata on infinite words. We just have to find the right $\omega\mathsf{AT}$ expression for **secureTraces** corresponding to the new kind of automaton. For instance, we can define $\omega\mathsf{AT}$ expressions for co-Büchi automata, generalized Büchi automata and Muller automata, but, for lack of space, we did not discuss them in this paper.

Our framework has a weakness for the verification of nondeterministic Büchi automata. In this case, we use Safra's construction to create a deterministic Rabin automaton, which yields an exponential blow-up of states. A simple solution might be to encode directly the nondeterministic Büchi automaton by changing Hypotheses (5) for $Up \leqslant p(\sum V)$ for $U \in Q$, $p \in \mathsf{P}$ and $V = \delta(U, p)$. Unfortunately, if we also keep our definition of respect (6), this leads to a subtle mistake. The language represented by the $\omega\mathsf{AT}$ expression for Büchi automata is not the language of an *existential* Büchi automaton (the automata defined in Sect. 3.1 are existential), but it represents the language of a *universal* Büchi automaton and that language is contained in the former.

At the moment, we have not worked on the practical implementation of the framework proposed here, but we plan to do it. We also want to adapt the method to handle state/event properties. In this paper, we only focus on event properties since they are simpler to handle in program verification, but state/event properties are useful for specifying program properties; experiments show that, in practice, this kind of property yields more efficient algorithms (spacewise and timewise) than a translation in a purely state or event-based formalism [2].

Acknowledgements

We thank the anonymous referees for their valuable comments.

References

1. Bolduc, C.: Oméga-algèbre — Théorie et application en vérification de programmes. Forthcoming M.Sc. thesis, Université Laval, Québec, Canada (2006)
2. Chaki, S., Clarke, E.M., Ouaknine, J., Sharygina, N., Sinha, N.: State/event-based software model checking. In: Fourth International Conference on Integrated Formal Methods (IFM) 2004. Volume 2999 of Lecture Notes in Computer Science, Springer-Verlag (2004) 128–147
3. Cohen, E.: Separation and reduction. In Backhouse, R., Oliveira, J.N., eds.: Proceedings of the 5th International Conference on Mathematics of Program Construction. Volume 1837 of Lecture Notes in Computer Science, Springer-Verlag (2000) 45–59
4. Cohen, E.: Omega algebra and concurrency control. Presentation made at the 56th meeting of the IFIP Working Group 2.1, Ameland, The Netherlands (2001)
5. Dwyer, M.B., Avrunin, G.S., Corbett, J.C.: Patterns in property specifications for finite-state verification. In: 21st International Conference on Software Engineering, IEEE Computer Society Press (1999) 411–420
6. Kozen, D.: Some results in dynamic model theory. Science of Computer Programming **51** (2004) 3–22
7. Kozen, D.: Kleene algebra with tests and the static analysis of programs. Technical report 2003-1915, Computer Science Department, Cornell University (2003)
8. Kozen, D., Smith, F.: Kleene algebra with tests: Completeness and decidability. In van Dalen, D., Bezem, M., eds.: 10th Int. Workshop on Computer Science Logic (CSL'96). Volume 1258 of Lecture Notes in Computer Science, Utrecht, The Netherlands, Springer-Verlag (1996) 244–259
9. Möller, B.: Lazy Kleene algebra. In Kozen, D., ed.: 7th International Conference on Mathematics of Program Construction. Volume 3125 of Lecture Notes in Computer Science, Springer-Verlag (2004) 252–273
10. Safra, S.: Complexity of Automata on Infinite Objects. Ph.D. thesis, Weizmann Institute of Science, Rehovot, Israel (1989)
11. Schneider, F.B.: Enforceable security policies. ACM Transactions on Information and System Security **3** (2000) 30–50
12. von Wright, J.: From Kleene algebra to refinement algebra. Volume 2385 of Lecture Notes in Computer Science, Springer-Verlag (2002) 233–262

Weak Contact Structures

Ivo Düntsch* and Michael Winter**

Department of Computer Science,
Brock University,
St. Catharines, Ontario, Canada, L2S 3A1
{duentsch, mwinter}@brocku.ca

Abstract. In this paper we investigate weak contact relations C on a lattice L, in particular, the relation between various axioms for contact, and their connection to the algebraic structure of the lattice. Furthermore, we will study a notion of orthogonality which is motivated by a weak contact relation in an inner product space. Although this is clearly a spatial application, we will show that, in case L is distributive and C satisfies the orthogonality condition, the only weak contact relation on L is the overlap relation; in particular no RCC model satisfies this condition.

1 Introduction

Various hybrid algebraic/relational systems have been proposed for reasoning about spatial regions, among them the Region Connection Calculus (RCC) [12], proximity structures [13], adjacency relations [9], and others. In most cases, the underlying algebraic structure is a Boolean algebra $\langle B, +, \cdot, ^*, 0, 1\rangle$ whose non-zero elements are called *regions*. One also has binary relations P and C, respectively called *part of relation* and *contact relation*. P is the underlying partial order of the algebra and constitutes the *mereological* part of the structure [11], and C is often regarded as its *topological* part [14]. C is related to P in varying degrees of strength. The part common to most axiomatizations for such structures consists of four axioms:

C0. $(\forall x)\neg 0Cx$
C1. $(\forall x)[x \neq 0 \Rightarrow xCx]$
C2. $(\forall x)(\forall y)[xCy \Rightarrow yCx]$
C3. $(\forall x)(\forall y)(\forall z)[xCy \text{ and } yPz \Rightarrow xCz]$.

Axiom C3 prescribes only a very weak connection between C and P, which holds in the most common interpretations: If a region x is in contact with a region y, and y is a part of z, then x is in contact with z.

A relation satisfying C0 – C3 will be called a *weak contact relation* (with respect to P). Note that these are indeed weaker than the contact relations of e.g. [2], which satisfy C0 – C3 and C5e below.

* The ordering of authors is alphabetical, and equal authorship is implied.

** Both authors gratefully acknowledge support from the Natural Sciences and Engineering Research Council of Canada.

W. MacCaull et al. (Eds.): RelMiCS 2005, LNCS 3929, pp. 73–82, 2006.

A *weak contact structure* is a tuple $\langle L, P, 0, 1, C\rangle$, where P is a bounded partial order on L (with smallest element 0 and largest element 1), and C is a weak contact relation (with respect to P). A *weak contact lattice* is a bounded lattice $\langle L, +, \cdot, 0, 1\rangle$ with a weak contact structure $\langle L, \leq, 0, 1, C\rangle$. We will usually denote a weak contact lattice by $\langle L, C\rangle$.

Some or all of the following additional axioms have appeared in various systems:

C4. $(\forall x)(\forall y)(\forall z)[xC(y+z) \Rightarrow (xCy \text{ or } xCz)]$ (The sum axiom)
C5c. $(\forall x)(\forall y)[(\forall z)(xCz \Rightarrow yCz) \Rightarrow xPy]$. (The compatibility axiom)
C5e. $(\forall x)(\forall y)[(\forall z)(xCz \Leftrightarrow yCz) \Rightarrow x = y]$. (The extensionality axiom)
C5d. $(\forall x \neq 1)(\exists y \neq 0)[x(-C)y]$. (The disconnection axiom)
C6. $(\forall x, y)[(x, y \neq 0 \wedge x + y = 1) \Rightarrow xCy]$ (The connection axiom)

If a weak contact relation satisfies one of the additional axioms, say C*x*, we will denote this by $C \models Cx$.

It may be worthy of mention that C5d is equivalent to

$$(\forall x)[(\forall z)(xCz \iff 1Cz) \Rightarrow x = 1] \tag{1.1}$$

C5e was one of the traditionally used axioms [14], and it is quite strong; it implies that each region is completely determined by those regions to which it is in contact. The difference between C5e and C5d is that in the former, C distinguishes any two non–zero elements, while with C5d, C distinguishes only 1 from any non–zero element, as can be seen from (1.1).

As we shall see below, C5c makes P definable from C, so that the primitive relation C suffices. Its strength also may be seen as a weakness, since on finite–cofinite Boolean algebras, and, in particular, finite Boolean algebras, the only relation C which satisfies C1 – C5c is the *overlap* relation, defined by

$$xOy \iff (\exists z)[z \neq 0 \wedge xP^{\smile} zPy]; \tag{1.2}$$

xOy means that x and y have a common non-zero part. If P is the underlying order $\leq$ of a lattice, then

$$xOy \iff (\exists z)[z \neq 0 \wedge z \leq x \wedge z \leq y], \tag{1.3}$$
$$\iff x \cdot y \neq 0. \tag{1.4}$$

If C satisfies, in addition, C6, then its underlying Boolean algebra is atomless [5]. This property makes, for example, the RCC [12] unsuitable for reasoning in finite structures. In order to remedy these effects, various measures have been proposed, all of which keep an underlying Boolean algebra, e.g. not demanding C5e, changing the basic relations, or employing second order structures [9, 7]. Another possibility is to relax the conditions on the underlying algebraic structure, e.g. not requiring that they be Boolean algebras; this seems sensible if our domain of interest does not include all possible Boolean combinations of regions.

Example 1. Consider the reference set

{North America, Canada,Mexico, Continental USA, ∅}

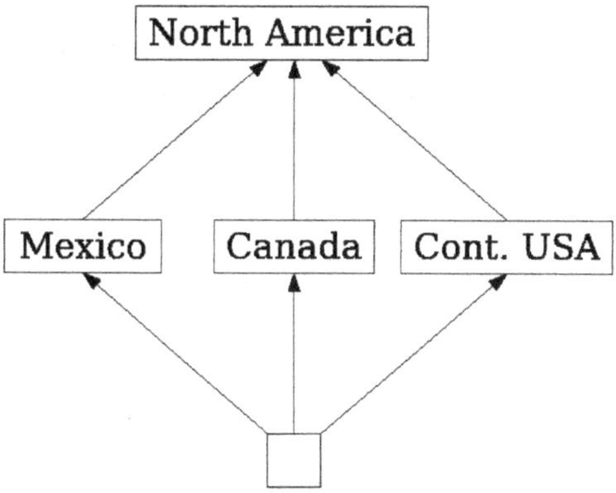

Fig. 1. Continental North America

with a "part–of" relation P. This relation generates the non–distributive lattice L shown in Figure 1. Weak contact in this example is defined by

$$xCy \iff x, y \neq 0 \text{ and } (xPy \text{ or } yPx).$$

Observe that here the overlap relation restricted to non-zero and non-universal regions, i.e. regions not equal to the empty region and 'North America', is just the identity.

Example 2. Another example[1] is the (non–distributive) lattice of linear subspaces of an inner product space where contact is given by $UCV \iff \neg(U \perp V)$, i.e., U and V are not orthogonal. We will return to this example in Section 5.

A related approach was put forward by Cohn and Varzi [1]. Motivated by topological considerations, they consider various types $\langle C, Q, \sigma \rangle$ of contact structures, where C is a reflexive and symmetric relation on a family of sets, a "parthood relation" Q is defined from C as

(1.5)
$$xQy \iff (\forall z)[xCz \Rightarrow yCz].$$

and a fusion operator σ, also defined from C. Loosely speaking, the fusion of a family of sets is its union. Q is clearly reflexive and transitive, but it need not be antisymmetric. It may be worthy to mention that Cohn and Varzi [1] do not, a priori, restrict the domain to algebraic or ordered structures.

In the present paper we investigate weak contact relations C on a lattice L, in particular, the interaction between the axioms C0 – C6 and the algebraic structure of the lattice. Our results are fairly basic, but we hope that they can assist the qualitative spatial reasoning community in choosing the axioms for (weak) contact relations appropriate for the domain under investigation.

2 Additional Definitions and Notation

If M is an ordered structure with smallest element 0, then $M^+ = \{y \in M : y \neq 0\}$. Furthermore, we shall usually identify a structure with its underlying set.

[1] We thank A. Urquhart for pointing this out.

Throughout, L will denote a bounded lattice. If $a \in L$, the *pseudocomplement* a^* of a is the largest $b \in L$ such that $a \cdot b = 0$. If every $a \in L$ has a pseudocomplement, then L is called *pseudocomplemented*. Observe that the lattice of Figure 3 shows that a pseudocomplemented lattice need not be distributive. On the other hand, it is well known that every complete distributive lattice is pseudocomplemented [10]. $a \in L^+$ is called *dense*, if $a \cdot b \neq 0$ for all $b \in L^+$. It is well known that $a + a^*$ is dense for each $a \in L$, and each dense element can be written in this form [10].

If $a \in L^+$, a pair $\langle b, c \rangle$ of non-zero elements of L is called a *partition of a* if $b + c = a$ and $b \cdot c = 0$.

For a set U, we denote by $\mathrm{Rel}(U)$ the set of all binary relations on U. If $R, S \in \mathrm{Rel}(U)$, then

$$R \,;\, S = \{\langle a, c \rangle : (\exists b)[aRb \text{ and } bRc]\}$$

is the *composition* of R and S, and $R^{\smile} = \{\langle a, b \rangle : bRa\}$ its *converse*. We also define $R(x) = \{y \in L : xRy\}$. The (right) *residual of R with respect to S* is the relation

$$R \setminus S = -(R^{\smile} \,;\, -S). \tag{2.1}$$

Here, complementation $-$ is taken in $\mathrm{Rel}(U)$. It is well known that

$$x(R \setminus S)y \iff R^{\smile}(x) \subseteq S^{\smile}(y), \tag{2.2}$$

and it is easy to see that $R \setminus R$ is a quasi order, i.e. reflexive and transitive. Furthermore, if $R, S, T \in \mathrm{Rel}(U)$, then the *de Morgan equivalences*

$$(R \,;\, S) \cap T = \emptyset \iff (R^{\smile} \,;\, T) \cap S = \emptyset \iff (T \,;\, S^{\smile}) \cap R = \emptyset \tag{2.3}$$

hold in $\mathrm{Rel}(U)$.

3 Mereology on Weak Algebraic Structures

Our first results concerns the algebraic structure of the collection of all contact relations on a lattice L:

Theorem 1. *The collection of weak contact relations is a complete lattice with smallest element O and largest element $L^+ \times L^+$.*

Proof. We only show that O is the smallest weak contact relation on L, and leave the rest to the reader. Clearly, O satisfies C0 – C3. Now, suppose that C is a weak contact relation on L. To show that $O \subseteq C$, let $x, y \in L^+$ such that xOy. By (1.4), there is some $z \neq 0$ with $z \leq x, y$. From C0 we obtain zCz which implies

$$zCz \overset{\text{C3}}{\Rightarrow} zCx \overset{\text{C2}}{\Rightarrow} xCz \overset{\text{C3}}{\Rightarrow} xCy. \qquad \square$$

It is known that for weak contact lattices which are Boolean algebras and satisfy C4, the axioms C5c, C5e, and C5d are equivalent. In our more general setting, we only have

Theorem 2. *C5c ⇒ C5e ⇒ C5d.*

Proof. The first implication follows immediately from the antisymmetry of $\subseteq$ and $\leq$. Suppose that C satisfies C5e. If xCz for all $z \neq 0$, then $x = 1$ by C5e which shows that C satisfies C5d. □

Let us next consider the case that C satisfies C4:

Theorem 3. *If $C \models C4$, then*

$$C \models C5c \iff C \models C5e.$$

Proof. "⇒" follows from Theorem 2. For the other direction, let $x, y \in L$ such that $C(x) \subseteq C(y)$; we will show that $x + y = y$: Suppose that $(x+y)Cz$; by C4, we have xCz or yCz, and from $C(x) \subseteq C(y)$, we obtain yCz in any case. Thus, $C(x+y) \subseteq C(y)$; since $C(y) \subseteq C(x+y)$ by C3, it follows that $C(x+y) = C(y)$, and C5e now implies that $x + y = y$, i.e. $x \leq y$. □

The following example shows that C4 and C5c are independent:

Example 3. There are weak contact lattices $\langle L_1, C\rangle, \langle L_2, C\rangle$ such that

1. $\langle L_1, C\rangle$ satisfies C5c and not C4.
2. $\langle L_2, C\rangle$ satisfies C4 and not C5e (and, hence, not C5c).

Indeed, for 1. consider Figure 2, and for 2. consider Figure 3.

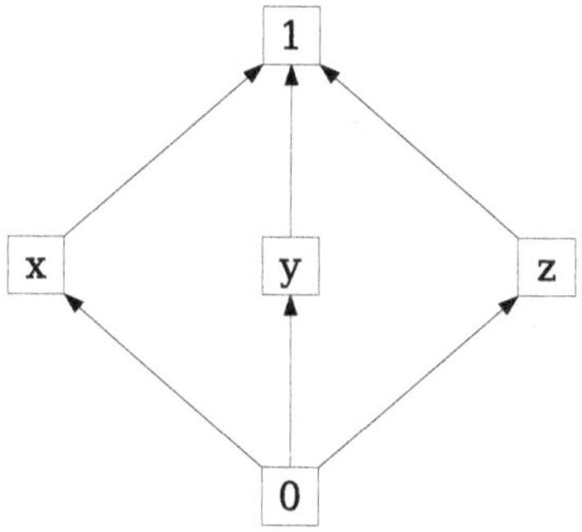

Fig. 2. $\langle L_1, O\rangle$

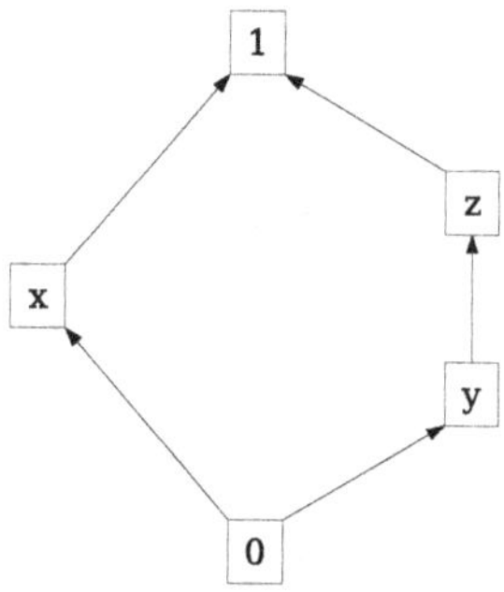

Fig. 3. $\langle L_2, O\rangle$

For the general case, we have

Theorem 4. *C5d ⇏ C5e ⇏ C5c.*

Proof. The lattice $\langle L_1, C\rangle$ of Figure 3 with $C = O$ satisfies C5d but not C5e: For each non-zero element there is a non-zero element disjoint to it, showing C5d. On the other hand, $O(y) = O(z)$, and $y \neq z$.

The lattice $\langle L_3, C\rangle$ shown in Figure 4 with $C = O \cup \{\langle x, z\rangle, \langle z, x\rangle\}$, satisfies C5e but not C5c: We have $C(x) \subseteq C(z)$, but $x \not\leq z$. □

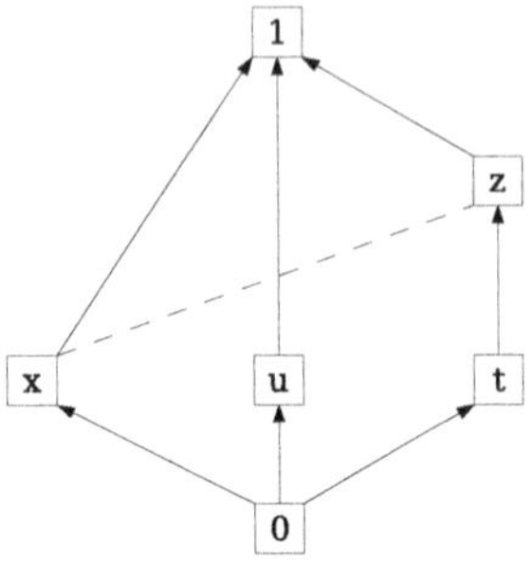

Fig. 4. $\langle L_3, C \rangle$

This leads to connections between C, $C \setminus C$, and P of different strengths:

Lemma 1. *1.* $C \,;\, (C \setminus C) \subseteq C$.
2. $C \setminus C$ *is antisymmetric if and only if* $C \models$ *C5e.*
3. $P \subseteq C \setminus C$.
4. $C \setminus C \subseteq P$ *if and only if* $C \models$ *C5c.*

Proof. 1. This follows immediately from the residual property of $C \setminus C$; for completeness, we give a short proof. Assume that $C \,;\, (C \setminus C) \not\subseteq C$, i.e. $[C \,;\, (C \setminus C)] \cap -C \neq \emptyset$. Then, the de Morgan equivalences (2.3), and the fact that C is symmetric show that

$$[C \,;\, (C \setminus C)] \cap -C \neq \emptyset \iff (C \,;\, -C) \cap (C \setminus C) \neq \emptyset$$
$$\iff (C \,;\, -C) \cap -(C \,;\, -C) \neq \emptyset,$$

a contradiction.

We observe that 1. is a compatibility condition such as C3 with respect to $C \setminus C$. Thus, it is implicitly valid in the setup of [1].

2. This was already observed without proof in [1], see also [4]. Let $C \setminus C$ be antisymmetric, and $a, b \in L$ such that $C(a) = C(b)$. From (2.2) and the fact that C is symmetric, we obtain that $a(C \setminus C)b$ and $b(C \setminus C)a$, and our hypothesis implies that $a = b$. Thus, $C \models$ C5e. Conversely, let $a(C \setminus C)b$ and $b(C \setminus C)a$. By (2.2), this implies $C(a) = C(b)$, and hence, $a = b$ by C5e.

3. Consider the following:

$$(x \leq y \wedge xCz) \overset{C2}{\Rightarrow} (zCx \wedge x \leq y) \overset{C3}{\Rightarrow} zCy \overset{C2}{\Rightarrow} yCz \Rightarrow C(x) \subseteq C(y) \Rightarrow x(C \setminus C)y.$$

4. This is just the definition. □

Observe that Figure 4 shows a weak contact lattice where $C \setminus C$ is a partial order, that strictly contains P; the additional pairs are $\langle x, z \rangle, \langle z, x \rangle$. It may be noted that, even if $C \setminus C$ is a partial order, it need not be a lattice order.

Corollary 1. *C5c implies C3.*

Proof. Since $C \models$ C5c, Lemma 1(3,4) give us $P = C \setminus C$. C3 now follows from Lemma 1(1). □

Already our weakest extensionality axiom may have an effect on the algebraic structure of L:

Theorem 5. *If L is a distributive bounded pseudocomplemented lattice and C satisfies C5d, then L is a Boolean algebra.*

Proof. Suppose that $x \neq 0, 1$, and assume that $x + x^* \neq 1$. Then, by C5d, there is some $z \neq 0$ such that $z(-C)(x + x^*)$. Hence, $z \cdot (x + x^*) = 0$, which contradicts the fact that $x + x^*$ is dense; therefore, $x + x^* = 1$, i.e. x^* is a complement of x. Since in distributive lattices complements are unique, it follows that L is a Boolean algebra. □

This strengthens a Theorem of [3].

In [6] we have shown that for a distributive L, $C \models$ C4 and $C \models$ C5e imply that $C = O$. Our next example shows that C4 is essential:

Example 4. Consider the sixteen element Boolean algebra shown in Figure 5. There, C is the smallest contact relation containing O and $\langle \{1,2\}, \{3,4\} \rangle$; $\uparrow M$ denotes the upset $\{x : (\exists y \in M) y \leq x\}$ induced by M. Clearly, C is a weak contact relation different from O, and we see from the table that it satisfies C5e, but not C4.

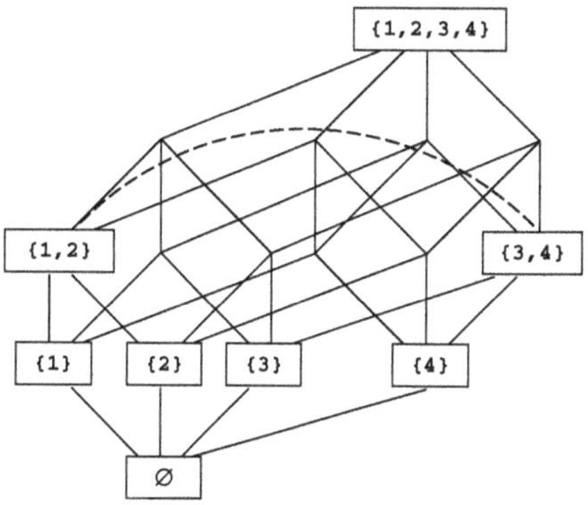

$C(\{1\})$	$= \uparrow\{1\}$
$C(\{2\})$	$= \uparrow\{2\}$
$C(\{3\})$	$= \uparrow\{3\}$
$C(\{4\})$	$= \uparrow\{4\}$
$C(\{1,2\})$	$= \uparrow\{3,4\} \cup \uparrow\{1\} \cup \uparrow\{2\}$
$C(\{1,3\})$	$= \uparrow\{1\} \cup \uparrow\{3\}$
$C(\{1,4\})$	$= \uparrow\{1\} \cup \uparrow\{4\}$
$C(\{2,3\})$	$= \uparrow\{2\} \cup \uparrow\{3\}$
$C(\{2,4\})$	$= \uparrow\{2\} \cup \uparrow\{4\}$
$C(\{3,4\})$	$= \uparrow\{1,2\} \cup \uparrow\{3\} \cup \uparrow\{4\}$
$C(\{1,2,3\})$	$= \uparrow\{1\} \cup \uparrow\{2\} \cup \uparrow\{3\}$
$C(\{1,3,4\})$	$= \uparrow\{1\} \cup \uparrow\{3\} \cup \uparrow\{4\}$
$C(\{1,2,4\})$	$= \uparrow\{1\} \cup \uparrow\{2\} \cup \uparrow\{4\}$
$C(\{2,3,4\})$	$= \uparrow\{2\} \cup \uparrow\{3\} \cup \uparrow\{4\}$

Fig. 5. A BA with $C \neq O$ satisfying C5c and not C4

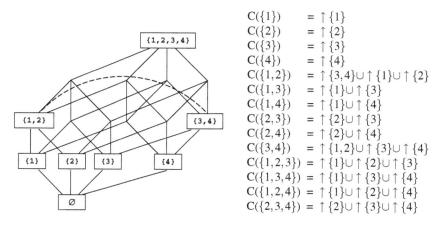

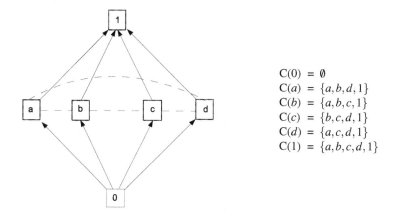

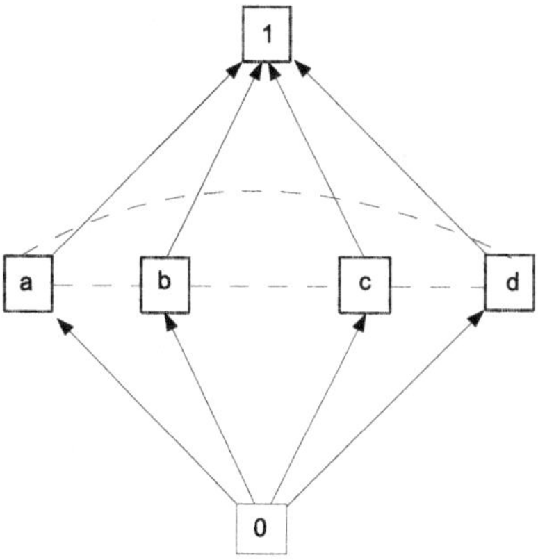

$C(0)$	$= \emptyset$
$C(a)$	$= \{a, b, d, 1\}$
$C(b)$	$= \{a, b, c, 1\}$
$C(c)$	$= \{b, c, d, 1\}$
$C(d)$	$= \{a, c, d, 1\}$
$C(1)$	$= \{a, b, c, d, 1\}$

Fig. 6. A non–distributive modular lattice with $C \neq O$ satisfying C0–C5d and N

An example of a weak contact relation satisfying all of C0–C5c with an underlying modular and non–distribute lattice is shown in Figure 6.

4 Overlap and the Lattice Structure

Since $xOy \iff x \cdot y \neq 0$, we can expect that additional properties of the weak contact relation O are strongly related to the lattice properties. In this section, we will explore this relationship.

First, observe that (1.4) implies that O is the universal relation on L^+ in case 0 is meet irreducible. Thus, in this case O is the only contact relation.

Theorem 6. $C = O \iff (\forall x)(\forall y)(\forall z)[xC(y \cdot z) \Leftrightarrow (x \cdot y)Cz]$.

Proof. "$\Rightarrow$": This follows immediately from (1.4) and the associativity of $\cdot$.

"$\Leftarrow$": By Theorem 1 it is sufficient to show $C \subseteq O$. Suppose xCy. Then, $xC(y \cdot 1)$, and hence $(x \cdot y)C1$ using the assumption on C. From C0 we conclude $x \cdot y \neq 0$. □

Lemma 2. *1. If L is distributive then O satisfies C4.*
2. If O satisfies C4 and C5e then L is distributive.
3. If L is a bounded pseudocomplemented lattice, then O satisfies C5c.
4. If L is a bounded pseudocomplemented lattice, then

$$L \text{ is a Boolean algebra} \iff O \text{ satisfies C4 and C5e.}$$

Proof. 1. Suppose that L is distributive, and let $aO(b+c)$; then $a \cdot (b+c) \neq 0$. Since L is distributive, this is equivalent to $a \cdot b + a \cdot c \neq 0$. It follows that $a \cdot b \neq 0$ or $a \cdot c \neq 0$, i.e. aOb or aOc.

2. Let $a, b, c, d \in L$. Then, we have

$$\begin{aligned} dO[a \cdot (b+c)] &\iff (d \cdot a)O(b+c) && \text{by Theorem 6} \\ &\iff (d \cdot a)Ob \text{ or } (d \cdot a)Oc && \text{by C3 and C4} \\ &\iff dO(b \cdot a) \text{ or } dO(c \cdot a) && \text{by Theorem 6} \\ &\iff dO(a \cdot b + a \cdot c) && \text{by C3 and C4} \end{aligned}$$

Thus, $O(a \cdot (b+c)) = O(a \cdot b + a \cdot c)$, and C5e implies $a \cdot (b+c) = a \cdot b + a \cdot c$.

3. Suppose $O(x) \subseteq O(y)$. If $x \cdot y^* \neq 0$ we conclude $y^* \in O(x) \subseteq O(y)$ and hence $y \cdot y^* \neq 0$, a contradiction. Therefore, $x \cdot y^* = 0$ which is equivalent to $x \leq y$.

4. "$\Rightarrow$": By 1. we know that O satisfies C4 and by 2. that O satisfies C5c. The latter implies C5e.

"$\Leftarrow$": By 2. we know that L is distributive. Since C5e implies C5d, we may conclude from Theorem 5 that L is a Boolean algebra . □

It can be seen from Figure 3 that the converse of 1. of Lemma 2 is not true: O satisfies C4, but the lattice is not distributive. Furthermore, we cannot replace O by an arbitrary contact relation C: If L is the eight element Boolean algebra with atoms x, y, z, let C be the smallest contact relation on L such that $xC(y+z)$. Then, $x(-C)y$ and $x(-C)z$, showing that $C \not\models$ C4.

The converse of 2. of Lemma 2 does not hold either: In any bounded chain we have $C(x) = C(1)$ for all $x \neq 0$, and thus, C5d is not satisfied.

Observe that in the proof of Lemma 2(3), C4 was only used to establish distributivity. Since every finite distributive bounded lattice is pseudocomplemented, the previous results show that, for finite L, we have to give up C4 or C5e or both not to result in a Boolean algebra, where O is the only contact relation which satisfies C0 – C5e. In other words, if L is a bounded distributive lattice which is not a Boolean algebra, then O does not satisfy C5d, see Lemma 2.

5 Orthogonality

Our notion of orthogonality is motivated by the weak contact relation in an inner product space given in Example 2. There, weak contact of two linear subspaces U and V was defined by $UCV \iff \neg(U \perp V)$, i.e. there are $u \in U$ and $v \in V$ with $\langle u, v\rangle \neq 0$; here, $\langle u, v\rangle$ is the inner product of u and v. It is easy to verify that this relation indeed satisfies C0-C3.

Example 2 suggests the following definition: A pair of non-zero elements $\langle y, z\rangle$ is called an *orthogonal partition* of an element x if $y(-C)z$ and $y + z = x$. Note, that $y(-C)z$ implies $y \cdot z = 0$, and that the definition is symmetric.

By Theorem 1 we have $O \subseteq C$ so that every orthogonal partition is a partition.

In the example, we are able to switch from a given partition to an orthogonal one using the orthonormalization procedure by Gram and Schmidt; this can be done by keeping one subspace and varying the other one. We use an abstract version of this technique as an additional property of weak contact providing orthogonalization:

N. $(\forall x)(\forall y)(\forall z)[\langle y, z\rangle$ is a partition of $x \Rightarrow$
$(\exists u)\langle y, u\rangle$ is an orthogonal partition of $x]$.

The weak contact relation in Example 2 has property N. Notice, that O always satisfies N.

If the underlying lattice is distributive, then N is very restrictive:

Lemma 3. *Let L be distributive.*

1. *If $\langle y, u\rangle$ and $\langle y, v\rangle$ are partitions of x then $u = v$.*
2. *If C satisfies N then $C = O$.*

Proof. 1. By definition we have $y \cdot u = 0 = y \cdot v$ and $y + u = x = y + v$ so that we conclude $u = v$ by the distributivity of L.

2. By Theorem 1 it remains to show that $C \subseteq O$. Suppose $y(-O)z$. If $y = 0$ or $z = 0$ we conclude $y(-C)z$ using C0. Otherwise, $\langle y, z\rangle$ is a partition of $y + z$. By property N there is a $u \in L$ so that $\langle y, u\rangle$ is an orthogonal partition of $y + z$. Since every orthogonal partition is a partition we conclude $z = u$ using 1. This implies $y(-C)z$. □

Corollary 2. *If L is a bounded pseudocomplemented lattice, then*

L is a Boolean algebra and C satisfies N $\iff C = O$ and C satisfies C4 and C5e.

Proof. The claim follows immediately from Lemma 3 and Lemma 2. □

The weak contact structure $\langle L, C \rangle$ given in Figure 6 satisfies C0–C5d as well as N; furthermore, $C \neq O$.

In any RCC model we have $C \neq O$ because of C6. Therefore, the previous theorem shows that there is no RCC model satisfying property N.

Finally, the weak contact structure of Figure 6 exhibits a weak contact relation $C \neq O$ on a non–distributive modular lattice which satisfies C0-C5e and N.

6 Conclusion and Outlook

We have looked at the "fine structure" of the interplay between properties of contact relations and those of the underlying algebra. In particular, we have shown that the various forms of extensionality – and their interplay with C4 – do not coincide when our basic structure is not a Boolean algebra. Our results are elementary and not difficult to prove; nevertheless, we hope that they are useful for the axiomatization for application domains where not all possible Boolean combinations of regions are required. We intend to continue our investigations by considering still weaker structures such as semi–lattices, and also topological domains in the spirit of [1]. Additionally, the topic of orthogonal partitions and weak contact relations satisfying N merits further attention.

References

1. Cohn, A.G., Varzi, A.: Connection relations in mereotopology. In Prade, H., ed.: Proceedings of the 13th European Conference on Artificial Intelligence (ECAI-98), Chichester, John Wiley & Sons (1998) 150–154
2. Düntsch, I., Orłowska, E.: A proof system for contact relation algebras. Journal of Philosophical Logic **29** (2000) 241–262
3. Düntsch, I., Orłowska, E., Wang, H.: Algebras of approximating regions. Fundamenta Informaticae **46** (2001) 71–82
4. Düntsch, I., Wang, H., McCloskey, S.: Relation algebras in qualitative spatial reasoning. Fundamenta Informaticae (2000) 229–248
5. Düntsch, I., Wang, H., McCloskey, S.: A relation algebraic approach to the Region Connection Calculus. Theoretical Computer Science **255** (2001) 63–83
6. Düntsch, I., Winter, M.: A representation theorem for Boolean contact algebras. Theoretical Computer Science (B) **347** (2005) 498–512
7. Eschenbach, C.: A predication calculus for qualitative spatial representation. [8] 157–172
8. Freksa, C., Mark, D.M., eds.: Spatial Information Theory, Proceedings of the International Conference COSIT '99. In Freksa, C., Mark, D.M., eds.: Spatial Information Theory, Proceedings of the International Conference COSIT '99. Lecture Notes in Computer Science, Springer–Verlag (1999)
9. Galton, A.: The mereotopology of discrete space. [8] 251–266
10. Grätzer, G.: General Lattice Theory. Birkhäuser, Basel (1978)
11. Leśniewski, S.: O podstawach matematyki. **30–34** (1927 – 1931)
12. Randell, D.A., Cohn, A.G., Cui, Z.: Computing transitivity tables: A challenge for automated theorem provers. In Kapur, D., ed.: Proceedings of the 11th International Conference on Automated Deduction (CADE-11). Volume 607 of LNAI., Saratoga Springs, NY, Springer (1992) 786–790
13. Vakarelov, D., Dimov, G., Düntsch, I., Bennett, B.: A proximity approach to some region–based theories of space. J. Appl. Non-Classical Logics **12** (2002) 527–529
14. Whitehead, A.N.: Process and reality. MacMillan, New York (1929)

On Relational Cycles

Alexander Fronk and Jörg Pleumann

Software Technology,
University of Dortmund, D-44221 Dortmund, Germany

Abstract. We provide relation-algebraic characterisations of elementary, ordinary, and maximal cycles in graphs. Relational specifications for the enumeration of cycles are provided. They are executable within the RELVIEW and RELCLIPSE tools and appear to be useful in various applications. Particularly, cycles offer a valuable instrument for analysing Petri Nets.

1 Introduction and Related Work

Cycles, as specific sequences of nodes and edges, play an important role in graph theory. In logistics, for instance, one likes to cover as much territory as possible by a course leading from a depot via all customers back to the depot without using any road in both directions. Eulerian cycles provide such courses in which every road is used exactly once. Hamiltonian cycles, as their dual, visit each customer exactly once. For bipartite graphs, specific assumptions regarding the existence of such cycles need to be made leading to specialised algorithms (cf. [1]). Considering relation algebra, there also exist algorithms to detect Eulerian paths [13] and Hamiltonian cycles in homogeneous relations [12]. Surprisingly, apart from these specific applications, hardly any work on paths and cycles exists except for a definition of cycle-freeness, the determination of nodes lying on a cycle, and criteria for the existence of cycles in directed and undirected graphs (cf. [15]). Particularly for the enumeration of cycles — to the best of our knowledge — no relational characterisation is given in the literature.

For many applications, it is sufficient to know the set S of nodes a cycle of a graph $\mathcal{G}$ consists of: From S, the (possibly infinite) uniquely determined set of all cycles of $\mathcal{G}$ can be deduced by assuming each node in S to be a starting node and only nodes from S to be traversed. We provide a relational characterisation of such node sets — instead of sequences of nodes and edges — and develop criteria for deciding whether a given subset S of nodes of a graph $\mathcal{G}$ is the node set of a maximal cycle, an ordinary cycle, or an elementary cycle in $\mathcal{G}$. The set S then allows for deducing sets of infinitely-many sequences of nodes and edges forming maximal or ordinary cycles over S, and sets of finitely-many finite sequences of nodes and edges forming elementary cycles over S.

In this paper we provide relation-algebraic characterisations of maximal, ordinary, and elementary cycles which we relate to strongly connected components: It turns out that the set of nodes of such a component forms the node set of a maximal cycle, and that other node sets can be characterised with respect

W. MacCaull et al. (Eds.): RelMiCS 2005, LNCS 3929, pp. 83–95, 2006.

to these components. Further, we develop relation-algebraic specifications for the enumeration of node sets of any cycles in homogeneous as well as in heterogeneous relations. Efficiency considerations are not a concern. Although the specifications are executable in RelView [3], RelClipse [11], or any other tool, such as PetRA [9], using the Kure-Java library [10] providing a mechanisation of relational algebra based on BDDs, they are still specifications. That is, they mathematically describe what needs to be considered to enumerate cycles but they do not claim to be efficiently executable. Thus, they serve as both a mathematical prototype and a test base for efficiently implemented algorithms. The latter will normally use appropriate and diverse data structures without any artificial reduction to relations only. A relational specification for the enumeration of elementary cycles that is of high-performance on medium-sized graphs is provided in the paper by Berghammer and Fronk on Feedback Vertex Sets [4].

The specifications provided here are found to be very useful in various applications. In particular, we apply them in our tool PetRA tailored for the relation-algebraic analysis of Petri Nets. Petri Nets are the most prominent representative of bipartite graphs in computer science. They are capable of and scalable to modelling many different systems and applications, for example, hardware layout or production lines in large factories. The enumeration of cycles does not only visually enhance the comprehension of the Petri Net topology and therefore of the system or application it models. Moreover, some Petri Net classes use cycles as a basic modelling concept. Our specifications thus allow to verify whether a Petri Net belongs to such a specific net class. Visual comprehension of a system complements its automated relational analysis as discussed in [5] for condition/event nets and in [8, 9] for place/transition systems.

The paper is organised as follows. Sect. 2 equips the reader with the relation-algebraic notation used in this paper; readers seeking a deeper understanding of relation algebra are referred to [15]. In Sect. 3, we provide the basic definitions needed to establish a characterisation of cycles and relate them to graph-theoretic ones. Sects. 4, 5, and 6 provide specifications for handling maximal, ordinary, and elementary cycles, respectively. We briefly discuss an application to Petri Nets in Sect. 7. The paper concludes in Sect. 8.

2 Preliminaries

A relation R between two sets X and Y is denoted by $R : X \leftrightarrow Y$. We write $[X \leftrightarrow Y]$ for the set of all relations over $X \times Y$ and always consider these sets to be nonempty. We write $R_{x,y}$ instead of $\langle x, y\rangle \in R$ since we consider relations as Boolean matrices.

For two relations R and S, $R \cup S$, $R \cap S$, and $R \subseteq S$ denote *union*, *intersection*, and *inclusion*, respectively, R^T denotes the *converse* of R, $\overline{R}$ its *negation*, and $R; S$ the *relational multiplication*. The *empty relation* is denoted by $\mathbb{O}$, the *universal relation* by $\mathbb{L}$, and the *identity relation* by $\mathbb{I}$.

A *vector* v on a set X, i.e., a relation v with $v; \mathbb{L} = v$, is denoted by $v : X \leftrightarrow \mathbb{1}$ for any singleton set $\mathbb{1} := \{\diamond\}$. We omit $\diamond$ as subscript and write v_x instead of

$v_{x,\diamond}$. A vector can be considered as a Boolean matrix with exactly one column, i.e., as a Boolean column vector, and represents the subset $\{x \in X \mid v_x\}$ of X. The vectors $(R; \mathbb{L}) : X \leftrightarrow \mathbb{1}$ and $(R^\mathsf{T}; \mathbb{L}) : Y \leftrightarrow \mathbb{1}$ are defined as the *domain* and *co-domain* of $R : X \leftrightarrow Y$, respectively. A *point* p on a set X is a vector $p : X \leftrightarrow \mathbb{1}$ with $p; p^\mathsf{T} \subseteq \mathbb{I}$ and $\mathbb{L}; p = \mathbb{L}$.

The *membership* relation $\in$ is modelled by a relation $\varepsilon : X \leftrightarrow 2^X$ on a set X and its powerset such that $\varepsilon_{x,X} :\Longleftrightarrow x \in X$. Each point $p : 2^X \leftrightarrow \mathbb{1}$ on a powerset can be transformed into a vector of type $[X \leftrightarrow \mathbb{1}]$ by $\varepsilon; p$. Each vector $v : 2^X \leftrightarrow \mathbb{1}$ modelling a subset S of 2^X can be transformed into a relation P of type $[X \leftrightarrow |S|]$, such that each column of P represents an element of S, by using the injection $inj(v) : S \leftrightarrow 2^X$ and calculating P through $\varepsilon; inj(v)^\mathsf{T}$.

We denote the *right residual*, defined through $\overline{R^\mathsf{T}\overline{S}}$, by $R\backslash S$. The *symmetric quotient*, defined through $(R\backslash S) \cap (\overline{R}\backslash \overline{S})$, is denoted by $\mathsf{syQ}(R, S)$.

R^{i+1} denotes $R; R^i$ for $i \geq 0$ with $R^0 = \mathbb{I}$. The relation $R^+ := \bigcup_{i \geq 1} R^i$ is called the *transitive closure* of R, and $R^* := \bigcup_{i \geq 0} R^i = R^+ \cup \mathbb{I}$ the *reflexive transitive closure* of R. A relation R is *acyclic* if and only if $R^+ \subseteq \overline{\mathbb{I}}$ holds.

3 Basic Definitions

Throughout this paper, we refer to $\mathcal{G}$ as any arbitrary but fixed directed graph of the form $\mathcal{G} = (V, R)$ with V a set of nodes and $R : V \leftrightarrow V$ a relation on these nodes, i.e. the set of edges of $\mathcal{G}$. In correspondence with relation algebra, we write $R_{x,y}$ if (x, y) is an edge in $\mathcal{G}$.

Definition 1 (Ordinary cycle). *An* ordinary cycle $\mathcal{C}$ *of* $\mathcal{G}$, *or* cycle *for short, is a finite sequence* $\langle n_0, \ldots, n_l \rangle$ *with* $l \geq 1$ *such that* $n_0, \ldots, n_l \in V$ *and both* $n_0 = n_l$ *and* $R_{n_i, n_{i+1}}$ *for each* $0 \leq i \leq l - 1$ *hold.*

Remark 1.

- In graph theory, such a sequence of nodes of a cycle $\mathcal{C}$ is called the *walk* of $\mathcal{C}$. It is uniquely defined by the nodes a cycle visits.
- In the remainder, we reasonably exclude cycles of infinite length, for example due to the repetition of loops, since it does not affect our characterisation but would unnecessarily complicate cycle treatment.

Definition 2 (Elementary cycle). *A cycle* $\mathcal{C}$ *of* $\mathcal{G}$ *is called* elementary *if*

1. *no node* $n \in V$ *occurs more than once in* $\mathcal{C}$ *except for* $n_0 = n_l$ *and if*
2. *there exist no two nodes* n_i *and* n_j *in* $\mathcal{C}$ *with* $j - i \geq 2$ *such that* R_{n_i, n_j} *holds.*

Remark 2. The second condition excludes such cycles $\mathcal{C}$ from being called elementary for which there exists an edge (x, y) in $\mathcal{G}$ between two non-neighbored nodes x and y in $\mathcal{C}$. Such an edge is called a *chord*. Elementary cycles, as we define them, are thus chordless w.r.t. $\mathcal{C}$. In graph theory, this condition is not required to hold for elementary cycles.

Definition 3 (Maximal cycle). *A cycle $\mathcal{C}$ of $\mathcal{G}$ is called* maximal *if there exists no nonempty sequence σ of nodes such that $\langle n_0, \ldots, n_i\rangle \circ \sigma \circ \langle n_{i+1} \ldots, n_l\rangle$ forms a cycle in $\mathcal{G}$, $0 \leq i < l$ ("$\circ$" denotes the usual concatenation of sequences).*

It is clear by these definitions that both elementary and maximal cycles are also ordinary ones, and that a cycle can be maximal and elementary simultaneously.

Definition 4 (Support). *The* support *of a cycle $\mathcal{C}$, $supp(\mathcal{C})$ for short, is defined as the set of nodes $\{n \in V \mid n \in \langle n_0, \ldots, n_l\rangle\}$.*

Example 1 (Supports). Figure 1 visualises supports of each kind of cycle in an arbitrary graph, shaded in gray. The node v_0 does not participate in any cycle. If $\mathcal{G}$ contained the edge $e = (v_7, v_4)$, then b) would not denote the support of an elementary cycle since e is then a chord w.r.t the cycle shown. In this case, $\{v_4, v_6, v_7\}$ would be a proper support.

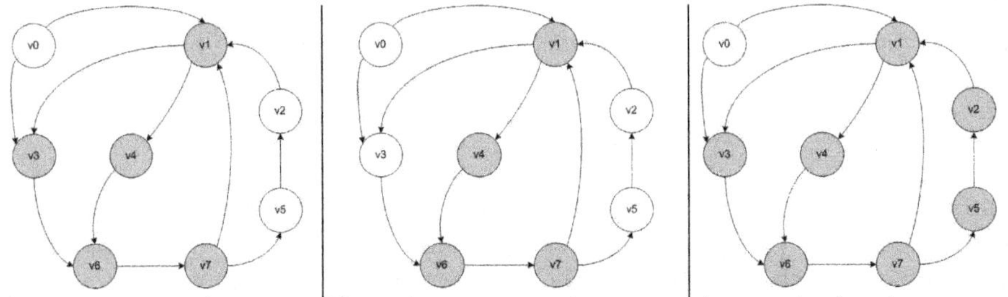

a) an ordinary cycle support | b) an elementary cycle support | c) a maximal cycle support

Fig. 1. Supports of cycles in a graph

In a support, the information on which node is adjacent to which other is lost. Nonetheless, from the support $supp(\mathcal{C})$ of a cycle $\mathcal{C}$ the set of all cycles visiting exactly the nodes in $supp(\mathcal{C})$ can uniquely be determined. If chords were permitted in elementary cycles, however, and if $supp(\mathcal{C})$ was the support of such a cycle, the set of cycles deduced from $supp(\mathcal{C})$ would contain non-elementary ones as well. Then, no distinction between ordinary and elementary cycle supports would be possible without checking the entire set of cycles deduced from a support. Our objective is to develop relation-algebraic specifications on how to enumerate supports of maximal cycles (Sec. 4), ordinary cycles (Sec. 5), and elementary cycles (Sec. 6). Therefore, we need sufficient characterisations to determine the kind of cycle a set of nodes is the support of. Hence, providing a relational characterisation for checking whether an arbitrary set of nodes is the support of a cycle or not is the crucial factor. Pairwise reachability will be the main criterion here. To see this, we first define the notion of relational reduction.

Definition 5 (Reduction). *Let $R : X \leftrightarrow X$ be a homogeneous relation, and let $v : X \leftrightarrow \mathbb{1}$ be a vector modelling any subset of X. The* reduction of R by v, *$R|_v : X \leftrightarrow X$ for short, is defined as $R \cap (v; v^{\mathsf{T}})$.*

Remark 3. It is easy to see that a pair (x, x') is contained in $R|_v : X \leftrightarrow X$ if and only if it is contained in R and both x and x' are contained in v.

Proposition 1. *A set of nodes $S \subseteq V$ modelled by a vector $v : V \leftrightarrow \mathbb{1}$ is the support of a cycle $\mathcal{C}$ if and only if the nodes in S are pairwise reachable with respect to $(R|_v)^*$.*

Proof. "⇒" From Def. 1 it clearly follows that the nodes forming a cycle are pairwise reachable, i.e. both ${R^*}_{x,y}$ and ${R^*}_{y,x}$ hold for any two nodes $x, y \in$ *supp*$(\mathcal{C})$. If we reduce R to the nodes contained in *supp*$(\mathcal{C})$ this property trivially holds as well.

"⇐" Let $v : V \leftrightarrow \mathbb{1}$ model a set S of pairwise reachable nodes with respect to $(R|_v)^*$. Further assume that S is not the support of a cycle. Then, there exists a node $x \in S$ with either no predecessor or no successor in S. Therefore, the nodes in S cannot be pairwise reachable with respect to $(R|_v)^*$. □

4 Maximal Cycles

The support of a maximal cycle can be characterised by means of strongly connected components as relation-algebraically defined in [15, Def. 6.1.6].

Definition 6 (Strongly connected component). *A set of nodes $S \subseteq V$ is called* strongly connected component *of $\mathcal{G}$ if S is an equivalence class of the equivalence relation $(R^* \cap R^{*\mathsf{T}})$, i.e., two nodes x and y are in S if and only if both ${R^*}_{x,y}$ and ${R^*}_{y,x}$ hold.*

Proposition 2 (Characterisation of maximal cycle supports). *A set of nodes $S \subseteq V$ is the support of a maximal cycle in $\mathcal{G}$ if and only if S is a strongly connected component of $\mathcal{G}$.*

Proof. "⇒" Let S be the support of a maximal cycle, and assume $x, y \in S$. By Prop. 1, this implies both ${R^*}_{x,y}$ and ${R^*}_{y,x}$ (resp. ${R^{*\mathsf{T}}}_{x,y}$). In converse, with Def. 3, ${R^*}_{x,y}$ and ${R^{*\mathsf{T}}}_{x,y}$ imply $x, y \in S$ since otherwise S is not the support of a *maximal* cycle. Together, x and y are in S if and only if both ${R^*}_{x,y}$ and ${R^{*\mathsf{T}}}_{x,y}$ hold, and therefore, together with Def. 6, S is a strongly connected component.

"⇐" Let S be a strongly connected component of $\mathcal{G}$. Assume S not to be the support of a cycle. Then, there exists a node $y \in S$ such that either ${R^*}_{x,y}$ or ${R^{*\mathsf{T}}}_{x,y}$ does not hold for some $x \in S$. Hence, S is not a strongly connected component.

Furthermore, S is the support of a maximal cycle: assume $\mathcal{C}'$ to be a cycle in $\mathcal{G}$ with $S \subset$ *supp*$(\mathcal{C}')$. Then, there exists a node $y \in$ *supp*$(\mathcal{C}') \setminus S$ with both ${R^*}_{x,y}$ and ${R^{*\mathsf{T}}}_{x,y}$ for some $x \in S$, and hence S is not a strongly connected component. □

The determination of all maximal cycle supports is therefore equivalent to the enumeration of all strongly connected components. In relation algebra, this task is now straightforward.

Proposition 3 (Enumeration of strongly connected components). *The set of all strongly connected components of $\mathcal{G}$ can be represented as a vector $SCC : 2^V \leftrightarrow \mathbb{1}$ defined as* $\mathsf{syQ}(\varepsilon, (R^* \cap R^{*\mathsf{T}})); \mathbb{L}$.

Proof. We proof the proposition by stepwise deriving Def. 6 as follows:

$$\begin{aligned} SCC_S :&\Longleftrightarrow (\mathsf{syQ}(\varepsilon, (R^* \cap R^{*\mathsf{T}})); \mathbb{L})_S \\ &\Longleftrightarrow \exists y : \mathsf{syQ}(\varepsilon, (R^* \cap R^{*\mathsf{T}}))_{S,y} \wedge \mathbb{L}_y \qquad \text{where } \mathbb{L} : V \leftrightarrow \mathbb{1} \\ &\Longleftrightarrow \exists y : \forall x : \varepsilon_{x,S} \leftrightarrow ((R^* \cap R^{*\mathsf{T}}))_{x,y} \qquad \text{where } \varepsilon : V \leftrightarrow 2^V \\ &\Longleftrightarrow \exists y : \forall x : x \in S \leftrightarrow R^*_{x,y} \wedge R^*_{y,x} \end{aligned}$$

□

Remark 4. To represent this enumeration as a list $MCList : V \leftrightarrow |SCC|$ of column vectors such that each vector models a maximal cycle of $\mathcal{G}$, we use the injection as shown in Sect. 2 and get $MCList := \varepsilon; inj(SCC)^{\mathsf{T}}$.

Given a set of nodes S, it is interesting to know whether there exists a maximal cycle in $\mathcal{G}$ completely containing S and, if so, which support it has. That is, we filter out the uniquely determined strongly connected component containing S. This task can be performed without enumerating all strongly connected components:

Proposition 4 (Maximal cycle support through given nodes). *Let $v : V \leftrightarrow \mathbb{1}$ be a vector modelling a set $S \subseteq V$ of nodes of $\mathcal{G}$. The support of the maximal cycle containing S, $MC(v) : V \leftrightarrow \mathbb{1}$, is given through $MC(v) := (v \backslash (R^* \cap R^{*\mathsf{T}}))^{\mathsf{T}}$.*

Proof. With Prop. 2, the support of the maximal cycle containing S is equal to the uniquely determined strongly connected component containing S. A vector $MC(v) : V \leftrightarrow \mathbb{1}$ modelling this support is thus the greatest vector containing nodes $y \in V$ such that $(R^* \cap R^{*\mathsf{T}})_{x,y}$ holds for each $x \in S$. Hence, $MC(v)^{\mathsf{T}}$ is the greatest solution X of $v; X \subseteq (R^* \cap R^{*\mathsf{T}})$. Therefore, $MC(v)$ is equal to the converse of the right-residual $v \backslash (R^* \cap R^{*\mathsf{T}})$. □

Remark 5. If this maximal cycle support does not exist, $MC(v)$ is empty.

5 Ordinary Cycles

In the light of Prop. 1 and Def. 6, the support of an ordinary cycle is either a maximal cycle support, or it is contained therein. As an immediate consequence of Prop. 1, the following characterisation of ordinary cycle supports is trivially true:

Proposition 5 (Characterisation of ordinary cycle supports). *A set of nodes $S \subseteq V$ is the support of an ordinary cycle in $\mathcal{G}$ if and only if S is a strongly connected subgraph of $\mathcal{G}$.*

Being a strongly connected subgraph means that each node in S has both a successor and a predecessor in S. With this characterisation, we can develop a relation-algebraic test for checking whether a given set of nodes is the support of an ordinary cycle:

Proposition 6 (Cycle test). *A set of nodes $S \subseteq V$ modelled by a vector $v : V \leftrightarrow \mathbb{1}$ is the support of an ordinary cycle if and only if* $\mathsf{syQ}(v, (R|_v)^* \cap (R|_v)^{*\mathsf{T}}) \neq \mathbb{O}$ *holds.*

Proof. With Prop. 5 and with $R|_v$ being the subgraph of $\mathcal{G}$ spanned by the nodes of S (Def. 5), we have for each set of nodes S modelled by a vector $v : V \leftrightarrow \mathbb{1}$:

$$\begin{aligned}
\mathit{isCycle}(v) :&\iff \mathsf{syQ}(v, (R|_v)^* \cap (R|_v)^{*\mathsf{T}}) \neq \mathbb{O}\\
&\iff \exists x : \mathsf{syQ}(v, (R|_v)^* \cap (R|_v)^{*\mathsf{T}})_{\diamond,x}\\
&\iff \exists x : \forall y : v_y \leftrightarrow ((R|_v)^* \cap (R|_v)^{*\mathsf{T}})_{y,x}\\
&\iff \exists x : \forall y : v_y \leftrightarrow (R|_v)^*{}_{y,x} \wedge (R|_v)^*{}_{x,y}
\end{aligned}$$

That is, S is the support of an ordinary cycle if and only if the nodes in S are strongly connected w.r.t. $(R|_v)^*$. □

Note that the union of ordinary cycle supports needs thus not necessarily be an ordinary cycle support.

For maximal cycles, the enumeration of all strongly connected components delivers the set of all supports of maximal cycles. For ordinary cycles, this enumeration is less easy. Proposition 5 qualifies ordinary cycle supports to be found as strongly connected subgraphs of strongly connected components. In a naïve approach, one could simply enumerate all subsets of strongly connected components and check with the above test for each subset whether it is the support of a cycle. As in the worst case there are $|2^V|$-many sets to test, this procedure will not be efficient. Since these sets need to be strongly connected with respect to R^* reduced to the nodes they contain, we can exclude weakly connected sets from being tested. The relation-algebraic enumeration of weakly connected sets is characterised as follows:

Proposition 7 (Enumeration of weakly connected node sets). *The set of all weakly connected node sets can be modelled by a vector $w : 2^V \leftrightarrow \mathbb{1}$ defined through* $(\varepsilon \cap \overline{(R^\mathsf{T};\varepsilon \cap R;\varepsilon)})^\mathsf{T};\mathbb{L}$.

Proof.

$$\begin{aligned}
w_W :&\iff ((\varepsilon \cap \overline{R^\mathsf{T};\varepsilon \cap R;\varepsilon})^\mathsf{T};\mathbb{L})_W\\
&\iff \exists x : (\varepsilon \cap \overline{R^\mathsf{T};\varepsilon \cap R;\varepsilon})_{x,W} \wedge \mathbb{L}_x \qquad \text{where } \mathbb{L} : V \leftrightarrow \mathbb{1}\\
&\iff \exists x : \varepsilon_{x,W} \wedge ((\overline{R^\mathsf{T};\varepsilon})_{x,W} \vee (\overline{R;\varepsilon})_{x,W})\\
&\iff \exists x : \varepsilon_{x,W} \wedge (\neg\exists y : (R_{y,x} \wedge \varepsilon_{y,W}) \vee \neg\exists y : (R_{x,y} \wedge \varepsilon_{y,W}))\\
&\iff \exists x : W_x \wedge (\neg\exists y : (R_{y,x} \wedge W_y) \vee \neg\exists y : (R_{x,y} \wedge W_y))
\end{aligned}$$

That is, the vector w represents the set of all node sets W in which there exists a node which has no successors or no predecessors in W. This is the definition of a set being weakly connected w.r.t. R. □

The relation-algebraic enumeration of the set of all subsets of strongly connected components of $\mathcal{G}$ is characterised as follows:

Proposition 8 (Enumeration of all subsets of strongly connected components). *The set of all subsets of strongly connected components of $\mathcal{G}$ can be modelled by a vector $subSCC : 2^V \leftrightarrow \mathbb{1}$ defined as $\overline{(\varepsilon \cap \overline{(R^* \cap R^{*\mathsf{T}})}; \varepsilon)}^{\mathsf{T}} ; \mathbb{L}$.*

Proof.

$$
\begin{aligned}
subSCC_S :&\iff (\overline{(\varepsilon \cap \overline{(R^* \cap R^{*\mathsf{T}})}; \varepsilon)}^{\mathsf{T}} ; \mathbb{L})_S \\
&\iff \neg\exists x : (\varepsilon \cap \overline{(R^* \cap R^{*\mathsf{T}})}; \varepsilon)_{x,S} \wedge \mathbb{L}_x \qquad \text{where } \mathbb{L} : V \leftrightarrow \mathbb{1} \\
&\iff \forall x : \overline{(\varepsilon \cap \overline{(R^* \cap R^{*\mathsf{T}})}; \varepsilon)}_{x,S} \\
&\iff \forall x : \overline{\varepsilon}_{x,S} \vee \overline{(\overline{(R^* \cap R^{*\mathsf{T}})}; \varepsilon)}_{x,S} \\
&\iff \forall x : \varepsilon_{x,S} \rightarrow \overline{(\overline{(R^* \cap R^{*\mathsf{T}})}; \varepsilon)}_{x,S} \\
&\iff \forall x : \varepsilon_{x,S} \rightarrow \neg(\exists y : \varepsilon_{y,S} \wedge \overline{(R^* \cap R^{*\mathsf{T}})}_{x,y}) \\
&\iff \forall x : x \in S \rightarrow \neg(\exists y : y \in S \wedge \neg(R^*{}_{x,y} \wedge R^*{}_{y,x})) \\
&\iff \neg(\exists x : x \in S \wedge \exists y : y \in S \wedge \neg(R^*{}_{x,y} \wedge R^*{}_{y,x})) \\
&\iff \neg(\exists x, y : x \in S \wedge y \in S \wedge \neg(R^*{}_{x,y} \wedge R^*{}_{y,x})) \\
&\iff \forall x, y : x \in S \wedge y \in S \rightarrow R^*{}_{x,y} \wedge R^*{}_{y,x}
\end{aligned}
$$

That is, the vector *subSCC* models the set of all sets S in which each pair of nodes in S is mutually reachable w.r.t. R^*, and hence, following Def. 6, which is a subset of a strongly connected component of $\mathcal{G}$. □

By eliminating all sets modelled by the vector w from the sets modelled by the vector *subSCC*, it is easy to see that only supports of cycles and supports of unions of finitely-many disjoint cycles remain; the latter occur in case a strongly connected component contains disjoint strongly connected subgraphs. Such unions are eliminated using the cycle test provided above. Thus, we can put things together and formulate a relation-algebraic specification for the columnwise enumeration (see Sect. 2) of the supports of all (ordinary) cycles. Since both elementary and maximal cycles are ordinary as well, their supports are also enumerated. Hence, the specification looks as follows. Note that a support candidate is a point of type $[2^V \leftrightarrow \mathbb{1}]$; to handle it appropriately, it needs to be converted into a vector of type $[V \leftrightarrow \mathbb{1}]$ by using the construction introduced in Sect. 2:

$supports = \emptyset$

$supportCandidates = subSCC(R) \cap \overline{w(R)}$

for each c in $supportCandidates$:

  if $isCycle(\varepsilon; c)$ then $supports = supports \cup c$

if $supports \neq \emptyset$ then $supportList = \varepsilon; inj(supports)^{\mathsf{T}}$

return $supportList$

Sometimes it is not necessary to enumerate all supports, particularly if one is interested in specific nodes being visited in a cycle. For any set of nodes S modelled by a vector $v : V \leftrightarrow \mathbb{1}$, the above specification can be extended to enumerate the set of all cycle supports that completely contain the nodes in S. This is trivially achieved by intersecting $subSCC \cap \overline{w}$ with the term $(v \backslash \varepsilon)^{\mathsf{T}}$ delivering – as is easy to see – the set of all sets of nodes completely containing S. This further restricts the set of cycle candidates to be tested.

6 Elementary Cycles

In an elementary cycle $\mathcal{C} = \langle n_0, \ldots, n_l \rangle$, the participating nodes are pairwise disjoint (Def. 2). That is, no node (except for $n_0 = n_l$) occurs more than once in the sequence $\langle n_0, \ldots, n_l \rangle$, and hence for each node both its predecessor and successor with respect to $supp(\mathcal{C})$ are uniquely determined. For the support of an elementary cycle $\mathcal{C}$, this means that $supp(\mathcal{C})$ does not contain branching and joining points in R reduced to $supp(\mathcal{C})$. It is also easy to see that branch-freeness excludes the existence of edges between non-neighbored nodes of $\mathcal{C}$, i.e., a node set without branching and joining points particularly spans a chordless subgraph (cf. Def. 2 and Rem. 2). A branching point of a relation R is an element in the domain of the multivalent part of R given by $mup(R) := R \cap (R; \overline{\mathbb{I}})$, i.e., the set of all pairs (x, y) with $R_{x,y}$ for which there exists a $z \neq y$ with $R_{x,z}$, whereas a joining point is an element in the domain of the multivalent part of the converse of R given by $R^{\mathsf{T}} \cap (R^{\mathsf{T}}; \overline{\mathbb{I}})$ (cf. [15]). We formalise this observation:

Proposition 9 (Branch-free cycles). *A set of nodes $S \subseteq V$ is the support of an elementary cycle if and only if it is the support of a cycle and it does neither contain branching nor joining points w.r.t. $R|_S$.*

To prove the above property, we use the observation that branching and joining points in cycles appear simultaneously:

Corollary 1 (Branches and joins occur simultaneously). *The support of a cycle $\mathcal{C}$ contains a branching point if and only if it contains a joining point.*

Proof. "⇒" Let $\mathcal{C}$ be a cycle the support S of which contains a branching point. Let $n_i, 0 \leq i \leq l$, be this branch. Then, there exist two nodes $x, y \in S$ with $(R|_S)_{n_i,x}$ and $(R|_S)_{n_i,y}$. By Prop. 1, there must exist a node $z \in S$ with both $(R|_S)^*{}_{x,z}$ and $(R|_S)^*{}_{y,z}$ must hold. Hence, z is a joining point in S.
"⇐" Analogously. □

To prove Prop. 9, it is sufficient to consider branching points.

Proof. "⇒" Let $\mathcal{C}$ be an elementary cycle. Then, $S := supp(\mathcal{C})$ is obviously the support of a cycle. Let us assume, S contains a branching point x, and let y be one of its successors in S. Then, $\mathcal{C}$ has the form $\langle n_0, \ldots, x, y, \ldots, n_l \rangle$. Since x is a branching point, there must exist a node $z \in S$ which also is a successor of x and different from y. Therefore, $\mathcal{C}$ must also have the form $\langle n_0, \ldots, x, z, \ldots, n_l \rangle$. Since y and z are distinct nodes, x must occur twice in $\mathcal{C}$, and hence $\mathcal{C}$ cannot be elementary.

"⇐" Let $S := supp(\mathcal{C})$ be the support of a cycle $\mathcal{C}$, and let it be free of branching points w.r.t. $R|_S$. Then, both the predecessor and successor of each node x in S is uniquely determined w.r.t. $R|_S$. Hence, each cycle $\mathcal{C}$' with $supp(\mathcal{C})' = S$ must be elementary. □

For any given support S of a cycle $\mathcal{C}$, we can now formulate a test for checking whether $\mathcal{C}$ is elementary. Since we need to check for branching points within S, we need to consider R reduced to S:

Proposition 10 (Elementary cycle test). *Let S be the support of a cycle. S is the support of an elementary cycle if and only if $S^\mathsf{T}; mup(R|_S) = \mathbb{O}$ holds.*

Proof.

$$\begin{aligned} isECycle(S) :&\iff S^\mathsf{T}; mup(R|_S) = \mathbb{O} \\ &\iff \neg\exists y : (S^\mathsf{T}; mup(R|_S))_{\diamond,y} \\ &\iff \neg\exists y : \exists x : {S^\mathsf{T}}_{\diamond,x} \wedge mup(R|_S)_{x,y} \\ &\iff \neg\exists x : S_x \wedge \exists y : mup(R|_S)_{x,y} \end{aligned}$$

That is, $S^\mathsf{T}; mup(R|_S) = \mathbb{O}$ holds if S does not contain a branching point. With Prop. 9, S then is the support of an elementary cycle. □

Each support of an elementary cycle is the support of a cycle. Therefore, we can extend the previously given specification for the enumeration of cycle supports in such a way that it only lists supports of elementary cycles. We proceed as follows: In addition to checking whether a set of nodes in *supportCandidates* is a cycle, we check the property formulated in Prop. 10 and get to column-wise enumerate the set of all elementary cycle supports in $\mathcal{G}$:

$supports = \emptyset$

$supportCandidates = subSCC(R) \cap \overline{w(R)}$

for each c in $supportCandidates$:

  if $isCycle(\varepsilon; c)$ and $isECycle(\varepsilon; c)$ then $supports = supports \cup c$

if $supports \neq \emptyset$ then $supportList = \varepsilon; inj(supports)^\mathsf{T}$

return $supportList$

It is easy to see that due to the absence of branching points elementary cycles are inclusion-minimal supports, i.e. there exists no cycle $\mathcal{C}'$ in $\mathcal{G}$ such that $supp(\mathcal{C}') \subset$

supp($\mathcal{C}$) holds for any elementary cycle $\mathcal{C}$, and hence the union of elementary cycle supports can never be an elementary cycle support.

7 Applications to Petri Nets

S-Systems as well as T-Systems are common classes of Petri Nets. Quite frequently, they consist of cycles as their main building blocks. A designer needs to identify these blocks visually since they form the processes the entire system consists of. Identifying the cycles of a Petri Net helps to understand the net topology and is thus vital for working with the underlying system.

On the level of machine-based analysis, as we proposed in [8, 9], the enumeration of cycles is of particular interest for strongly connected conservative Petri Nets. In such nets, the supports of elementary cycles provide the supports of minimal semi-positive place invariants [6, Chap. 5]. Cycle enumeration is thus an alternative to calculating the supports of such place invariants by solving specific linear equations.

A theory considering so called *handles* is elaborated in [7]. It relies on the absence of certain cyclic structures and then allows to deduce interesting net properties. For a brief explanation, let an elementary cycle $\mathcal{C} = \langle n_0, \dots, p, \dots, t, \dots, n_l \rangle$ be given with p and t any place and transition of the net, respectively. Further, let p be a branching point and t be a joining point with successors of p and predecessors of t not in *supp*($\mathcal{C}$). A path from p to t using only elements not in the support of $\mathcal{C}$ is called a PT-handle (the dual is called TP-handle). Then, a strongly connected Petri Net in which no elementary cycle has a PT- or TP-handle is structurally live, consistent, and conservative. With our specification of elementary cycle enumeration, the detection of such structures is feasible, and again we provide an alternative to costly proving the latter properties individually.

We applied the specifications developed here to the analysis of Petri Nets as well as to the enumeration of Feedback Vertex Sets [4]. Based on our object-oriented Java library, KURE-Java [16, 14, 10], which mechanises the calculus of relations and is thus capable of executing the relational specifications, we implemented a Petri Net tool, PETRA [9] (see Fig. 2), for both editing and relation-algebraically analysing Petri Nets.

We integrated the specifications described above into PETRA. KURE-Java allows for the integration of relation algebra with arbitrary visualisations and thereby offers a flexible way to graphically model any system, to transform it into suitable relations, and to illustrate the results an executed relational specification delivers when the system is analysed. These executable relational specifications are written in the RELVIEW language [3] supported by KURE-Java and thus executable in PETRA. More generally, they are executable within any application using RELVIEW or RELCLIPSE [11], a re-implementation of RELVIEW as a Plug-In for Eclipse under Windows, as GUI-based applications, or KURE-Java directly. The specifications developed here consider homogeneous relations in general and thus comprise bipartite graphs as a special case since they can easily be represented as such. PETRA, KURE-Java and RELCLIPSE can be downloaded at `http://ls10-www.cs.uni-dortmund.de`.

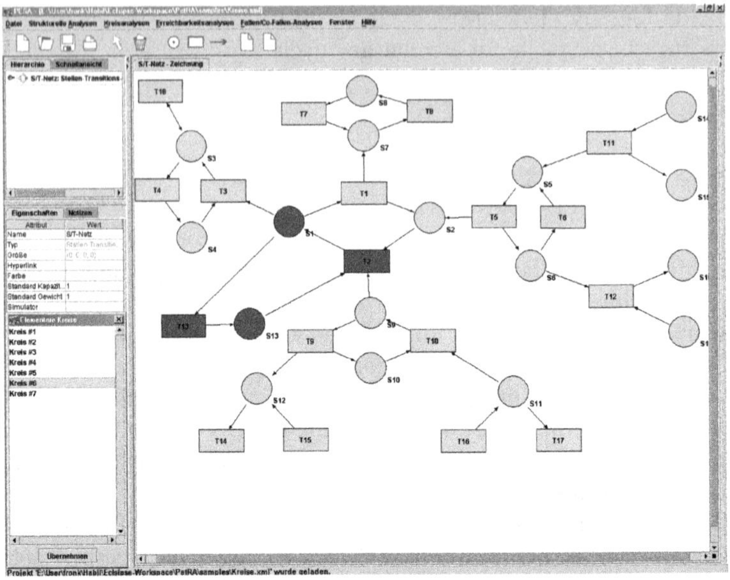

Fig. 2. A screenshot of the Petri Net tool PETRA. A list of all elementary cycle supports of the net is provided in the lower left window; selecting an entry from the list displays the participating places and transition in red.

8 Conclusion

We developed a relation-algebraic characterisation of ordinary, elementary, and maximal cycles in graphs. Further, we presented specifications for testing whether a given set of nodes is the support of any such cycle as well as for enumerating all their supports. In summary:

- the set of all supports of all maximal cycles of a graph $\mathcal{G}$ is identical to the set of all strongly connected components of $\mathcal{G}$ and hence inclusion-maximal in the set of all supports;
- the support of an ordinary cycle forms a strongly connected subgraph of a strongly connected component of $\mathcal{G}$;
- an elementary cycle is an inclusion-minimal ordinary cycle.

Supports were consequently characterised through pairwise reachability for ordinary cycles and additionally through the absence of branching points for elementary ones.

References

[1] J. Bang-Jensen and G. Gutin. *Digraphs: Theory, Algorithms and Applications.* Monographs in Mathematics. Springer, 2000.

[2] B. Baumgarten. *Petri-Netze: Grundlagen und Anwendungen.* BI-Wissenschafts-Verlag, 1990.

[3] R. Behnke, R. Berghammer, E. Meyer, and P. Schneider. RELVIEW – A system for calculating with relations and relational programming. In E. Astesiano, editor, *Proceedings of the 1st International Conference on Fundamental Approaches to Software Engineering*, volume 1382 of *Lecture Notes in Computer Science (LNCS)*, pages 318 – 321. Springer, 1996.
[4] R. Berghammer and A. Fronk. Exact computation of minimum feedback vertex sets with relational algebra. *Fundamenta Informaticae*, 70:1–16, 2006. To appear.
[5] R. Berghammer, B. Karger, and C. Ulke. Relational-algebraic analysis of Petri Nets with RELVIEW. In T. Margaria and B. Steffen, editors, *Proceedings of the 2nd Workshop on Tools and Applications for the Construction and Analysis of Systems (TACAS '96)*, volume 1055 of *Lecture Notes in Computer Science (LNCS)*, pages 49–69. Springer, 1996.
[6] J. Desel and J. Esparza. *Free Choice Petri Nets*, volume 40 of *Cambridge Tracts in Theoretical Computer Science*. Cambridge University Press, 1995.
[7] J. Esparza and M. Silva. Circuits, Handles, Bridges and Nets. In G. Rozenberg, editor, *Advances in Petri Nets*, Lecture Notes in Computer Science (LNCS), 1990.
[8] A. Fronk. Using relation algebra for the analysis of Petri Nets in a CASE tool based approach. In *2nd IEEE International Conference on Software Engineering and Formal Methods (SEFM), Beijing*, pages 396–405. IEEE, 2004.
[9] A. Fronk and J. Pleumann. Relationenalgebraische Analyse von Petri-Netzen: Konzepte und Implementierung. In Ekkart Kindler, editor, *Proceedings of the 11th Workshop on Algorithms and Tools for Petri Nets (AWPN)*, pages 61–68. Universität Paderborn, Germany, September 2004. English version appeared in *Petri Net News Letters*, April 2005.
[10] A. Fronk, J. Pleumann, R. Schönlein, and O. Szymanski. KURE-Java. `http://ls10-www.cs.uni-dortmund.de/index.php?id=136`, 2005.
[11] A. Fronk and R. Schönlein. RelClipse. `http://ls10-www.cs.uni-dortmund.de/index.php?id=137`.
[12] Th. Hoffmann. *Fallstudien relationaler Programmentwicklung am Beispiel ausgewählter Graphdurchlaufstrategien.* PhD thesis, Universität Kiel, 2002.
[13] Institute of Computer Science and Applied Mathematics, Faculty of Engineering, Christian-Albrechts-University of Kiel, Germany. RelView Examples - Graph-theoretic Algorithms. `http://www.informatik.uni-kiel.de/~progsys/relview/Graph`.
[14] U. Milanese. KURE: Kiel University Relation Package. `http://cvs.informatik.uni-kiel.de/~kure/`, 2003.
[15] G. Schmidt and Th. Ströhlein. *Relations and graphs.* EATCS Monographs on Theoretical Computer Science. Spinger, 1993.
[16] O. Szymanski. Relationale Algebra im dreidimensionalen Software-Entwurf – ein werkzeugbasierter Ansatz. Master's thesis, Universität Dortmund, 2003.

A Framework for Kleene Algebra with an Embedded Structure

Hitoshi Furusawa*

Research Center for Verification and Semantics, AIST,
3-11-46 Nakoji, Amagasaki, Hyogo 661-0974, Japan
hitoshi.furusawa@aist.go.jp

Abstract. This paper proposes a framework for Kleene algebras with embedded structures that enables different kinds of Kleene algebras such as a Kleene algebra with tests and a Kleene algebra with relations to be handled uniformly. This framework guarantees the existence of free algebra if the embedded structures satisfy certain conditions.

1 Introduction

Kozen [10] defined a Kleene algebra with tests to be a Kleene algebra that has an embedded Boolean algebra. Desharnais [3] defined a Kleene algebra with relations to be a Kleene algebra that has an embedded relation algebra. It may also be necessary to consider Kleene algebras with other embedded structures. For instance, it has been shown that Turing computability can be captured in terms of a Kleene category (heterogeneous or typed Kleene algebra) equipped with an embedded lattice structure in each homset of endomorphisms for each object [14, Section 5].

The common feature among Kleene algebras with tests and Kleene algebras with relations is the fact that their underlying idempotent semiring structure is shared. A *slightly* more general definition of a Kleene algebra with tests, by comparison with that of Kozen, was introduced by paying attention to this feature [6, 7]. In this definition, a Kleene algebra with tests is defined as a triple consisting of a Boolean algebra, a Kleene algebra, and a function from the carrier set of the Boolean algebra to the carrier set of the Kleene algebra that is a homomorphism between their underlying idempotent semirings rather than an exact embedding. Hence, the category of Kleene algebras with tests is the comma category (U_{BI}, U_{KI}) of forgetful functors U_{BI} from the category **Bool** of Boolean algebras to the category **ISR** of idempotent semirings, and U_{KI} from the category **Kleene** of Kleene algebra to the category **ISR**. A systematic free construction of a Kleene algebra with tests from a pair of sets was given in [5] by means of this definition.

While the free construction in [5] is captured using adjunction between the category of sets, **Set**, and the category **Bool**, adjunction between **Set** and the

* This work was done in part while the author was visiting McMaster University under the Japan Society for the Promotion of Science (JSPS) Bilateral Programs for Scientist Exchanges.

W. MacCaull et al. (Eds.): RelMiCS 2005, LNCS 3929, pp. 96–107, 2006.

category **Kleene**, the forgetful functor from the category **Bool** to the category **ISR**, adjunction between the category **ISR** and the category **Kleene**, and coproducts in the category **Kleene**, we can observe that the key element of the construction is an adjunction $F_{IK} \dashv U_{KI}$ between the category **ISR** and the category **Kleene**, and coproducts in the category **Kleene**. It does not seem to be very important for the free construction that the embedded structure is a Boolean algebra.

Parameterising the category **Bool** in the setting of [5], we provide a framework for Kleene algebras with embedded structure $\mathcal{F}_{-}$. For example, by replacing the category **Bool** with the category **RA** of relation algebras, we obtain the category $\mathcal{F}_{\mathbf{RA}}$ of slightly generalised Kleene algebras with relations. We will show that a similar free construction is also available under a few conditions. Generally, the existence of free algebras ensures that structures have canonical models called initial semantics. The conditions provide sufficient requirements for the existence of free Kleene algebras with an embedded structure. The construction makes clear that the free Kleene algebra with an embedded structure generated by a pair of sets can be represented by a coproduct in the category **Kleene**. This fact is independent from embedded structures.

2 Comma Categories

A framework for a Kleene algebra with an embedded structure may be provided using the notion of comma categories. We therefore recall some basics of comma categories. For more details of category theory we refer to [1, 12].

Definition 1 (Comma category). Given categories and functors

$$E \xrightarrow{T} C \xleftarrow{S} D \ ,$$

the *comma category* (T, S) has as objects all triples $\langle e, d, f \rangle$, with an object e in E, an object d in D, and an arrow $f\colon T(e) \to S(d)$ in C, and as arrows from $\langle e, d, f \rangle$ to $\langle e', d', f' \rangle$ all pairs $\langle k, h \rangle$ of arrows $k\colon e \to e'$ in E and $h\colon d \to d'$ in D such that $f' \circ T(k) = S(h) \circ f$.

For two objects d_1 and d_2 in D, where D is a category with coproducts, $d_1 + d_2$ denotes the coproduct in D of d_1 and d_2. For two arrows $f\colon d_1 \to d'_1$ and $g\colon d_2 \to d'_2$, $f + g$ denotes the unique arrow in D such that the following two squares commute:

$$\begin{array}{ccccc}
d_1 & \xrightarrow{i} & d_1 + d_2 & \xleftarrow{j} & d_2 \\
{\scriptstyle f}\downarrow & & \downarrow{\scriptstyle f+g} & & \downarrow{\scriptstyle g} \\
d'_1 & \xrightarrow[i']{} & d'_1 + d'_2 & \xleftarrow[j']{} & d'_2
\end{array}$$

i and j are injections of coproduct $d_1 + d_2$, and i' and j' likewise relate to $d'_1 + d'_2$. Furthermore, the mappings

$$(d_1, d_2) \mapsto d_1 + d_2, \quad (f, g) \mapsto f + g$$

determine a functor $+$ from $D \times D$ to D.

The following property plays a key rôle in this paper.

Theorem 1. *Given categories and functors*

$$E \xrightarrow{T} C \xleftarrow{S} D \ ,$$

assuming that S *has a left adjoint* V *and the category* D *has coproducts, then the functor* $\Upsilon\colon (T,S) \to E \times D$ *which is defined by the mappings*

$$\langle e,d,f\rangle \mapsto (e,d), \quad \langle k,h\rangle \mapsto (k,h)$$

has a left adjoint.

For each object c in C and d in D, the bijection from $D(V(c),d)$ to $C(c,S(d))$, constituting the adjunction $V \dashv S$, is denoted by $\rho_{c,d}$. The subscript c,d will be omitted. The functor Θ from $E \times D$ to (T,S) which is defined by the mappings

$$(e,d) \mapsto \langle e, V(T(e))+d, \rho(i)\rangle, \quad (k,h) \mapsto \langle k, V(T(k))+h\rangle \ ,$$

where i is the first injection of the coproduct $V(T(e))+d$, is a left adjoint to Υ. See the Appendix for further details of the proof.

3 Kleene Algebras

In this section we recall some basics of Kleene algebras [8, 9] and related structures. Several examples of Kleene algebras are contained in [2, 4].

Definition 2 (Kleene algebra). A *Kleene algebra* is a set K equipped with nullary operators 0, 1 and binary operators $+$, $\cdot$, and a unary operator *, where the tuple $(K,+,\cdot,0,1)$ is an idempotent semiring and these data satisfy the following properties:

$$\begin{aligned} 1+(p\cdot p^*) &= p^* \\ 1+(p^*\cdot p) &= p^* \\ p\cdot r \le r &\implies p^*\cdot r \le r \\ r\cdot p \le r &\implies r\cdot p^* \le r \end{aligned}$$

where $\le$ refers to the natural partial order

$$p \le q \overset{\text{def}}{\iff} p+q=q \ .$$

A Kleene algebra will be called *trivial* if $0=1$, and called *non-trivial* otherwise. A *Kleene algebra homomorphism* is a function between the carrier sets of Kleene algebras which preserves nullary operators 0, 1, binary operators $+$, $\cdot$, and a unary operator *. The category of Kleene algebras and homomorphisms between them will be denoted by **Kleene**.

Remark 1. **Kleene** has binary coproducts.

Given two Kleene algebras, a quotient of the set of regular languages over disjoint union of the carrier sets of the given Kleene algebras with respect to the equivalence relation on the set of regular languages which identifies corresponding original operators of the given two Kleene algebras and the regular sets forms a coproduct of the two Kleene algebras.

The injections of a coproduct in **Kleene** are not always one-to-one. Trivial Kleene algebras have only one element since, for each a,

$$a = a \cdot 1 = a \cdot 0 = 0.$$

For each Kleene algebra K, there exists a unique Kleene algebra homomorphism from K to the trivial Kleene algebra. From a trivial Kleene algebra, there exists a Kleene algebra homomorphism if the target is also trivial. So, the coproduct of a trivial Kleene algebra and a non-trivial Kleene algebra is trivial. Thus, we have an injection which is not one-to-one. This example is due to Wolfram Kahl.

A Kleene algebra $\mathbf{K}$ is called *integral* if it has no zero divisors, that is,

$$a \neq 0 \ \wedge \ b \neq 0 \implies a \cdot b \neq 0$$

holds for all $a, b \in K$. This notion was introduced in [4].

Proposition 1. *Let* $\mathbf{J} = (J, +_J, \cdot_J, {}^{*_J}, 0_J, 1_J)$ *and* $\mathbf{K} = (K, +_K, \cdot_K, {}^{*_K}, 0_K, 1_K)$ *be non-trivial Kleene algebras. If* $\mathbf{K}$ *is integral, then the following holds.*

(i) The mapping $f \colon K \to J$ *defined to be* $f(a) = 0_J$ *if* $a = 0_K$*, and otherwise* $f(a) = 1_J$*, is a Kleene algebra homomorphism.*
(ii) The first injection $j \colon \mathbf{J} \to \mathbf{J} + \mathbf{K}$ *is one-to-one.*

Proof. For each $a, b \in K$, if $a \neq 0_K$ and $b \neq 0_K$, then $a +_K b \neq 0_K$ and $a \cdot_K b \neq 0_K$ since $\mathbf{K}$ is integral. So, *(i)* follows from

$$\begin{aligned} f(a) +_J f(b) &= 1_J +_J 1_J = 1_J = f(a +_K b) \\ f(a) \cdot_J f(b) &= 1_J \cdot_J 1_J = 1_J = f(a \cdot_K b) \\ f(a)^{*_J} &= 1_J^{*_J} = 1_J = f(a^{*_K}) \ . \end{aligned}$$

The case of $a = 0_K$ or $b = 0_K$ is immediate from $0_K^{*_K} = 1_K$, $0_J^{*_J} = 1_J$, and $f(0_K) = 0_J$.

(ii) will be proved using f as given in *(i)*. Taking $\mathrm{id}_{\mathbf{J}}$ and f, a unique intermediating arrow $h \colon \mathbf{J} + \mathbf{K} \to \mathbf{J}$ with respect to them exists. By the definition of coproducts, h satisfies $\mathrm{id}_{\mathbf{J}} = h \circ j$. Thus j is one-to-one.

Set and **ISR** denote the categories of sets and functions, idempotent semirings and their homomorphisms respectively. $U_K \colon$ **Kleene** $\to$ **Set** denotes the forgetful functor which takes a Kleene algebra to its carrier set. The functor U_K is decomposed by functors $U_{KI} \colon$ **Kleene** $\to$ **ISR** and $U_I \colon$ **ISR** $\to$ **Set**, where $U_{KI}(\mathbf{K})$ is an idempotent semiring obtained by forgetting the * operator and U_I takes an idempotent semiring to its carrier set. Since idempotent semirings are axiomatised by equational theory, the functor U_I has a left adjoint F_I. Also, since the structure of Kleene algebras is obtained by adding the * operation to the structure of idempotent semirings and axiomatised by equational Horn theory, the functor U_{KI} has a left adjoint F_{IK}. Thus, $F_K \stackrel{\text{def}}{=} F_{IK} \circ F_I$ is a left adjoint to U_K.

Remark 2. For a set Σ, $\mathbf{Reg}(\Sigma)$ denotes the Kleene algebra consisting of the set of regular sets over Σ together with the standard operations on regular sets. Clearly, $\mathbf{Reg}(\Sigma)$ is integral. Moreover, Kozen showed that $\mathbf{Reg}(\Sigma) \cong F_K(\Sigma)$ in [9].

The situation we state above is as follows:

$$\mathbf{Set} \underset{U_I}{\overset{F_I}{\rightleftarrows}} \mathbf{ISR} \underset{U_{KI}}{\overset{F_{IK}}{\rightleftarrows}} \mathbf{Kleene} \qquad (F_I \dashv U_I,\ F_{IK} \dashv U_{KI})$$

$$\begin{array}{ll} F_B \stackrel{\text{def}}{=} F_{IB} \circ F_I & F_K \stackrel{\text{def}}{=} F_{IK} \circ F_I \\ U_B = U_I \circ U_{BI} & U_K = U_I \circ U_{KI} \end{array}$$

For each idempotent semiring $\mathbf{S}$ and Kleene algebra $\mathbf{K}$, the bijection from $\mathbf{Kleene}(F_{IK}(\mathbf{S}), \mathbf{K})$ to $\mathbf{ISR}(\mathbf{S}, U_{KI}(\mathbf{K}))$, constituting the adjunction $F_{IK} \dashv U_{KI}$, is denoted by $\varphi_{\mathbf{S},\mathbf{K}}$. The subscript $\mathbf{S}, \mathbf{K}$ will be omitted unless its absence causes confusion.

A Kleene algebra will be called **-continuous* [8] if it holds that

$$p \cdot q^* \cdot r = \sup_n p \cdot q^n \cdot r$$

where $q^0 = 1$, $q^{n+1} = q \cdot q^n$, and the supremum respects the ordering defined by $\leq$. An idempotent semiring equipped with an arbitrary join will be called a *standard Kleene algebra* ($\mathbf{S}$-algebra [2], or quantale [13]).

Let $\mathbf{S} = (S, +, \cdot, 0, 1)$ be an idempotent semiring. A non-empty subset $A \subseteq S$ closed under $+$ and closed downward under the ordering $\leq$ with respect to $+$ is called an ideal of $\mathbf{S}$. As in the case of *-continuous Kleene algebras [2, 8], the set $\mathcal{I}_S$ of ideals forms an $\mathbf{S}$-algebra. In an $\mathbf{S}$-algebra, the * operation can be defined as reflexive transitive closure, yielding a Kleene algebra $\mathbf{K}_{\mathcal{I}_S}$. The mapping $\langle \_ \rangle$ which takes $a \in S$ to the least ideal which contains a, *i.e.*, $\langle a \rangle = \{b \in S \mid b \leq a\}$, determines a one-to-one idempotent semiring homomorphism $\langle \_ \rangle \colon \mathbf{S} \to U_{KI}(\mathbf{K}_{\mathcal{I}_S})$.

For an idempotent semiring $\mathbf{S}$, the arrow $\varphi(\mathrm{id}_{F_{IK}(\mathbf{S})})$ in $\mathbf{ISR}$ is a component of the unit of $F_{IK} \dashv U_{KI}$ with respect to $\mathbf{S}$, that is, for each Kleene algebra $\mathbf{K}$ and an arrow $f \colon \mathbf{S} \to U_{KI}(\mathbf{K})$ in $\mathbf{ISR}$, there is a unique arrow $\overline{f} \colon F_{IK}(\mathbf{S}) \to \mathbf{K}$ in $\mathbf{Kleene}$ such that $f = U_{KI}(\overline{f}) \circ \varphi(\mathrm{id}_{F_{IK}(\mathbf{S})})$.

Proposition 2. *For each idempotent semiring* $\mathbf{S}$, $\varphi(\mathrm{id}_{F_{IK}(\mathbf{S})})$ *is one-to-one.*

Proof. Replacing $f \colon \mathbf{S} \to U_{KI}(\mathbf{K})$ with $\langle \_ \rangle \colon \mathbf{S} \to U_{KI}(\mathbf{K}_{\mathcal{I}_S})$, we have an arrow $\overline{\langle \_ \rangle} \colon F_{IK}(\mathbf{S}) \to \mathbf{K}_{\mathcal{I}_S}$ satisfying $\langle \_ \rangle = U_{KI}(\overline{\langle \_ \rangle}) \circ \varphi(\mathrm{id}_{F_{IK}(\mathbf{S})})$. Since $\langle \_ \rangle$ is one-to-one, $\varphi(\mathrm{id}_{F_{IK}(\mathbf{S})})$ is also one-to-one.

This fact induces the following property:

Proposition 3. *φ preserves one-to-one mappings.*

Proof. Let $m\colon F_{IK}(\mathbf{S}) \to \mathbf{K}$ be a one-to-one Kleene algebra homomorphism. Since $\varphi(\mathrm{id}_{F_{IK}(\mathbf{S})})$ is a component of the unit of $F_{IK} \dashv U_{KI}$ with respect to $\mathbf{S}$, we have $\varphi(m) = U_{KI}(m) \circ \varphi(\mathrm{id}_{F_{IK}(\mathbf{S})})$. Also, since $U_{KI}(m)$ is m itself and both of m and $\varphi(\mathrm{id}_{F_{IK}(\mathbf{S})})$ are one-to-one, $\varphi(m)$ is also one-to-one.

4 Framework

This section provides a framework for Kleene algebras with an embedded structure.

Let $\mathcal{X}$ be the category of a certain kind of algebras which have an underlying idempotent semiring structure and homomorphisms. The forgetful functor from $\mathcal{X}$ to **ISR** will be denoted by $U_{XI}\colon \mathcal{X} \to \mathbf{ISR}$.

Definition 3 (Framework). A *framework* $\mathcal{F}_{-}$ for Kleene algebra with an embedded structure is the mapping which takes the category $\mathcal{X}$ of a certain kind of algebras which have an underlying idempotent semiring structure and homomorphisms to the comma category (U_{XI}, U_{KI}).

Note that the third component of an object $\langle \mathbf{X}, \mathbf{K}, i\rangle$ in $\mathcal{F}_{\mathcal{X}}$ is not necessarily an inclusion mapping.

Bool denotes the category of Boolean algebras and their homomorphisms. An object $\langle \mathbf{B}, \mathbf{K}, i\rangle$ in $\mathcal{F}_{\mathbf{Bool}}$ may not be a Kleene algebra with tests in the sense of [10, 11] but in the sense of [5]. Let $\mathcal{F}^{\subseteq}_{\mathbf{Bool}}$ be the full subcategory of $\mathcal{F}_{\mathbf{Bool}}$ such that the third component of each object is an inclusion. Then an object in $\mathcal{F}^{\subseteq}_{\mathbf{Bool}}$ is a Kleene algebra with tests in the sense of [10, 11].

It is known that the image of $U_{BI}(\mathbf{B})$ under i forms Boolean algebra again for each object $\langle \mathbf{B}, \mathbf{K}, i\rangle$ in $\mathcal{F}_{\mathbf{Bool}}$. However, if we consider $\mathcal{F}_{\mathbf{BM}}$, where **BM** denotes the category of Boolean monoids and their homomorphisms, the third component of an object may not preserve the structure of Boolean monoids.

Example 1.

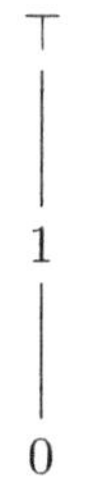

Consider the Kleene algebra $K = (\{0, 1, \top\}, +, \cdot, {}^*, 0, 1)$ with $+$ defined by the least upper bound with respect to the ordering, $\cdot$ defined by the table

$\cdot$	0	1	$\top$
0	0	0	0
1	0	1	$\top$
$\top$	0	$\top$	$\top$

and * defined by $0^* = 1^* = 1$ and $\top^* = \top$. This Kleene algebra appeared in [2, 4].

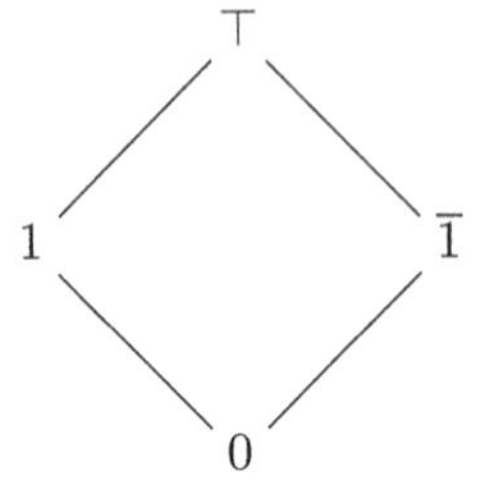

Also consider the Boolean monoid $M = (\{0, 1, \bar{1}, \top\}, +, \sqcap, ^-, \cdot, 0, \top, 1)$ with operators $+, \sqcap, ^-$ defined by the least upper bound, the greatest lower bound, and the complement with respect to the ordering, and $\cdot$ defined by the table

$\cdot$	0	1	$\bar{1}$	$\top$
0	0	0	0	0
1	0	1	$\bar{1}$	$\top$
$\bar{1}$	0	$\bar{1}$	$\bar{1}$	$\bar{1}$
$\top$	0	$\top$	$\bar{1}$	$\top$

Define the mapping $f\colon \{0, 1, \bar{1}, \top\} \to \{0, 1, \top\}$ by

$$0 \mapsto 0, \quad 1 \mapsto 1, \quad \bar{1} \mapsto \top, \quad \top \mapsto \top\ .$$

Then f is an idempotent semiring homomorphism from $U_{MI}(M)$ to $U_{KI}(K)$, where U_{MI} is the forgetful functor from **BM** to **ISR**. Thus, $\langle M, K, f\rangle$ is an object of $\mathcal{F}_{\mathbf{BM}}$ whose third component does not preserve Boolean monoid structure.

Let **RA** be the category of relation algebras and their homomorphisms. The full subcategory $\mathcal{F}^{\subseteq}_{\mathbf{RA}}$ of $\mathcal{F}_{\mathbf{RA}}$ such that the third component i of each object $\langle \mathbf{R}, \mathbf{K}, i\rangle$ is an inclusion is the category of Kleene algebras with relations in the sense of Desharnais [3].

Let $\langle \mathbf{X}, \mathbf{K}, i\rangle$ be an object of $\mathcal{F}_{\mathcal{X}}$. If i is a one-to-one mapping, then the image $\mathsf{im}(i)$ of the carrier set X of $\mathbf{X}$ forms an object $i[\mathbf{X}]$ in $\mathcal{X}$, and $\langle \mathbf{X}, \mathbf{K}, i\rangle$ is isomorphic to $\langle i[\mathbf{X}], \mathbf{K}, \subseteq\rangle$, where $i[\mathbf{X}]$ is exactly embedded in the Kleene algebra $\mathbf{K}$.

For a category $\mathcal{X}$, define $\Psi_{\mathcal{X}}$ to be the functor from $\mathcal{F}_{\mathcal{X}}$ to $\mathcal{X} \times \mathbf{Kleene}$ which takes an object $\langle \mathbf{X}, \mathbf{K}, f\rangle$ to the pair $(\mathbf{X}, \mathbf{K})$ and an arrow $\langle h, k\rangle$ to the pair (h, k).

Also, for a category $\mathcal{X}$, define $\Phi_{\mathcal{X}}$ to be the functor from $\mathcal{X} \times \mathbf{Kleene}$ to $\mathcal{F}_{\mathcal{X}}$ which takes $(\mathbf{X}, \mathbf{K})$ to $\langle \mathbf{X}, F_{IK}(U_{XI}(\mathbf{X})) + \mathbf{K}, \varphi(i)\rangle$, where i is the first injection of the coproduct $F_{IK}(U_{XI}(\mathbf{X})) + \mathbf{K}$, and (f, g) to $\langle f, F_{IK}(U_{XI}(f)) + g\rangle$ in $\mathcal{F}_{\mathcal{X}}$.

The next property follows from Theorem 1.

Theorem 2. *$\Phi_{\mathcal{X}}$ is a left adjoint to $\Psi_{\mathcal{X}}$.*

Let U_X be the forgetful functor from $\mathcal{X}$ to **Set** which takes an object $\mathbf{X}$ to the carrier set X of $\mathbf{X}$ and an arrow $f\colon \mathbf{X} \to \mathbf{X}'$ to a function $f\colon X \to X'$ from the carrier set X of $\mathbf{X}$ to the carrier set X' of $\mathbf{X}'$. If U_X has a left adjoint $F_X\colon \mathbf{Set} \to \mathcal{X}$, then the functor $U_X \times U_K\colon \mathcal{X} \times \mathbf{Kleene} \to \mathbf{Set} \times \mathbf{Set}$ has a left adjoint $F_X \times F_K$. In this case, we have the following sequence of adjunctions:

$$\mathbf{Set} \times \mathbf{Set} \underset{U_X \times U_K}{\overset{F_X \times F_K}{\rightleftarrows}} \mathcal{X} \times \mathbf{Kleene} \underset{\Psi_{\mathcal{X}}}{\overset{\Phi_{\mathcal{X}}}{\rightleftarrows}} \mathcal{F}_{\mathcal{X}}$$

5 Observation

$\mathcal{F}_{\mathcal{X}}^{\subseteq}$ denotes the full subcategory of $\mathcal{F}_{\mathcal{X}}$ such that the third component of each object is an inclusion. It was shown in [5] that, for each object $\langle \mathbf{B}, \mathbf{K}, i\rangle$ in $\mathcal{F}_{\mathbf{Bool}}$, we have an object $\langle i[\mathbf{B}], \mathbf{K}, \subseteq\rangle$ in the category $\mathcal{F}_{\mathbf{Bool}}^{\subseteq}$ of Kleene algebras with tests in the sense of Kozen since the image $\mathsf{im}(i)$ of $U_{BI}(\mathbf{B})$ under the idempotent semiring homomorphism i forms a Boolean algebra $i[\mathbf{B}]$. This fact induces the functor from $\mathcal{F}_{\mathbf{Bool}}$ to $\mathcal{F}_{\mathbf{Bool}}^{\subseteq}$, and, moreover, the functor is a left adjoint to the forgetful functor from $\mathcal{F}_{\mathbf{Bool}}^{\subseteq}$ to $\mathcal{F}_{\mathbf{Bool}}$.

On the other hand, Example 1 shows that the third component f of $\langle B, K, f\rangle$ does not preserve Boolean monoid structure. Thus, we cannot have adjunction between $\mathcal{F}_{\mathbf{BM}}$ and $\mathcal{F}_{\mathbf{BM}}^{\subseteq}$ in a similar way to the case of $\mathcal{F}_{\mathbf{Bool}}$ and $\mathcal{F}_{\mathbf{Bool}}^{\subseteq}$. But while we may not have adjunction between $\mathcal{F}_{\mathcal{X}}^{\subseteq}$ and $\mathcal{F}_{\mathcal{X}}$ for each $\mathcal{X}$, the adjunction between $\mathbf{Set} \times \mathbf{Set}$ and $\mathcal{F}_{\mathcal{X}}^{\subseteq}$ still can be obtained.

Assume that $U_X\colon \mathcal{X} \to \mathbf{Set}$ has a left adjoint $F_X\colon \mathbf{Set} \to \mathcal{X}$. For a set A, the carrier of $F_X(A)$ should have at least two distinct elements which may be the zero element and the identity element in $U_{XI}(F_X(A))$ because $F_X(A)$ is a free algebra which has an underlying idempotent semiring structure. So, $F_{IK}(U_{XI}(F_X(A)))$ is a non-trivial Kleene algebra since a component of the unit of $F_{IK} \dashv U_{KI}$ with respect to each idempotent semiring is one-to-one by Proposition 2. Also, by Remark 2, $F_K(\Sigma)$ is integral. Thus we have the following property.

Theorem 3. *Let i be the first injection of $F_{IK}(U_{XI}(F_X(A))) + F_K(B)$. Then, the third component $\varphi(i)$ of an object $\langle F_X(A), F_{IK}(U_{XI}(F_X(A))) + F_K(B), \varphi(i)\rangle$ in $\mathcal{F}_{\mathcal{X}}$ is one-to-one for each pair (A, B) of sets if $U_X\colon \mathcal{X} \to \mathbf{Set}$ has a left adjoint F_X.*

Proof. Take the object $\langle F_X(A), F_{IK}(U_{XI}(F_X(A))) + F_K(B), \varphi(i)\rangle$ in $\mathcal{F}_{\mathcal{X}}$. Then i is one-to-one by Proposition 1. Therefore, by Proposition 3, $\varphi(i)$ is one-to-one.

So, as we mentioned in Section 4, $\mathsf{im}(\varphi(i))$ forms an object $\varphi(i)[F_X(A)]$ in $\mathcal{X}$, and

$$\begin{aligned}&\langle F_X(A), F_{IK}(U_{XI}(F_X(A))) + F_K(B), \varphi(i)\rangle\\ &\quad\cong \langle \varphi(i)[F_X(A)], F_{IK}(U_{XI}(F_X(A))) + F_K(B), \subseteq\rangle\ .\end{aligned}$$

The mapping

$$(A, B) \mapsto \langle \varphi(i)[F_X(A)], F_{IK}(U_{XI}(F_X(A))) + F_K(B), \subseteq\rangle$$

determines a functor $\Xi_{\mathcal{X}}$ from $\mathbf{Set} \times \mathbf{Set}$ to $\mathcal{F}_{\mathcal{X}}^{\subseteq}$ and the following holds:

Theorem 4. *The functor $\Xi_{\mathcal{X}}$ is a left adjoint to the forgetful functor from $\mathcal{F}^{\subseteq}_{\mathcal{X}}$ to* **Set** $\times$ **Set** *if $U_X\colon \mathcal{X} \to$* **Set** *has a left adjoint F_X.*

The following is immediate since $U_B\colon$ **Bool** $\to$ **Set** has a left adjoint F_B.

Corollary 1. *The functor $\Xi_{\mathbf{Bool}}$ which is determined by the mapping*

$$(A, B) \mapsto \langle \varphi(i)[F_B(A)], F_{IK}(U_{BI}(F_B(A))) + F_K(B), \subseteq\rangle$$

is a left adjoint to the forgetful functor from $\mathcal{F}^{\subseteq}_{\mathbf{Bool}}$ to **Set** $\times$ **Set**.

Hence, the triple $\langle \varphi(i)[F_B(A)], F_{IK}(U_{BI}(F_B(A)))+F_K(B), \subseteq\rangle$ is the free Kleene algebra with tests, in the sense of [10, 11], generated by a pair of sets A and B.

Since the structure of relation algebras is obtained by adding composition, the identity element with respect to composition, and the converse to Boolean algebra structure, and is axiomatised by equations, the forgetful functor $U_R\colon$ **RA** $\to$ **Set** also has a left adjoint $F_R\colon$ **Set** $\to$ **RA**. Note that the statement is not true if we require that relation algebras are complete, that is, relation algebras have an arbitrary join.

Corollary 2. *The functor $\Xi_{\mathbf{RA}}$ which is determined by the mapping*

$$(A, B) \mapsto \langle \varphi(i)[F_R(A)], F_{IK}(U_{RI}(F_R(A))) + F_K(B), \subseteq\rangle$$

is a left adjoint to the forgetful functor from $\mathcal{F}^{\subseteq}_{\mathbf{RA}}$ to **Set** $\times$ **Set**.

Hence, the triple $\langle \varphi(i)[F_R(A)], F_{IK}(U_{RI}(F_R(A)))+F_K(B), \subseteq\rangle$ is the free Kleene algebra with relations, in the sense of [3], generated by a pair of sets A and B if we do not require the completeness of relation algebras.

6 Conclusion

We proposed a general framework for a Kleene algebra with an embedded structure. If the embedded structure satisfies the following conditions we have a free construction from a pair of sets.

- $\mathcal{X}$ is the category of a certain kind of algebras which have underlying idempotent semiring structure and homomorphisms,
- $U_X\colon \mathcal{X} \to$ **Set** has a left adjoint.

Considering that the embedding should share the underlying idempotent semiring structure, the first condition is quite natural. The second condition is also acceptable if we consider the fact that the forgetful functor from the category of an (essentially) algebraic structure on the carrier sets of their objects consisting of a number of finitary operations which are required to satisfy a number of Horn clauses has a left adjoint. The construction of a left adjoint from **Set** $\times$ **Set** to $\mathcal{F}_{\mathcal{X}}$ provides a presentation of free algebras as coproducts in **Kleene**.

Acknowledgements. The author would like to thank Xiaoheng Ji, Wolfram Kahl, Dexter Kozen, Hiroshi Watanabe, and the anonymous referees for discussions and helpful comments.

References

1. M. Barr and C. Wells. Category Theory for Computing Science, Third Edition, Les Publications CRM, (1999).
2. John Horton Conway. Regular Algebra and Finite Machines. Chapman and Hall, London, 1971.
3. Jules Desharnais. Kleene Algebras with Relations. In R. Berghammer and B. Möller, editors, *Relational and Kleene-Algebraic Methods in Computer Science: 7th International Seminar on Relational Methods in Computer Science and 2nd International Workshop on Applications of Kleene Algebra*, volume 3051 of Springer Lecture Notes in Computer Science, 8–20, 2004. (Invited Talk).
4. J. Desharnais, B. Möller, and G. Struth. Kleene Algebra with Domain, *Technical Report 2003-07, Institut für Informatik, Universität Augsburg* (2003).
5. Hitoshi Furusawa. A Free Construction of Kleene Algebra with Tests. In Kozen, editor, *Mathematics of Program Construction, 7th International Conference, MPC 2004*, volume 3125 of *Springer Lecture Notes in Computer Science*, 129–141, 2004.
6. H. Furusawa and Y. Kinoshita. Essentially Algebraic Structure for Kleene Algebra with Tests and Its Application to Semantics of **While** Programs, *IPSJ Transactions on Programming*, Vol. 44, No. SIG 4 (PRO 17), 47–53 (2003).
7. Y. Kinoshita and H. Furusawa. Essentially Algebraic Structure for Kleene Algebra with Tests, *Computer Software*, Vol. 20, No. 2, 47–53 (2003). In Japanese.
8. Dexter Kozen. On Kleene Algebras and Closed Semirings. In Rovan, editor, *Proceedings of Mathematical Foundations of Computer Science*, volume 452 of *Springer Lecture Notes in Computer Science*, 26–47, 1990.
9. Dexter Kozen. A Completeness Theorem for Kleene Algebras and the Algebra of Regular Events. *Information and Computation*, 110:366–390 (1994).
10. Dexter Kozen. Kleene Algebra with Tests. *ACM Transactions on Programming Languages and Systems*, Vol. 19, No. 3, 427–443 (1997).
11. D. Kozen and F. Smith. Kleene Algebra with Tests: Completeness and Decidability. In D. van Delen and M. Bezem, editors, *Proc 10th Int. Workshop Computer Science Logic (CSL'96)*, volume 1258 of *Springer Lecture Notes in Computer Science*, 244–259 (1996).
12. Saunders Mac Lane. Categories for the Working Mathematician, Second Edition, Springer, (1998).
13. C. Mulvey and J. Pelletier. A Quantisation of the Calculus of Relations, *Proc. of Category Theory 1991*, CMS Conference Proceedings, No. 13, Amer. Math. Soc., 345–360 (1992).
14. Izumi Takeuti. Kleene Category as a Model of Calculation, *Proc 2nd Symposium on Science and Technology for System Verification, Programming Science Group Technical Report, Research Center for Verification and Semantics, National Institute of Advanced Industrial Science and Technology (AIST)*, PS-2005-017, 151–160, October 2005.

Appendix (Proof of Theorem 1)

First, we prove that Θ defined in Section 2 is a functor from $E \times D$ to (T, S). The image $\langle e, V(T(e)) + d, \rho(i) \rangle$ of an object (e, d) in $E \times D$ under Θ is an object in (T, S) since e is an object in E, $V(T(e)) + d$ is an object in D, and $\rho(i)$ is an arrow in C whose source is $T(e)$ and target is $S(V(T(e)) + d)$. By the definition of $V(T(k)) + h$, for a pair (k, h) of an arrow $k\colon\ e \to e'$ in E and an arrow $h\colon\ d \to d'$ in D, the following diagram commutes:

$$\begin{array}{ccc} V(T(e)) & \xrightarrow{\ i\ } & V(T(e)) + d \\ {\scriptstyle V(T(k))}\Big\downarrow & & \Big\downarrow{\scriptstyle V(T(k)) + h} \\ V(T(e')) & \xrightarrow[\ i'\ ]{} & V(T(e')) + d' \end{array}$$

where i and i' are the first injections of the coproducts $V(T(e))+d$ and $V(T(e'))+d'$ respectively. Hence, by the naturality of ρ, the following diagram commutes:

$$\begin{array}{ccc} T(e) & \xrightarrow{\ \rho(i)\ } & S(V(T(e)) + d) \\ {\scriptstyle T(k)}\Big\downarrow & & \Big\downarrow{\scriptstyle S(V(T(k)) + h)} \\ T(e') & \xrightarrow[\ \rho(i')\ ]{} & S(V(T(e')) + d') \end{array}$$

Thus, the image $\langle k, V(T(k)) + h\rangle$ of each arrow $(k, h)\colon\ (e, d) \to (e', d')$ in $E \times D$ under Θ is an arrow form $\langle e, V(T(e))+d, \rho(i)\rangle$ to $\langle e, V(T(e'))+d', \rho(i')\rangle$ in (T, S). Since $V\colon\ C \to D$, $T\colon\ E \to C$, and $+\colon\ D \times D \to D$ are functors, Θ preserves identities and compositions. Therefore, we have proved that Θ is a functor from $E \times D$ to (T, S).

Next, we prove that Θ is a left adjoint to Υ. Define the mapping $\xi_{(e,d),\langle e',d',l'\rangle}$ from the set $(T, S)(\Theta(e, d), \langle e', d', l'\rangle)$ of arrows from $\Theta(e, d)$ to $\langle e', d', l'\rangle$ in (T, S) to the set $E \times D((e, d), \Upsilon\langle e', d', l'\rangle)$ of arrows from (e, d) to $\Upsilon\langle e', d', l'\rangle$ in $E \times D$ for each object (e, d) in $E \times D$ and $\langle e', d', l'\rangle$ in (T, S) as follows:

$$\left\langle \begin{array}{c} e \\ {\scriptstyle f}\Big\downarrow \\ e' \end{array} ,\ \begin{array}{c} V(T(e)) + d \\ \Big\downarrow{\scriptstyle g} \\ d' \end{array} \right\rangle \mapsto \left(\begin{array}{c} e \\ {\scriptstyle f}\Big\downarrow \\ e' \end{array} ,\ \begin{array}{c} d \\ \Big\downarrow{\scriptstyle g \circ j} \\ d' \end{array} \right)$$

where j is the second injection of $V(T(e))+d$. In the sequel, ξ means $\xi_{(e,d),\langle e',d',l'\rangle}$. It is sufficient to show that ξ is bijective. The bijectivity immediately follows from verifying that the arrow $\langle f, g\rangle$ satisfies

$$g \circ i = V(T(f)) \circ \rho^{-1}(l')$$

since intermediating arrow g is a unique arrow which commutes the following two squares

$$\begin{array}{ccccc} V(T(e)) & \xrightarrow{\ i\ } & V(T(e)) + d & \xleftarrow{\ j\ } & d \\ {\scriptstyle V(T(f))}\Big\downarrow & & \vdots\,{\scriptstyle g} & & \Big\downarrow{\scriptstyle j} \\ V(T(e')) & \xrightarrow[\ \rho^{-1}(l')\ ]{} & d' & \xleftarrow[\ g\ ]{} & V(T(e)) + d \end{array}$$

Indeed, this equation holds since the following square commutes by the definition of $\langle f, g \rangle$.

$$\begin{array}{ccc} T(e) & \xrightarrow{\rho(i)} & S(V(T(e)) + d) \\ {\scriptstyle T(f)}\downarrow & & \downarrow{\scriptstyle S(g)} \\ T(e') & \xrightarrow[l']{} & S(d') \end{array}$$

Non-termination in Unifying Theories of Programming

Walter Guttmann

Abteilung Programmiermethodik und Compilerbau,
Universität Ulm, 89069 Ulm, Germany
walter.guttmann@uni-ulm.de

Abstract. Within the *Unifying Theories of Programming* framework, program initiation and termination has been modelled by introducing a pair of variables in order to satisfy the required algebraic properties. We replace these variables with the improper value $\bot$ that is frequently used to denote undefinedness. Both approaches are proved isomorphic using the relation calculus, and the existing operations and laws are carried over. We split the isomorphism by interposing "intuitive" relations.

1 Introduction

The *Unifying Theories of Programming* framework [1], hereafter abbreviated as UTP, takes a relational view on semantics: The meaning of a non-deterministic, imperative program is described by a predicate relating the initial and final values of its observable variables. This intuitive view, however, turns out to be too simplistic. Certain laws observed in practice for programs are not satisfied by arbitrary predicates, e.g., the zero laws of sequential composition:

$$\mathit{true}; P \;=\; \mathit{true} \;=\; P; \mathit{true},$$

where *true* is the meaning of the totally unpredictable program. To rectify the situation one might, e.g., redefine the sequential composition operator, but we appreciate a solution that retains its meaning as relational composition. The class of predicates employed to describe programs is instead restricted by so-called healthiness conditions.

The definition of the proper subclass depends on the introduction of a free variable – actually a pair *ok* and ok' denoting the initial and final value – to model that a program has been started and terminated. Such information about the definedness of a program is often denoted by the distinguished, improper value $\bot$ that, depending on the context, represents an undefined variable, expression, or state of execution.

In Sect. 3 we show how these two views coincide. To this end, we replace *ok* and ok' with $\bot$ to deal with non-termination in UTP and prove that the new and original approaches are isomorphic. This alleviates the objection to introduce the improper value put forward by [2]. The formal development is carried out in the relation calculus [3].

W. MacCaull et al. (Eds.): RelMiCS 2005, LNCS 3929, pp. 108–120, 2006.

We continue the investigation in Sect. 4 by proving that both views are in one-to-one correspondence with the intuitive relational reading of predicates mentioned above. The latter class is installed as a seamless intermediate, thus splitting the isomorphism.

First of all, we describe the principles of [1] we will need in the rest of the paper. We also state several structural properties.

2 Basics of Unifying Theories of Programming

In practice one can observe about the execution of a program the initial and final values of its variables. UTP therefore models a program as a predicate P whose free variables come from a set αP, its alphabet, that is partitioned according to

$$\alpha P \;=\; in\alpha P \cup out\alpha P$$

into a set of undashed variables $in\alpha P$ standing for initial values and a set of dashed variables $out\alpha P$ standing for final values. We will focus on the case $out\alpha P = (in\alpha P)'$, the extension to the inhomogeneous case being uncomplicated. Unless stated otherwise we will use $v = in\alpha P$ and $v' = out\alpha P$ treating both v and v' as a single variable rather than a set of variables. Thus, v takes as its value a tuple of values, one for each element of $in\alpha P$, and similarly does v'.

Every predicate P may be identified with the set $\{(v, v') : P(v, v')\}$ of all pairs of observations that satisfy it [1]. We will adopt this relational attitude at the end of this section.

2.1 Combining Predicates

The *non-deterministic choice* between predicates P and Q is just

$$P \vee Q,$$

so we do not have to introduce a new notation.

The *conditional* uses a predicate without dashed variables b, to choose between predicates P and Q as per

$$P \lhd b \rhd Q \;=_{\mathsf{def}}\; (b \wedge P) \vee (\neg b \wedge Q).$$

The *sequential composition* of predicates P and Q with a common alphabet is defined by

$$P;Q \;=_{\mathsf{def}}\; \exists v_0 : P[v_0/v'] \wedge Q[v_0/v],$$

where $P[v_0/v']$ denotes the substitution of v_0 for v' in P. Attention has to be paid when a substitution is applied to $P;Q$ since v' in P and v in Q are replaced by v_0 which is bound by the existential quantifier. For example, if e does not depend on v and v_0, $(P;Q)[e/v'] = P;(Q[e/v'])$.

2.2 Designs

Given predicates P_1 and P_2 that do not contain the auxiliary variables *ok* and *ok'* a *design* is defined as

$$P_1 \vdash P_2 \;=_{\text{def}}\; ok \wedge P_1 \implies ok' \wedge P_2 .$$

Its informal reading is that "if the program starts in a state satisfying P_1, it will terminate, and on termination P_2 will be true".

The *no-operation* program is the design

$$\mathbb{I}_D \;=_{\text{def}}\; true \vdash v = v' ,$$

where v and v' vary depending on the context.

The *assignment* of the value of an expression e to a variable x is the design

$$x := e \;=_{\text{def}}\; true \vdash x' = e \wedge w' = w$$

where w denotes all variables except x.

2.3 Healthiness Conditions

The restriction to designs does not yield all required algebraic properties. These are enforced by the four *healthiness conditions* imposed on a predicate P:

H1. $P = (ok \implies P)$.
H2. $\left[P[false/ok'] \implies P[true/ok']\right]$.
H3. $P = P; \mathbb{I}_D$.
H4. $P; true = true$.

The outer brackets in H2 denote the universal closure. A characterisation of H1 is given by [1, Theorem 3.2.2] stating

$$P \text{ is H1} \iff (true; P = true) \wedge (\mathbb{I}_D; P = P),$$

only H2 lacks a convincing algebraic formulation, as already remarked by [2]. We solve this problem by restricting our notion of design.

Proposition 1. *Every predicate that is H3 is also H2.*

Proof. Let P be a predicate that is H3 and has v and ok as its variables, then $P[false/ok'] = (P; \mathbb{I}_D)[false/ok'] = P; (\mathbb{I}_D[false/ok']) \implies P; (\mathbb{I}_D[true/ok']) = (P; \mathbb{I}_D)[true/ok'] = P[true/ok']$ since composition is monotonic and $\mathbb{I}_D$ is H2.

Designs are characterised by [1, Theorem 3.2.3] as just those predicates that are H1 and H2, thus by Proposition 1 a predicate that is H1 and H3 is a design, too. Following the terminology of [2] we call such a predicate a *normal* design. Finally, a predicate that is H4 is called *feasible*. The relations of predicates satisfying the various laws are displayed in Fig. 1.

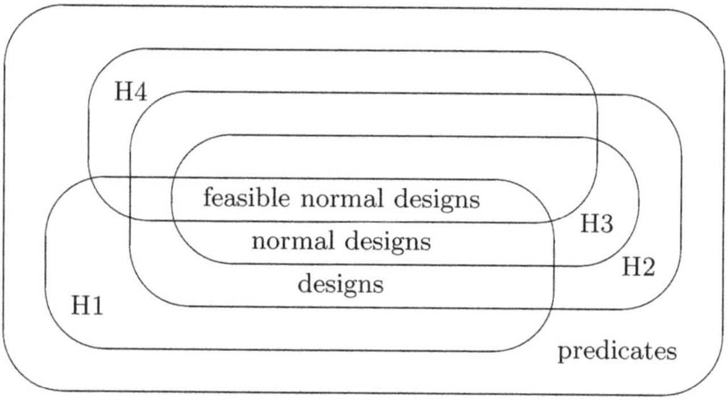

Fig. 1. Predicates satisfying different healthiness conditions

2.4 The Complete Lattice of Predicates

Using conjunction and disjunction as the lattice operators, predicates form a complete lattice. The predicates satisfying any combination of H1, H2, and H3 (but not H4) form a complete sub-lattice [4]. Rather unfortunately, the underlying order employed by UTP is the *reverse* implication ordering where $P \sqsubseteq Q$ if and only if $[Q \implies P]$. Thus, the weakest fixed point μF of a monotonic function F from predicates to predicates exists and it is used to define *recursion*:

$$\mu F =_{\text{def}} \bigvee \{X : [X \implies F(X)]\}.$$

With recursion, all constructs are in place to define the programming language of UTP. Following [1, Table 5.0.1], a program may be constructed from the constants *true* and assignment, the combinators introduced in Sect. 2.1, and recursion. From [1, Chapter 5.4] it follows that F is continuous if it is composed from these program constructs.

The program constructs preserve H1–H4 with one exception: In general, μF need not be feasible even if F is continuous and preserves all healthiness conditions, including feasibility. A counterexample is given by $\mu F_0 = \neg ok$ for

$$F_0 =_{\text{def}} P_1 \vdash P_2 \mapsto \mathit{true} \vdash x + x' > \min\{x + x' : P_1 \implies P_2\},$$

with the natural numbers as the domain of x and x'. Nevertheless, if F is composed from program constructs, μF is feasible.

2.5 Predicates and Relations

Up to now we have used the language of predicates since this is the formalism employed by [1]. We will retain these terms whenever we discuss the connection to UTP but will switch to relational terms [3] for the formal development.

To establish a common notation we will denote, for relations R and S, union as $R \vee S$, intersection as $R \wedge S$, inclusion as $R \leq S$, composition as $R ; S$, transposition as R^{T}, and complement as $\overline{R}$, and the constants empty relation as $\perp\!\!\!\perp$, identity relation as $\mathbb{I}$, and universal relation as $\mathbb{T}$. Unary operators have highest precedence, followed by composition, union and intersection, conditional, the formation of designs, finally equality and inclusion with lowest precedence.

A relation R is a *vector* if $R = R;\mathbb{T}$. The design $P_1 \vdash P_2$ is a normal design if and only if P_1 is a vector [1, Theorem 3.2.4].

3 Representing Non-termination with Improper Values

Given a predicate P, for any choice of values for v either there exists a choice for v' that satisfies P or not. We may say accordingly that, for the given assignment to v, P is defined or not. Recall that in UTP v' actually represents a set of variables, yet there is no notion of separate definedness for each variable. One cannot state that, for the given assignment to v, certain variables represented by v' are defined whereas others are not. When we introduce the improper value $\perp$ we therefore do not extend the range of every variable by distinct instances of $\perp$ but add just one value, respectively, to the range of v and v'. This corresponds to the construction of the *smash-product* of semantic domains.

3.1 Bottom Predicates

In the following we need to distinguish two kinds of predicates.

1. Presented in Sects. 2.2 and 2.3, (feasible) normal designs contain the auxiliary variables *ok* and *ok*′. We assume that $\perp$ denotes a value that is not in the range of the variables of designs.
2. We call a predicate that does not contain *ok* and *ok*′ and has the improper value $\perp$ in the range of its variables a $\perp$*-predicate*. Two special $\perp$-predicates are the vector $\mathbb{V} =_{\mathsf{def}} (v = \perp)$ and $\mathbb{V}^{\mathsf{T}} = (v' = \perp)$.

We will use the same symbols to denote the operations on $\perp$-predicates and designs, relying on the context for disambiguation.

A design $P_1 \vdash P_2$ is composed of predicates P_1 and P_2 that neither contain *ok* and *ok*′, nor $\perp$. Let us call predicates of this type *basic relations*. They play a formal role in the following development, but will receive additional interpretation in Sect. 4 as "intuitive" relations. To facilitate the construction of $\perp$-predicates we introduce two operations that convert to and from basic relations.

Definition 2. *The operations $\cdot^+$ from basic relations to $\perp$-predicates and $\cdot^-$ from $\perp$-predicates to basic relations are lattice homomorphisms and satisfy the axioms*

$$(P;Q)^+ = P^+;Q^+ \qquad P^{+-} = P$$
$$\overline{P}^- = \overline{P^-} \qquad P^{-+} = P \wedge \overline{\mathbb{V}} \wedge \overline{\mathbb{V}^{\mathsf{T}}}$$

When a relation is viewed as a set of pairs, the intention for $\cdot^+$ is the identity, and $\cdot^-$ removes all pairs having $\perp$ as one component. The following development nevertheless uses just the stated axioms and the homomorphism qualities, namely monotonicity and distributivity over union and intersection. Lemma 3 provides derived properties of the operations for subsequent use.

Lemma 3. *1. The lower adjoint* $\cdot^+$ *and the upper adjoint* $\cdot^-$ *form a Galois connection,* $\bot^+ = \bot$, $\bot^- = \bot$, $\top^- = \top$, $\top^+ = \overline{\mathbb{V}} \wedge \overline{\mathbb{V}^\mathsf{T}}$, $\overline{\mathbb{V}}^- = \top$, $\mathbb{I}^- = \mathbb{I}$, *and* $\mathbb{I}^+ = \mathbb{I} \wedge \overline{\mathbb{V}} \wedge \overline{\mathbb{V}^\mathsf{T}}$.

2. $P^+ = P^+ \wedge \overline{\mathbb{V}} \wedge \overline{\mathbb{V}^\mathsf{T}}$ *and* $P^+ \wedge Q^{-+} = P^+ \wedge Q$.
3. $\overline{P^+}; \top = \top$ *and* $\overline{P^+; \top} \geq \mathbb{V}$.
4. $(P^+; Q)^- = P; Q^-$ *and* $(P^+; \top)^- = P; \top$.
5. $P^+ \wedge (Q; R^-)^+ = P^+ \wedge Q^+; R$ *and* $P^+ \wedge \overline{Q; R^-}^+ = P^+ \wedge \overline{Q^+; R}$.
6. *If* B *is a vector,* $P^+ \wedge B^+ = P^+ \wedge B^+; \top$ *and* $P^+ \wedge \overline{B}^+ = P^+ \wedge \overline{B^+; \top}$.
7. *If* B *is a vector,* $(P \lhd B \rhd Q)^+ = P^+ \lhd B^+; \top \rhd Q^+$.
8. *If* P *is a vector,* $\overline{P^{-+}; \top} = \overline{P} \vee \mathbb{V}$.
9. $(P; \top)^+; \top = P^+; \top$.

Proof.

1. $P^+ \leq Q \implies P = P^{+-} \leq Q^- \implies P^+ \leq Q^{-+} = Q \wedge \overline{\mathbb{V}} \wedge \overline{\mathbb{V}^\mathsf{T}} \leq Q$.
$\bot \leq \bot^- \implies \bot^+ \leq \bot \implies \bot^+ = \bot$, hence $\bot^- = \bot^{+-} = \bot$.
$\top^+ \leq \top \implies \top \leq \top^- \implies \top = \top^-$, hence $\top^+ = \top^{-+} = \top \wedge \overline{\mathbb{V}} \wedge \overline{\mathbb{V}^\mathsf{T}} = \overline{\mathbb{V}} \wedge \overline{\mathbb{V}^\mathsf{T}}$, hence $\top = \top^{+-} = (\overline{\mathbb{V}} \wedge \overline{\mathbb{V}^\mathsf{T}})^- \leq \overline{\mathbb{V}}^-$.
$\mathbb{I}^- = (\mathbb{I}^-; \mathbb{I})^{+-} = (\mathbb{I}^{-+}; \mathbb{I}^+)^- = ((\mathbb{I} \wedge \overline{\mathbb{V}} \wedge \overline{\mathbb{V}^\mathsf{T}}); \mathbb{I}^+)^- = (\overline{\mathbb{V}} \wedge \mathbb{I}; (\overline{\mathbb{V}} \wedge \mathbb{I}^+))^- = (\overline{\mathbb{V}} \wedge \mathbb{I}^+)^- = \overline{\mathbb{V}}^- \wedge \mathbb{I}^{+-} = \mathbb{I}$, hence $\mathbb{I}^+ = \mathbb{I}^{-+} = \mathbb{I} \wedge \overline{\mathbb{V}} \wedge \overline{\mathbb{V}^\mathsf{T}}$.
2. $P^+ = P^{+-+} = P^+ \wedge \overline{\mathbb{V}} \wedge \overline{\mathbb{V}^\mathsf{T}}$, hence $P^+ \wedge Q^{-+} = P^+ \wedge Q \wedge \overline{\mathbb{V}} \wedge \overline{\mathbb{V}^\mathsf{T}} = P^+ \wedge Q$.
3. By (2), $\overline{P^+}; \top = \overline{P^+ \wedge \overline{\mathbb{V}} \wedge \overline{\mathbb{V}^\mathsf{T}}}; \top \geq \overline{\overline{\mathbb{V}^\mathsf{T}}}; \top = \mathbb{V}^\mathsf{T}; \top = \top$ and
$\overline{P^+; \top} = \overline{(P^+ \wedge \overline{\mathbb{V}} \wedge \overline{\mathbb{V}^\mathsf{T}}); \top} \geq \overline{\overline{\overline{\mathbb{V}}}; \top} = \overline{\overline{\mathbb{V}}} = \mathbb{V}$.
4. $(P^+; Q)^- = (P^+; Q)^{-+-} = (P^+; Q \wedge \overline{\mathbb{V}} \wedge \overline{\mathbb{V}^\mathsf{T}})^- = ((\overline{\mathbb{V}} \wedge P^+); (Q \wedge \overline{\mathbb{V}^\mathsf{T}}))^- = ((P^+ \wedge \overline{\mathbb{V}^\mathsf{T}}); (Q \wedge \overline{\mathbb{V}^\mathsf{T}}))^- = (P^+; (Q \wedge \overline{\mathbb{V}} \wedge \overline{\mathbb{V}^\mathsf{T}}))^- = (P^+; Q^{-+})^- = P; Q^-$ by (2), hence $(P^+; \top)^- = P; \top^- = P; \top$ by (1).
5. By (4) and (2), $P^+ \wedge (Q; R^-)^+ = P^+ \wedge (Q^+; R)^{-+} = P^+ \wedge Q^+; R$ and $P^+ \wedge \overline{Q; R^-}^+ = P^+ \wedge \overline{(Q^+; R)^-}^+ = P^+ \wedge \overline{Q^+; R}^{-+} = P^+ \wedge \overline{Q^+; R}$.
6. By (1) and (5), $P^+ \wedge B^+ = P^+ \wedge (B; \top)^+ = P^+ \wedge (B; \top^-)^+ = P^+ \wedge B^+; \top$ and $P^+ \wedge \overline{B}^+ = P^+ \wedge \overline{B; \top}^+ = P^+ \wedge \overline{B; \top^-}^+ = P^+ \wedge \overline{B^+; \top}$.
7. $(P \lhd B \rhd Q)^+ = (B^+ \wedge P^+) \vee (\overline{B}^+ \wedge Q^+) = (B^+; \top \wedge P^+) \vee (\overline{B^+; \top} \wedge Q^+) = P^+ \lhd B^+; \top \rhd Q^+$ by (6).
8. $\overline{P^{-+}; \top} = \overline{(P \wedge \overline{\mathbb{V}} \wedge \overline{\mathbb{V}^\mathsf{T}}); \top} = \overline{P \wedge \overline{\mathbb{V}} \wedge \overline{\mathbb{V}^\mathsf{T}}; \top} = \overline{P} \vee \mathbb{V} \vee \overline{\overline{\mathbb{V}^\mathsf{T}}; \top} = \overline{P} \vee \mathbb{V}$.
9. $P^+; \top \leq (P; \top)^+; \top = P^+; \top^+; \top \leq P^+; \top$.

3.2 From Auxiliary Variables to Improper Values

The transition from normal designs to $\bot$-predicates is accomplished by the function $\mathcal{I}$, whose effect on the constructs of UTP introduced in Sect. 2 is given by Lemma 5.

Definition 4. *The mapping* $\mathcal{I}$ *from normal designs to* $\bot$*-predicates is*

$$\mathcal{I} =_{\text{def}} P_1 \vdash P_2 \;\mapsto\; \overline{{P_1}^+; \top} \vee {P_2}^+ .$$

The no-operation program is transformed to $\mathbb{I}_\bot =_{\text{def}} \mathcal{I}(\mathbb{I}_D)$.

Lemma 5. *For a vector B and normal designs P and Q,*

1. $\mathcal{I}(P \vee Q) = \mathcal{I}(P) \vee \mathcal{I}(Q)$,
2. $\mathcal{I}(P \lhd B \rhd Q) = \mathcal{I}(P) \lhd B^+;\mathbb{T} \rhd \mathcal{I}(Q)$,
3. $\mathcal{I}(P;Q) = \mathcal{I}(P);\mathcal{I}(Q)$,
4. $\mathbb{I}_\bot = \mathbb{I} \vee \mathbb{V}$, *and*
5. $\mathcal{I}(\mathbb{T}) = \mathbb{T}$.

Proof. We will use [1, Theorem 3.1.4] that describes non-deterministic choice, conditional, and sequential composition of two designs as a design. Let $P_1 \vdash P_2$ and $Q_1 \vdash Q_2$ be normal designs.

1. By Lemma 3(6), $\mathcal{I}((P_1 \vdash P_2) \vee (Q_1 \vdash Q_2)) = \mathcal{I}(P_1 \wedge P_2 \vdash Q_1 \vee Q_2) =$
   $\overline{(P_1^+ \wedge Q_1^+)};\mathbb{T} \vee (P_2 \vee Q_2)^+ = \overline{(P_1^+ \wedge Q_1^+;\mathbb{T})};\mathbb{T} \vee P_2^+ \vee Q_2^+ =$
   $\overline{P_1^+;\mathbb{T} \wedge Q_1^+;\mathbb{T}} \vee P_2^+ \vee Q_2^+ = \overline{P_1^+;\mathbb{T}} \vee P_2^+ \vee \overline{Q_1^+;\mathbb{T}} \vee Q_2^+ =$
   $\mathcal{I}(P_1 \vdash P_2) \vee \mathcal{I}(Q_1 \vdash Q_2)$.
2. By Lemma 3(7), $\mathcal{I}((P_1 \vdash P_2) \lhd B \rhd (Q_1 \vdash Q_2)) =$
   $\mathcal{I}(P_1 \lhd B \rhd Q_1 \vdash P_2 \lhd B \rhd Q_2) = \overline{(P_1 \lhd B \rhd Q_1)^+};\mathbb{T} \vee (P_2 \lhd B \rhd Q_2)^+ =$
   $\overline{(P_1^+ \lhd B^+;\mathbb{T} \rhd Q_1^+)};\mathbb{T} \vee (P_2^+ \lhd B^+;\mathbb{T} \rhd Q_2^+) =$
   $(\overline{P_1^+;\mathbb{T}} \lhd B^+;\mathbb{T} \rhd \overline{Q_1^+;\mathbb{T}}) \vee (P_2^+ \lhd B^+;\mathbb{T} \rhd Q_2^+) =$
   $\overline{P_1^+;\mathbb{T}} \vee P_2^+ \lhd B^+;\mathbb{T} \rhd \overline{Q_1^+;\mathbb{T}} \vee Q_2^+ = \mathcal{I}(P_1 \vdash P_2) \lhd B^+;\mathbb{T} \rhd \mathcal{I}(Q_1 \vdash Q_2)$.
3. $\overline{P_1^+;\mathbb{T}} = \overline{P_1^+;\mathbb{T}};\mathbb{T};\mathbb{V} \le \overline{P_1^+;\mathbb{T}};\overline{Q_1^+;\mathbb{T}} \le \overline{P_1^+;\mathbb{T}};(\overline{Q_1^+;\mathbb{T}} \vee Q_2^+) \le$
   $\overline{P_1^+;\mathbb{T}};\mathbb{T} = \overline{P_1^+;\mathbb{T}}$ by Lemma 3(3).

   $P_1^+ \wedge \overline{P_2^+;\overline{\overline{Q_1^+;\mathbb{T}}}} = P_1^+ \wedge \overline{P_2;\overline{Q_1^+;\mathbb{T}}}^{+} = P_1^+ \wedge \overline{P_2;\overline{\overline{Q_1};\mathbb{T}}}^{+} =$
   $P_1^+ \wedge \overline{P_2;\overline{Q_1}}^{+}$ by Lemma 3(5&4).
   Therefore, $\mathcal{I}(P_1 \vdash P_2);\mathcal{I}(Q_1 \vdash Q_2) = (\overline{P_1^+;\mathbb{T}} \vee P_2^+);(\overline{Q_1^+;\mathbb{T}} \vee Q_2^+) =$
   $\overline{P_1^+;\mathbb{T}};(\overline{Q_1^+;\mathbb{T}} \vee Q_2^+) \vee P_2^+;(\overline{Q_1^+;\mathbb{T}} \vee Q_2^+) =$
   $\overline{P_1^+;\mathbb{T}} \vee P_2^+;\overline{Q_1^+;\mathbb{T}} \vee P_2^+;Q_2^+ = \overline{P_1^+;\mathbb{T} \wedge \overline{P_2^+;\overline{\overline{Q_1^+;\mathbb{T}}}}} \vee (P_2;Q_2)^+ =$
   $\overline{(P_1^+ \wedge \overline{P_2^+;\overline{\overline{Q_1^+;\mathbb{T}}}})};\mathbb{T} \vee (P_2;Q_2)^+ = \overline{(P_1^+ \wedge \overline{P_2;\overline{Q_1}}^{+})};\mathbb{T} \vee (P_2;Q_2)^+ =$
   $\mathcal{I}(P_1 \wedge \overline{P_2;\overline{Q_1}} \vdash P_2;Q_2) = \mathcal{I}((P_1 \vdash P_2);(Q_1 \vdash Q_2))$.
4. By Lemma 3(1&8), $\mathbb{I}_\bot = \mathcal{I}(\mathbb{I}_D) = \mathcal{I}(\mathbb{T} \vdash \mathbb{I}) = \overline{\mathbb{T}^+;\mathbb{T}} \vee \mathbb{I}^+ = \overline{\mathbb{T}^{-+};\mathbb{T}} \vee \mathbb{I}^+ =$
   $\overline{\mathbb{T}} \vee \mathbb{V} \vee (\mathbb{I} \wedge \overline{\mathbb{V}} \wedge \overline{\mathbb{V}^\mathsf{T}}) = \mathbb{V} \vee (\mathbb{I} \wedge \overline{\mathbb{V}^\mathsf{T}}) = \mathbb{V} \vee (\mathbb{I} \wedge \overline{\mathbb{V}^\mathsf{T}})^\mathsf{T} = \mathbb{V} \vee (\mathbb{I} \wedge \overline{\mathbb{V}}) = \mathbb{V} \vee \mathbb{I}$.
5. By Lemma 3(1), $\mathcal{I}(\mathbb{T}) = \mathcal{I}(\bot\!\!\!\bot \vdash \bot\!\!\!\bot) = \overline{\bot\!\!\!\bot^+;\mathbb{T}} \vee \bot\!\!\!\bot^+ = \overline{\bot\!\!\!\bot;\mathbb{T}} \vee \bot\!\!\!\bot = \overline{\bot\!\!\!\bot} = \mathbb{T}$.

Corollary 6. *$\mathcal{I}$ is monotonic.*

Proof. $\mathcal{I}$ is a join homomorphism by Lemma 5(1), therefore an order homomorphism [4].

3.3 Healthiness Conditions

We introduce healthiness conditions similar to those presented in Sect. 2.3 for $\bot$-predicates. The predicates in the image of $\mathcal{I}$ satisfy these laws. Lemma 9 states consequences of various combinations of the healthiness conditions.

Definition 7. *A $\perp$-predicate P is a normal $\perp$-predicate if it satisfies the left zero law $\mathbb{T};P = \mathbb{T}$, and the unit laws $\mathbb{I}_\perp;P = P = P;\mathbb{I}_\perp$. P is called feasible if it satisfies the right zero law $P;\mathbb{T} = \mathbb{T}$.*

Lemma 8. *$\mathcal{I}(P)$ is a (feasible) normal $\perp$-predicate for every (feasible) normal design P.*

Proof. $\mathbb{I}_\perp;\mathcal{I}(P) = \mathcal{I}(\mathbb{I}_D);\mathcal{I}(P) = \mathcal{I}(\mathbb{I}_D;P) = \mathcal{I}(P) = \mathcal{I}(P;\mathbb{I}_D) = \mathcal{I}(P);\mathcal{I}(\mathbb{I}_D) = \mathcal{I}(P);\mathbb{I}_\perp$ by Lemma 5(3). $\mathbb{T};\mathcal{I}(P) = \mathcal{I}(\mathbb{T});\mathcal{I}(P) = \mathcal{I}(\mathbb{T};P) = \mathcal{I}(\mathbb{T}) = \mathbb{T} = \mathcal{I}(\mathbb{T}) = \mathcal{I}(P;\mathbb{T}) = \mathcal{I}(P);\mathcal{I}(\mathbb{T}) = \mathcal{I}(P);\mathbb{T}$ by Lemma 5(5&3).

Lemma 9. *Let P be a $\perp$-predicate.*

1. *If P satisfies the left zero and left unit laws, $\mathbb{V} \le P$.*
2. *If P satisfies the right unit law, $\overline{P};\mathbb{T} \wedge P \le \overline{\mathbb{V}^\mathsf{T}}$.*
3. *If P is a vector or P satisfies the right unit law, $P^-;\mathbb{T} = (P;\mathbb{T})^-$.*
4. *If P satisfies the right unit and right zero laws, $P^-;\mathbb{T} = \mathbb{T}$.*

Proof. We will use Lemma 5(4) in the first three parts of the proof.

1. $\mathbb{V} = \mathbb{V} \wedge \mathbb{T} = \mathbb{V} \wedge \mathbb{T};P = \mathbb{V};P \le (\mathbb{I} \vee \mathbb{V});P = \mathbb{I}_\perp;P = P$.
2. $\overline{P};\mathbb{T} \wedge P \le (\overline{P} \wedge P;\mathbb{T}^\mathsf{T});(\mathbb{T} \wedge \overline{P^\mathsf{T}};P) \le \mathbb{T};\overline{P^\mathsf{T}};P \le \mathbb{T};\overline{\mathbb{I}_\perp{}^\mathsf{T}} \le \mathbb{T};\overline{\mathbb{V}^\mathsf{T}} = \overline{\mathbb{V}^\mathsf{T}}$ by the Dedekind and Schröder rules [3].
3. If P is a vector, $P;\mathbb{I}_\perp = P;(\mathbb{I} \vee \mathbb{V}) = P;\mathbb{I} \vee P;\mathbb{T};\mathbb{V} = P \vee P;\mathbb{T} = P$, thus P satisfies the right unit law. $\mathbb{I}_\perp;\overline{\mathbb{V}} = (\mathbb{I} \vee \mathbb{V});\overline{\mathbb{V}} = \mathbb{I};\overline{\mathbb{V}} \vee \mathbb{V};\overline{\mathbb{V}} = \overline{\mathbb{V}};\overline{\mathbb{V}} \vee \mathbb{V};\overline{\mathbb{V}} = \mathbb{T};\overline{\mathbb{V}} = \mathbb{T}$, hence, $(P;\mathbb{T})^- = (P;\mathbb{I}_\perp;\overline{\mathbb{V}})^- = (P;\overline{\mathbb{V}})^- \wedge \overline{\mathbb{V}}^- = ((P \wedge \overline{\mathbb{V}} \wedge \overline{\mathbb{V}^\mathsf{T}});\mathbb{T})^- = (P^{-+};\mathbb{T})^- = P^-;\mathbb{T}$ by Lemma 3(1&4).
4. By (3) and Lemma 3(1), $P^-;\mathbb{T} = (P;\mathbb{T})^- = \mathbb{T}^- = \mathbb{T}$.

3.4 From Improper Values to Auxiliary Variables

We now move forward to reverse the mapping to $\perp$-predicates by the function $\mathcal{H}$. The predicates in the image of $\mathcal{H}$ satisfy the healthiness conditions of Sect. 2.3.

Definition 10. *The mapping $\mathcal{H}$ from normal $\perp$-predicates to designs is*

$$\mathcal{H} \ =_{\mathrm{def}} \ P \ \mapsto \ (\overline{P};\mathbb{T})^- \vdash P^- .$$

Lemma 11. *$\mathcal{H}$ is monotonic.*

Proof. $P \le Q$ implies both $(\overline{Q};\mathbb{T})^- \le (\overline{P};\mathbb{T})^-$ and $(\overline{Q};\mathbb{T})^- \wedge P^- \le P^- \le Q^-$. By [1, Theorem 3.1.2], $(\overline{P};\mathbb{T})^- \vdash P^- \le (\overline{Q};\mathbb{T})^- \vdash Q^-$, thus $\mathcal{H}(P) \le \mathcal{H}(Q)$.

Lemma 12. *$\mathcal{H}(P)$ is a (feasible) normal design for every (feasible) normal $\perp$-predicate P.*

Proof. By Lemma 9(3), $(\overline{P};\mathbb{T})^-;\mathbb{T} = (\overline{P};\mathbb{T};\mathbb{T})^- = (\overline{P};\mathbb{T})^-$, thus the design $\mathcal{H}(P)$ is a normal design by [1, Theorem 3.2.4]. If P is a feasible normal $\perp$-predicate, $\mathcal{H}(P);\mathbb{T} = ((\overline{P};\mathbb{T})^- \vdash P^-);(\perp\!\!\!\perp \vdash \perp\!\!\!\perp) = (\overline{P};\mathbb{T})^- \wedge \overline{P^-;\mathbb{T}} \vdash P^-;\perp\!\!\!\perp = (\overline{P};\mathbb{T})^- \wedge \overline{\mathbb{T}} \vdash \perp\!\!\!\perp = \perp\!\!\!\perp \vdash \perp\!\!\!\perp = \mathbb{T}$ by [1, Theorem 3.1.4] and Lemma 9(4), hence $\mathcal{H}(P)$ is a feasible normal design.

3.5 Isomorphism

The main result of this section shows that $\mathcal{H}$ exactly undoes the effect of $\mathcal{I}$. Therefore, sharpening Lemma 8, the image of $\mathcal{I}$ for (feasible) normal designs consists precisely of the (feasible) normal $\perp$-predicates. Analogously, sharpening Lemma 12, the image of $\mathcal{H}$ for (feasible) normal $\perp$-predicates consists precisely of the (feasible) normal designs. The situation is displayed in Fig. 2.

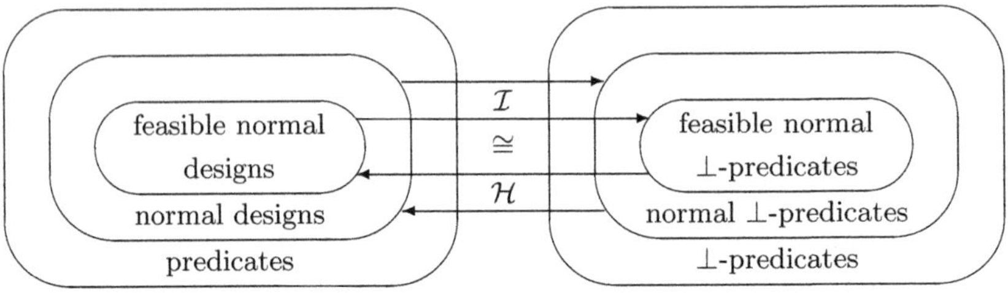

Fig. 2. Isomorphism mapping between (feasible) normal designs and (feasible) normal $\perp$-predicates

Theorem 13. *$\mathcal{I}$ and $\mathcal{H}$ are inverse to each other.*

Proof. By Lemmas 8 and 12 the compositions are well-defined.
Let $P_1 \vdash P_2$ be a normal design, then $\overline{\mathcal{I}(P_1 \vdash P_2)};\mathbb{T} = \overline{\overline{\overline{P_1{}^+;\mathbb{T}}} \vee P_2{}^+;\mathbb{T}} = (P_1{}^+;\mathbb{T} \wedge \overline{P_2{}^+});\mathbb{T} = P_1{}^+;\mathbb{T} \wedge \overline{P_2{}^+};\mathbb{T} = P_1{}^+;\mathbb{T}$ by Lemma 3(3).
Therefore, by Lemma 3(4), $\mathcal{H}(\mathcal{I}(P_1 \vdash P_2)) = (\overline{\mathcal{I}(P_1 \vdash P_2)};\mathbb{T})^- \vdash \mathcal{I}(P_1 \vdash P_2)^- = (P_1{}^+;\mathbb{T})^- \vdash \overline{P_1{}^+;\mathbb{T}}^{\,-} \vee P_2{}^{+-} = P_1;\mathbb{T} \vdash \overline{P_1;\mathbb{T}} \vee P_2 = P_1 \vdash \overline{P_1} \vee P_2 = P_1 \vdash P_2$.
Conversely, let P be a normal $\perp$-predicate, then, by Lemmas 3(8) and 9(2&1), $\mathcal{I}(\mathcal{H}(P)) = \mathcal{I}((\overline{P};\mathbb{T})^- \vdash P^-) = \overline{(\overline{P};\mathbb{T})^{-+}};\mathbb{T} \vee P^{-+} = \overline{\overline{P};\mathbb{T}} \vee \mathbb{V} \vee (P \wedge \overline{\mathbb{V}} \wedge \overline{\mathbb{V}^\mathsf{T}}) = \overline{\overline{P};\mathbb{T}} \vee \mathbb{V} \vee (P \wedge \overline{\mathbb{V}^\mathsf{T}}) = \overline{\overline{P};\mathbb{T}} \vee \mathbb{V} \vee P = \overline{\overline{\overline{P};\mathbb{T}} \wedge \overline{P}} = P$.

Corollary 14. *The complete lattice of normal designs is isomorphic to the complete lattice of normal $\perp$-predicates. Non-deterministic choice, conditional, sequential composition, and recursion are simulated as*

$$\begin{aligned} \mathcal{I}(\mathcal{H}(P) \vee \mathcal{H}(Q)) &= P \vee Q, \\ \mathcal{I}(\mathcal{H}(P) \lhd B \rhd \mathcal{H}(Q)) &= P \lhd B^+;\mathbb{T} \rhd Q, \\ \mathcal{I}(\mathcal{H}(P);\mathcal{H}(Q)) &= P;Q, \\ \mathcal{I}(\mu F) &= \mu F_\perp, \end{aligned}$$

where $F_\perp =_{\mathrm{def}} \mathcal{I} \circ F \circ \mathcal{H}$.

Proof. Together with Corollary 6 and Lemma 11 follows the isomorphism [4]. The simulation of non-deterministic choice, conditional, and sequential composition is a consequence along with the distributivity of $\mathcal{I}$ granted by Lemma 5.

Being an isomorphism of complete lattices, $\mathcal{I}$ even distributes over arbitrary disjunction. For the simulation of recursion, we therefore have

$$\begin{aligned}\mathcal{I}(\mu F) &= \mathcal{I}(\bigvee \{X : [X \le F(X)]\}) = \bigvee \{\mathcal{I}(X) : [X \le F(X)]\} \\ &= \bigvee \{\mathcal{I}(X) : [\mathcal{H}(\mathcal{I}(X)) \le F(\mathcal{H}(\mathcal{I}(X)))]\} \\ &= \bigvee \{Y : [\mathcal{H}(Y) \le F(\mathcal{H}(Y))]\} = \bigvee \{Y : [Y \le \mathcal{I}(F(\mathcal{H}(Y)))]\} \\ &= \bigvee \{Y : [Y \le F_\perp(Y)]\} = \mu F_\perp .\end{aligned}$$

by the isomorphism, the surjectivity of $\mathcal{I}$, Corollary 6 and Lemma 11.

4 Representing Non-termination Intuitively

To every predicate P modelling a program there is a corresponding relation R consisting of just those tuples (v, v') with $P(v, v')$. The intuitive reading of R is that $(v, v') \in R$ if and only if it is possible to observe the final values v' for the variables of the program, when they have been initialised with v. For a given initial assignment v, non-determinism is thus modelled by having more than one v' with $(v, v') \in R$, and non-termination by having no such v'.

In this section we show that these "intuitive" relations are in one-to-one correspondence with feasible normal designs and feasible normal $\perp$-predicates. To this end, we will split the isomorphisms $\mathcal{I}$ and $\mathcal{H}$ in two. Although intuitive relations form a complete lattice and the resulting functions are bijective, they do not preserve the lattice structure of normal designs, nor the structure of feasible normal designs.

In the following, an *intuitive relation* is a predicate that does not contain *ok* and *ok'* and does not have $\perp$ in the range of its variables. This distinguishes intuitive relations from both designs and $\perp$-predicates.

4.1 Eliminating Auxiliary Variables

The transition from feasible normal designs to intuitive relations and back is accomplished by the functions $\mathcal{I}_d$ and $\mathcal{H}_d$. Although they are inverse to each other, they do not preserve the structure since they are not even monotonic.

Definition 15. *The mapping $\mathcal{I}_d$ from feasible normal designs to intuitive relations is*

$$\mathcal{I}_d \;=_{\text{def}}\; P_1 \vdash P_2 \;\mapsto\; P_1 \wedge P_2 .$$

The mapping $\mathcal{H}_d$ from intuitive relations to feasible normal designs is

$$\mathcal{H}_d \;=_{\text{def}}\; P \;\mapsto\; P;\mathbb{T} \vdash P .$$

The normal design $\mathcal{H}_d(P)$ is feasible since $\mathcal{H}_d(P);\mathbb{T} = (P;\mathbb{T} \vdash P);(\perp\!\!\!\perp \vdash \perp\!\!\!\perp) = P;\mathbb{T} \wedge \overline{P;\mathbb{T}} \vdash P;\perp\!\!\!\perp = \perp\!\!\!\perp \vdash \perp\!\!\!\perp = \mathbb{T}$ by [1, Theorem 3.1.4].

Lemma 16. *$\mathcal{I}_d$ and $\mathcal{H}_d$ are inverse to each other.*

Proof. If P is intuitive relation, $\mathcal{I}_d(\mathcal{H}_d(P)) = \mathcal{I}_d(P;\top\!\!\!\top \vdash P) = P;\top\!\!\!\top \wedge P = P$. Let $P_1 \vdash P_2$ be a feasible normal design, then $\perp\!\!\!\perp \vdash \perp\!\!\!\perp = \top\!\!\!\top = (P_1 \vdash P_2);\top\!\!\!\top = (P_1 \vdash P_2);(\perp\!\!\!\perp \vdash \perp\!\!\!\perp) = P_1 \wedge \overline{P_2;\top\!\!\!\top} \vdash P_2;\perp\!\!\!\perp$ by [1, Theorem 3.1.4]. This implies $P_1 \wedge \overline{P_2;\top\!\!\!\top} \leq \perp\!\!\!\perp$, therefore $\mathcal{H}_d(\mathcal{I}_d(P_1 \vdash P_2)) = (P_1 \wedge P_2);\top\!\!\!\top \vdash P_1 \wedge P_2 = P_1 \wedge P_2;\top\!\!\!\top \vdash P_1 \wedge P_2 = P_1 \vdash P_1 \wedge P_2 = P_1 \vdash P_2$, both by [1, Theorem 3.1.2].

4.2 Eliminating Improper Values

To complete the picture, the transition from intuitive relations to feasible normal $\perp$-predicates and back is accomplished by the functions $\mathcal{I}_b$ and $\mathcal{H}_b$. Although they are inverse to each other, they too do not preserve the structure since they are not even monotonic. Their definition combines the bijections of Sects. 3 and 4.1. The structure-preserving mappings $\mathcal{I}$ and $\mathcal{H}$ are thus split in two, with intuitive relations as an intermediate lattice whose structure is different from that of normal designs and normal $\perp$-predicates. The result is displayed in Fig. 3.

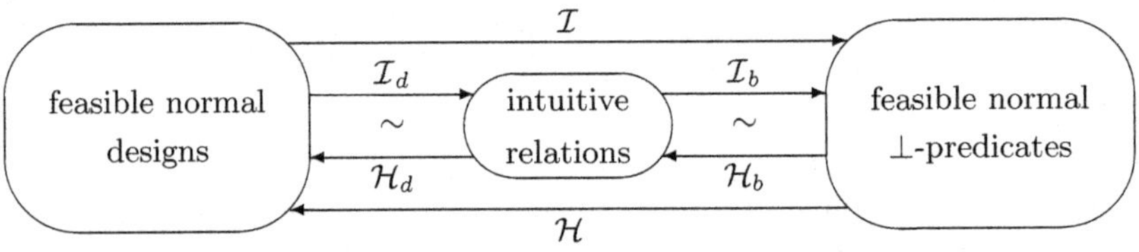

Fig. 3. Intuitive relations as an intermediate

Definition 17. *The mapping $\mathcal{I}_b$ from intuitive relations to $\perp$-predicates is given by $\mathcal{I}_b =_{\text{def}} \mathcal{I} \circ \mathcal{H}_d$. It maps to feasible normal $\perp$-predicates by Lemma 8, since $\mathcal{H}_d$ maps to feasible normal designs.*
The mapping $\mathcal{H}_b$ from feasible normal $\perp$-predicates to intuitive relations is given by $\mathcal{H}_b =_{\text{def}} \mathcal{I}_d \circ \mathcal{H}$. The composition is well-defined by Lemma 12.

Lemma 18. *$\mathcal{I}_b(P) = \overline{P^+;\top\!\!\!\top} \vee P^+$ and $\mathcal{H}_b(P) = (\overline{P};\top\!\!\!\top)^{-} \wedge P^-$.*

Proof. $\mathcal{I}_b(P) = \mathcal{I}(\mathcal{H}_d(P)) = \mathcal{I}(P;\top\!\!\!\top \vdash P) = \overline{(P;\top\!\!\!\top)^+;\top\!\!\!\top} \vee P^+ = \overline{P^+;\top\!\!\!\top} \vee P^+$ by Lemma 3(9), and $\mathcal{H}_b(P) = \mathcal{I}_d(\mathcal{H}(P)) = \mathcal{I}_d((\overline{P};\top\!\!\!\top)^{-} \vdash P^-) = (\overline{P};\top\!\!\!\top)^{-} \wedge P^-$.

Theorem 19. *$\mathcal{I}_b \circ \mathcal{I}_d = \mathcal{I}$, $\mathcal{H}_d \circ \mathcal{H}_b = \mathcal{H}$, $\mathcal{I}_b$ and $\mathcal{H}_b$ are inverse to each other.*

Proof. By Lemma 16, $\mathcal{I}_b \circ \mathcal{I}_d = \mathcal{I} \circ \mathcal{H}_d \circ \mathcal{I}_d = \mathcal{I}$, and $\mathcal{H}_d \circ \mathcal{H}_b = \mathcal{H}_d \circ \mathcal{I}_d \circ \mathcal{H} = \mathcal{H}$. By Theorem 13 and Lemma 16, $\mathcal{H}_b \circ \mathcal{I}_b = \mathcal{I}_d \circ \mathcal{H} \circ \mathcal{I} \circ \mathcal{H}_d = \mathcal{I}_d \circ \mathcal{H}_d = 1$ and $\mathcal{I}_b \circ \mathcal{H}_b = \mathcal{I} \circ \mathcal{H}_d \circ \mathcal{I}_d \circ \mathcal{H} = \mathcal{I} \circ \mathcal{H} = 1$.

5 Conclusion

Let us summarise the three contributions of our paper. In Sect. 3 we have defined an alternative basis to model non-termination in UTP, and proved both views isomorphic. This provides further confidence into UTP as an adequate tool for modelling program semantics. In Sect. 4 we have set up a one-to-one correspondence to the intuitive relational reading of predicates. A minor contribution, Sect. 2 is a concise presentation of the building blocks of UTP and its structural properties.

Our results are in line with [5] who discuss the need for a stability convention, and argue that "it is really of no interest which convention we choose". The two examples of conventions presented in [5] correspond to our normal designs and normal $\perp$-predicates, and so do the "partial relations" of [5] to the intuitive reading, but their connections are not investigated further there.

In terms of the classification of [6], $\perp$-predicates and intuitive relations correspond to the general and partial semantic models, respectively. Normal $\perp$-predicates and feasible normal $\perp$-predicates, however, differ from the general and total semantic models by relating a state to all proper outcomes whenever it is related to the looping outcome.

Relations augmented by $\perp$ subjected to the Egli-Milner ordering are used to define the semantics of an imperative language in [7]. The same ordering on relations extended with a "definedness predicate" is used for an applicative language in [8], then addressed by relation algebraic means in [9]. Such relations are shown isomorphic to state transformers by [10] who also discuss the Smyth ordering. For further pointers in connection with UTP we refer to the bibliography of [1].

The technique used in Sect. 3 can be applied analogously to a modified notion of designs, called *prescriptions*, defined in [2] to deal with general correctness, as opposed to the goal of total correctness pursued by UTP. The results from Sect. 4, however, do not carry over, for prescriptions provide a strictly finer description of program behaviour than designs.

Demonic operators are defined by [11] for modelling total correctness in modal semirings. Using abstract versions of the mappings defined in Sect. 4.1 they can be derived from the corresponding operators on designs. The development is given in [12] where the mappings are also used to calculate the semantics of the demonic while loop.

Further work is concerned with the links of UTP to functional programming, where the introduction of $\perp$ is conventional.

References

1. Hoare, C.A.R., He, J.: Unifying theories of programming. Prentice Hall Europe (1998)
2. Dunne, S.: Recasting Hoare and He's unifying theory of programs in the context of general correctness. In Butterfield, A., Strong, G., Pahl, C., eds.: 5th Irish Workshop on Formal Methods. EWiC, The British Computer Society (2001)
3. Schmidt, G., Ströhlein, T.: Relationen und Graphen. Springer-Verlag (1989)
4. Szász, G.: Introduction to Lattice Theory. 3rd edn. Academic Press (1963)
5. Hehner, E.C.R., Malton, A.J.: Termination conventions and comparative semantics. Acta Informatica **25**(1) (1988) 1–14
6. Nelson, G.: A generalization of Dijkstra's calculus. ACM Transactions on Programming Languages and Systems **11**(4) (1989) 517–561

7. de Bakker, J.W.: Semantics and termination of nondeterministic recursive programs. In Michaelson, S., Milner, R., eds.: (Third International Colloquium on) Automata, Languages and Programming, Edinburgh University Press (1976) 435–477
8. Broy, M., Gnatz, R., Wirsing, M.: Semantics of nondeterministic and noncontinuous constructs. In Bauer, F., Broy, M., eds.: Program Construction. Number 69 in Lecture Notes in Computer Science, Springer-Verlag (1979) 553–592
9. Berghammer, R., Zierer, H.: Relational algebraic semantics of deterministic and nondeterministic programs. Theoretical Computer Science **43**(2-3) (1986) 123–147
10. Apt, K.R., Plotkin, G.D.: Countable nondeterminism and random assignment. Journal of the ACM **33**(4) (1986) 724–767
11. Desharnais, J., Möller, B., Tchier, F.: Kleene under a modal demonic star. Journal of Logic and Algebraic Programming, special issue on Relation Algebra and Kleene Algebra (in press 2005)
12. Guttmann, W., Möller, B.: Modal design algebra. First International Symposium on Unifying Theories of Programming (to appear 2006)

Towards an Algebra of Hybrid Systems

Peter Höfner and Bernhard Möller

Institut für Informatik, Universität Augsburg, D-86135 Augsburg, Germany
{hoefner, moeller}@informatik.uni-augsburg.de

Abstract. We present a trajectory-based model for describing hybrid systems. For this we use left quantales and left semirings, thus providing a new application for these algebraic structures. Furthermore, we sketch a connection between game theory and hybrid systems.

1 Introduction

Hybrid systems are heterogeneous systems characterised by the interaction of discrete and continuous dynamics. They are an effective tool for modelling, design and analysis of a large number of technical systems. Such models are used for example in (air-)traffic controls, car-locating systems, chemical and biological processes and automated manufacturing.

This paper is based on work about hybrid systems by Sintzoff [16], Davoren/Nerode [4], Henzinger [7] and Lynch et al. [13]. In the latter two cases, the authors present two different ways to encode hybrid systems in a kind of finite state machines. These descriptions are very unhandy in calculations concerning liveness and safety properties.

The paper shows how a number of concepts can be recast and thus be made more workable in the setting of (left) semirings and (lazy) Kleene algebras [5, 15] and other algebras (e.g [2, 11]), thus providing an interesting application for them. Furthermore, we show how to express and calculate properties of hybrid systems and, more generally, of Boolean left test quantales, using some temporal operators. Finally, we sketch a connection to game theory to show how to adapt results from that area (see e.g. [1]) to an algebra of hybrid systems.

2 Trajectory-Based Model

We motivate the coming definitions by an example.

Running Example. (Temperature Control)
The hybrid automaton of Figure 1, adapted from [7], models a thermostat. The variable x represents the temperature. Initially, it is equal to 20 degrees and the heater is off (control mode *Off*). The temperature falls according to the flow condition $\dot{x} = -0.1x$. If the jump condition $x < 19$ is reached, the heater may start. The invariant condition $x \geq 18$ ensures that the heater will start at the latest when the temperature is equal to 18 degrees. In control mode

W. MacCaull et al. (Eds.): RelMiCS 2005, LNCS 3929, pp. 121–133, 2006.

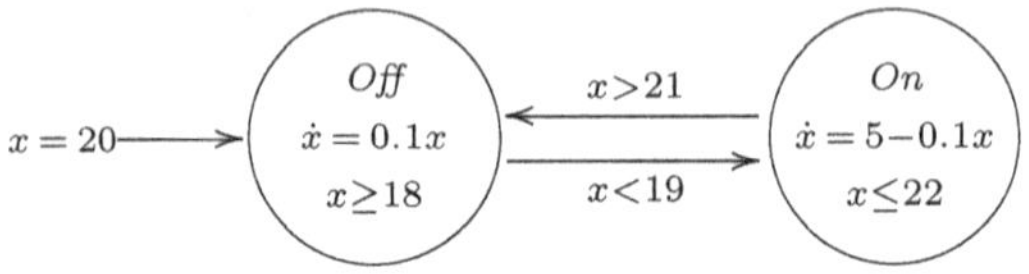

Fig. 1. Thermostat automaton

On, the temperature rises according to the flow condition $\dot{x} = 5 - 0.1x$. If the temperature reaches the second jump condition, the heater is switched off and the procedure starts again (with another initial value). □

For modelling this kind of system, we use trajectories (cf. e.g. [16]) that reflect the variation of the values of the variables over time. Let V be a set of *values* and D a set of *durations* (e.g. $\mathbb{N}$, $\mathbb{Q}$, $\mathbb{R}$, ...). We assume a cancellative addition $+$ on D and an element $0 \in D$ such that $(D, +, 0)$ is a commutative monoid and the relation $x \le y \overset{\text{def}}{\Leftrightarrow} \exists\, z \,.\, x + z = y$ is a linear order on D. Then 0 is the least element and $+$ is isotone w.r.t. $\le$. Moreover, 0 is indivisible, i.e., $x + y = 0 \Leftrightarrow x = y = 0$. D may include the special value ∞. If so, ∞ is required to be an annihilator w.r.t. $+$ and hence the greatest element of D (and cancellativity of $+$ is restricted to elements in $D - \{\infty\}$). For $d \in D$ we define the interval $\mathsf{tim}\, d$ of admissible times as

$$\mathsf{tim}\, d \overset{\text{def}}{=} \begin{cases} [0, d] & \text{if } d \neq \infty \\ [0, d[& \text{otherwise} \end{cases}$$

A *trajectory* t is a pair (d, g), where $d \in D$ and $g : \mathsf{tim}\, d \to V$. Then d is the *duration* of the trajectory, the image of $\mathsf{tim}\, d$ under g is its *range* $\mathsf{ran}\,(d, g)$. The set of all trajectories is denoted by TRA.

We define composition of trajectories (d_1, g_1) and (d_2, g_2) as

$$(d_1, g_1) \cdot (d_2, g_2) \overset{\text{def}}{=} \begin{cases} (d_1 + d_2, g) & \text{if } d_1 \neq \infty \wedge g_1(d_1) = g_2(0) \\ (d_1, g_1) & \text{if } d_1 = \infty \\ \text{undefined} & \text{otherwise} \end{cases}$$

with $g(x) = g_1(x)$ for all $x \in [0, d_1]$ and $g(x + d_1) = g_2(x)$ for all $x \in \mathsf{tim}\, d_2$.

For a zero-length trajectory $(0, g_1)$ we have $(0, g_1) \cdot (d_2, g_2) = (d_2, g_2)$ if $g_1(0) = g_2(0)$; otherwise the composition is undefined. Likewise, $(d_2, g_2) \cdot (0, g_1) = (d_2, g_2)$ if $g_1(0) = g_2(d_2)$ or $d_2 = \infty$. For a value $v \in V$, let $\underline{v} \overset{\text{def}}{=} (0, g)$ with $g(0) = v$ be the corresponding zero-length trajectory.

A *process* is a set of trajectories, consisting of possible behaviours of a hybrid system. The finite and infinite parts of a process A are defined as

$$\mathsf{inf}\, A \overset{\text{def}}{=} \{(d, g) \in A \mid d = \infty\} \qquad \mathsf{fin}\, A \overset{\text{def}}{=} A - \mathsf{inf}\, A$$

Composition is lifted to processes as follows:

$$A \cdot B \overset{\text{def}}{=} \mathsf{inf}\, A \cup \{a \cdot b \mid a \in \mathsf{fin}\, A, b \in B\}$$

The set I of all zero-length trajectories is the neutral element. A restricted form of composition, the *chop* $A^\frown B$, yields only trajectories that, after a finite trajectory of A, actually enter the second process. It is defined as $A^\frown B \stackrel{\text{def}}{=} (\mathsf{fin}\, A) \cdot B$, which implies $A \cdot B = \mathsf{inf}\, A \cup A^\frown B$.

Running Example. To use trajectories, we first set $V = D = \mathbb{R}$. Now we define two processes, one for each control mode:

$$A_{\mathrm{Off}} \stackrel{\text{def}}{=} \{(d,g) \mid d \in D,\, \dot{g}(t) = 0.1t\}\,,$$
$$A_{\mathrm{On}} \stackrel{\text{def}}{=} \{(d,g) \mid d \in D,\, \dot{g}(t) = 5 - 0.1t\}\,.$$

A_{Off} models all possible behaviours when the heater is off, whereas A_{On} describes the thermostat when the heater is on. The initial state is $R_{20} \stackrel{\text{def}}{=} \{\underline{20}\}$ $(= \{(0,g) \mid g(0) = 20\})$. Hence, we can formalise the starting sequence of the thermostat described above as

$$R_{20} \cdot A_{\mathrm{Off}} \cdot A_{\mathrm{On}}\,.$$

Since we want to describe the whole behaviour of the thermostat, we need the possibility for iteration. Let * be an operator for finite iteration (we will show the existence of * in Section 3). Then we can describe the system as

$$R_{20} \cdot (A_{\mathrm{Off}} \cdot A_{\mathrm{On}})^*\,.$$

In this way, the automaton is replaced by a corresponding regular expression. In Section 4 we show how to model jump and invariant conditions by restricting the ranges of trajectories. □

3 Left Semirings and Domain

Now, let's have a closer look at the algebraic structure of the trajectory-based model.

A *left semiring* is a quintuple $(S, +, 0, \cdot, 1)$ such that $(S, +, 0)$ is a commutative monoid and $(S, \cdot, 1)$ is a monoid such that $\cdot$ is left-distributive over $+$ and *left-strict*, i.e., $0 \cdot a = 0$. The left semiring is *idempotent* if $+$ is idempotent and $\cdot$ is right-isotone, i.e., $b \leq c \Rightarrow a \cdot b \leq a \cdot c$, where the *natural order* $\leq$ on S is given by $a \leq b \stackrel{\text{def}}{\Leftrightarrow} a + b = b$. Left-isotony of $\cdot$ follows from its left-distributivity. Moreover, 0 is the $\leq$-least element. A *semiring* is a left semiring in which $\cdot$ is also right-distributive and right-strict.

A left idempotent semiring S is called a *left quantale* if S is a complete lattice under the natural order and $\cdot$ is universally disjunctive in its left argument. Following [3], one might also call a left quantale a *left standard Kleene algebra*. A left quantale is *Boolean* if its underlying lattice is a completely distributive Boolean algebra.

An important left semiring (that is even a semiring and a left quantale) is REL, the algebra of binary relations over a set under relational composition.

A *left test semiring (quantale)* is a pair $(S, \mathsf{test}(S))$, where S is an idempotent left semiring (a left quantale) and $\mathsf{test}(S) \subseteq [0,1]$ is a Boolean subalgebra of the set $[0,1]$ of S such that $0, 1 \in \mathsf{test}(S)$ and join and meet in $\mathsf{test}(S)$ coincide with $+$ and $\cdot$, respectively. This definition corresponds to the one given in [12]. We will use $a, b, c \ldots$ for arbitrary S-elements and $p, q, r, \ldots$ for tests. By $\neg$ we denote complementation in $\mathsf{test}(S)$.

An important property of left test semirings is distribution of test multiplication over meet [15]: if $a \sqcap b$ exists then

$$p \cdot (a \sqcap b) \;=\; p \cdot a \;\sqcap\; b \;=\; p \cdot a \;\sqcap\; p \cdot b\;.$$

A *left domain semiring (quantale)* is a pair $(S, \ulcorner)$, where S is a left test semiring (quantale) and the *domain* operation $\ulcorner : S \to \mathsf{test}(S)$ satisfies

$$a \leq \ulcorner a \cdot a \quad (\text{d1}), \qquad \ulcorner(p \cdot a) \leq p \quad (\text{d2}), \qquad \ulcorner(a \cdot \ulcorner b) \leq \ulcorner(a \cdot b) \quad (\text{d3}).$$

The axioms are the same as in [5]; their relevant consequences can still be proved over left semirings (quantales) (see [15]). In particular, $\ulcorner$ is universally disjunctive and hence $\ulcorner 0 = 0$. In contrast to arbitrary complete Boolean test semirings [14], the domain operation is guaranteed to exist in left test quantales.

Checking all the axioms for the case of processes, we get

Lemma 3.1 *1. The processes form a Boolean left domain quantale*

$$\text{PRO} \overset{\text{def}}{=} (\mathcal{P}(\text{TRA}), \cup, \emptyset, \cdot, I, \ulcorner)$$

with $\mathsf{test}(\text{PRO}) = \mathcal{P}(\{\underline{v} \mid v \in V\})$ *and* $\ulcorner A = \{\underline{g(0)} \mid (d, g) \in A\}$.

2. *Additionally,* $\cdot$ *is positively disjunctive in its right argument, and chop inherits the disjunctivity properties from* $\cdot$ *and is associative, too.*
3. *Since* 0 *is indivisible, the meet with a test distributes over composition:*

$$P \in \mathsf{test}(\text{PRO}) \;\Rightarrow\; P \cap A \cdot B \;=\; (P \cap A) \cdot (P \cap B)$$

As in [15], we can extend an idempotent left semiring by finite and infinite iteration. A *left Kleene algebra* is a structure $(S, {}^*)$ consisting of an idempotent semiring S and an operation * that satisfies the left *unfold* and *induction* axioms

$$1 + a \cdot a^* \leq a^*\;, \qquad b + a \cdot c \leq c \Rightarrow a^* \cdot b \leq c\;.$$

To express infinite iteration we axiomatise an ω-operator over a left Kleene algebra. A *left* ω *algebra* [2] is a pair $(S, {}^\omega)$ such that S is a left Kleene algebra and ${}^\omega$ satisfies the *unfold* and *coinduction* axioms

$$a^\omega = a \cdot a^\omega\;, \qquad c \leq a \cdot c + b \Rightarrow c \leq a^\omega + a^* \cdot b\;.$$

Lemma 3.2

1. *Every left quantale can be extended to a left Kleene algebra by defining* $a^* \stackrel{\text{def}}{=} \mu x \,.\, a \cdot x + 1$.
2. *If the left quantale is a completely distributive lattice then it can be extended to a left* ω *algebra by setting* $a^\omega \stackrel{\text{def}}{=} \nu x \,.\, a \cdot x$. *In this case,*

$$\nu x \,.\, a \cdot x + b \quad = \quad a^\omega + a^* \cdot b \,.$$

The proof uses fixpoint fusion.

Since by Lemma 3.1 PRO forms a left quantale, we also have finite iteration * and infinite iteration $^\omega$ with all their laws available. Moreover, being Boolean, the quantale is separated, which provides a number of useful laws about the interaction of inf and fin with the semiring and iteration operations [15].

4 Range Assertions, Safety and Liveness

Often, it is necessary to restrict the range of a process A. Here, the range $\mathsf{ran}\, A$ is defined as $\mathsf{ran}\, A \stackrel{\text{def}}{=} \bigcup_{t \in A} \mathsf{ran}\, t$.

Running Example. We model the jump and invariant conditions for the transition from *Off* to *On*. First, we generally set

$$R_{[l,u]} \stackrel{\text{def}}{=} \{\underline{x} \mid x \in [l,u]\} \,.$$

Then the sequence "Off–jump–On" equals $A_{\mathrm{Off}} \cdot R_{[18,19]} \cdot A_{\mathrm{On}}$. As a safety condition for the thermostat of Figure 1 we want to guarantee the temperature to be between 18 and 22 degrees, i.e., we want to restrict the range of $A_{\mathrm{Off}} \cdot A_{\mathrm{On}}$ and $(A_{\mathrm{Off}} \cdot A_{\mathrm{On}})^*$. Thus we need to define a process containing all trajectories that never leave the range $[18, 22]$. □

We do this by observing that every test $P \in \mathsf{test}(\mathrm{PRO})$ is isomorphic to a subset of the value set V of the trajectories.

With $\top \stackrel{\text{def}}{=} \mathrm{TRA}$ and $\mathsf{F} \stackrel{\text{def}}{=} \mathsf{fin}\,(\mathrm{TRA})$ we define, for $P \in \mathsf{test}(\mathrm{PRO})$,

$$\Diamond P \stackrel{\text{def}}{=} \mathsf{F} \cdot P \cdot \top \,, \qquad \Box P \stackrel{\text{def}}{=} \overline{\Diamond \neg P}, .$$

Hence, $\Box P$ describes a safety aspect, viz. the set of all trajectories whose range satisfies the "invariant" P, i.e., $\Box P = \{t \in \mathrm{TRA} \mid \mathsf{ran}\, t \subseteq P\}$. Thus, the requested safety condition for the thermostat can be modelled as $\Box R_{[18,22]}$. Dually, $\Diamond P$ can be used to describe liveness aspects.

We now generalise these operators to an arbitrary general Boolean left test quantale S. Let $\top$ be the greatest element of S and set $\mathsf{F} \stackrel{\text{def}}{=} \mathsf{fin}\, \top$ and $\mathsf{N} \stackrel{\text{def}}{=} \mathsf{inf}\, \top$. By general results in [15] we have $F \cdot 0 = 0$, $\mathsf{N} = \top \cdot 0 = \mathsf{N} \cdot a$ for all a and $\mathsf{F} = \overline{\mathsf{N}}$. Moreover, F is downward closed and $1 \leq \mathsf{F}$, so that also $p \leq \mathsf{F}$ for all $p \in \mathsf{test}(S)$. Finally, $\mathsf{F} \cdot \mathsf{F} \leq \mathsf{F}$. Let now, for $p \in \mathsf{test}(S)$,

$$\Diamond p \stackrel{\text{def}}{=} \mathsf{F} \cdot p \cdot \top \,, \qquad \Box p \stackrel{\text{def}}{=} \overline{\Diamond \neg p}.$$

Thus, $\Box p$ corresponds to the "always p" operator of von Karger [11], whence the notation. Since $\Diamond$ and $\Box$ do not yield tests as their results, they cannot be nested. This does no harm, since nested safety requirements do not seem to be useful anyway. Moreover, all other algebraic operations are available for them. Our goal is now to derive a number of useful algebraic laws for $\Diamond$ and $\Box$.

Lemma 4.1 *Assume a left test quantale in which $\cdot$ is also positively right-disjunctive. Then $\Diamond$ is universally disjunctive and $\Box$ is universally conjunctive. In particular, both operators are isotone.*

Therefore we can define a general operator $\mathsf{ran} : S \to \mathsf{test}(S)$ by the Galois connection

$$\mathsf{ran}\, a \leq p \overset{\text{def}}{\Leftrightarrow} a \leq \Box p \,. \tag{1}$$

Running Example. Looking again at the safety requirement of the thermostat we see that by the condition $A_{\mathrm{Off}} \cdot A_{\mathrm{On}} \leq \Box R_{[18,22]}$ we indeed restrict the range of $A_{\mathrm{Off}} \cdot A_{\mathrm{On}}$ as claimed in the beginning of this section. Using the meet

$$A_{\mathrm{Off}} \cdot A_{\mathrm{On}} \sqcap \Box R_{[18,22]} \tag{th-rest}$$

is another way to enforce the restriction. □

By (1), ran is universally disjunctive. Moreover, we obtain

$$a \leq \Box(\mathsf{ran}\, a)\,, \qquad \mathsf{ran}\,(\Box p) \leq p\,, \qquad p \leq \Box p \Rightarrow \mathsf{ran}\, p \leq p\,.$$

For the following proofs and properties we introduce shorthands for the finite and infinite parts of boxes:

$$f_p \overset{\text{def}}{=} \mathsf{fin}\,(\Box p) = \mathsf{F} \sqcap \Box p\,, \qquad i_p \overset{\text{def}}{=} \mathsf{inf}\,(\Box p) = \mathsf{N} \sqcap \Box p\,.$$

Now we can show

Lemma 4.2 *Assume a right-distributive left test quantale S and $p \in \mathsf{test}(S)$.*

1. *$\Box p = p \cdot (\Box p) = (\Box p) \cdot p$.*
2. *If additionally $p \leq \Box p$ then $\ulcorner(\Box p) = p$.*

Proof. 1. We first show $\Box p = p \cdot (\Box p)$.
($\geq$) is clear by $p \leq 1$ and isotony.
($\leq$) We first show $\Box p \leq p \cdot \top$. By shunting this is equivalent to $\top \leq \overline{\Box p} + p \cdot \top$, i.e., to $\top \leq \mathsf{F} \cdot \neg p \cdot \top + p \cdot \top$, which holds by $1 \leq \mathsf{F}$, distributivity and Boolean algebra. Now we obtain $\neg p \cdot \Box p \leq 0$ and hence $\Box p = p \cdot \Box p + \neg p \cdot \Box p = p \cdot \Box p$.
Next, we show $\Box p = (\Box p) \cdot p$.
($\geq$) follows as above.
($\leq$) Splitting $\Box p$ into its finite and infinite parts and using distributivity, we get the equivalent claim $f_p + i_p \leq f_p \cdot p + i_p \cdot p = f_p \cdot p + i_p$. Since finite and infinite elements have empty intersection, this reduces to $f_p \leq f_p \cdot p$. For this we first show $f_p \leq F \cdot p$. By shunting, this is equivalent to $\top \leq \overline{f_p} + \mathsf{F} \cdot p$, i.e., to $\top \leq \mathsf{N} + \mathsf{F} \cdot \neg p \cdot \top + \mathsf{F} \cdot p$, which holds by $1 \leq \top$, distributivity, Boolean algebra and $\top = \mathsf{N} + \mathsf{F}$. Now we obtain $f_p \cdot \neg p \leq 0$ and hence $f_p = f_p \cdot p + f_p \cdot \neg p = f_p \cdot p$.

2. Axiom (d2) and 1. imply $\ulcorner(\Box p) \leq p$. The reverse inequation follows from the assumption $p \leq \Box p$, isotony of domain and $\ulcorner p = p$. $\Box$

Some of the following properties are satisfied only in a special kind of left semirings. Since elements of the form $\Box p$ correspond to safety properties, we call a left semiring (quantale) S *safety-closed* if $(\Box p) \cdot (\Box p) \leq \Box p$. In a safety-closed left semiring, $(\Box p)^+ = \Box p$ and

$$a \leq \Box p \Leftrightarrow a^+ \leq \Box p \Leftrightarrow a^+ \leq (\Box p)^+ , \tag{2}$$

where $b^+ \stackrel{\text{def}}{=} b \cdot b^*$. In Section 5 we will present a sufficient condition for safety-closedness. By that result, PRO is safety-closed.

Lemma 4.3 *Suppose that S is right-distributive and safety-closed.*

1. $\Box p \sqcap a \cdot b = (\Box p \sqcap a) \cdot (\Box p \sqcap b)$.
2. $\Diamond p \sqcap a \cdot b = (\Diamond p \sqcap a) \cdot b + \mathsf{fin}\, a \cdot (\Diamond p \sqcap b)$.
3. The box is multiplicatively idempotent, i.e., $(\Box p) \cdot (\Box p) = \Box p$.

Proof. 1. We show the claim first for finite a, i.e., for $a \leq \mathsf{F}$.
Let, for abbreviation, $s \stackrel{\text{def}}{=} \Box p$ and $d \stackrel{\text{def}}{=} \overline{s} = \Diamond\neg p$. By Boolean algebra and distributivity,

$$a \cdot b = (a \sqcap s) \cdot (b \sqcap s) + (a \sqcap s) \cdot (b \sqcap d) + (a \sqcap d) \cdot b$$

Now we observe that, by definition of d, we have $\mathsf{F} \cdot d \leq d$ and $d \cdot \top \leq d$, so that the last two summands are $\leq d$ by isotony. Hence,

$$a \cdot b \sqcap s = (a \sqcap s) \cdot (b \sqcap s) \sqcap s \leq (a \sqcap s) \cdot (b \sqcap s) .$$

The converse inequation holds by isotony and safety-closedness.
For arbitrary a we calculate, using $\mathsf{fin}/\mathsf{inf}$ decomposition, Boolean algebra and the claim for $\mathsf{fin}\, a \leq \mathsf{F}$,

$$\begin{aligned} a \cdot b \sqcap s &= (\mathsf{inf}\, a + \mathsf{fin}\, a \cdot b) \sqcap s \\ &= (\mathsf{inf}\, a \sqcap s) + ((\mathsf{fin}\, a \cdot b) \sqcap s) \\ &= \mathsf{inf}\, (a \sqcap s) + (\mathsf{fin}\, a \sqcap s) \cdot (b \sqcap s) \\ &= \mathsf{inf}\, (a \sqcap s) + \mathsf{fin}\, (a \sqcap s) \cdot (b \sqcap s) \\ &= (a \sqcap s) \cdot (b \sqcap s) . \end{aligned}$$

2. We show the claim for finite a; for infinite a the proof proceeds analogously to that of 1. Set $d \stackrel{\text{def}}{=} \Diamond p$ and $s \stackrel{\text{def}}{=} \overline{d} = \Box\neg p$. By Boolean algebra and distributivity,

$$d \sqcap a \cdot b = d \sqcap (d \sqcap a) \cdot b + d \sqcap (s \sqcap a) \cdot (d \sqcap b) + d \sqcap (s \sqcap a) \cdot (s \sqcap b) .$$

The first of these summands is below $(d \sqcap a) \cdot b$, the second one is below $a \cdot (d \sqcap b)$ and the third one is 0 by isotony, safety-closedness and $d \sqcap s = 0$. Hence, the sum is below $(d \sqcap a) \cdot b + a \cdot (d \sqcap b)$.
The converse inequation follows by $d \cdot b \leq d$, $a \leq \mathsf{F}$, $\mathsf{F} \cdot d \leq d$ and isotony.
3. This is a consequence of 1., since

$$\Box p = \Box p \sqcap \top = \Box p \sqcap \top \cdot \top = (\Box p \sqcap \top) \cdot (\Box p \sqcap \top) = \Box p \cdot \Box p .$$

$\Box$

Running Example. Returning to requirement (th-rest), we can transform the safety requirement $R_{20} \cdot (A_{\mathrm{Off}} \cdot A_{\mathrm{On}})^* \sqcap \Box p$ into $R_{20} \cdot ((A_{\mathrm{Off}} \sqcap \Box p) \cdot (A_{\mathrm{On}} \sqcap \Box p))^*$ by (2) and Lemma 4.3.1. Hence, it suffices to guarantee the safety requirement for the two processes A_{Off} and A_{On}. □

Lemma 4.4 *Assume a right-distributive and safety-closed left test quantale* S*, in which* $p \sqcap a \cdot b = (p \sqcap a) \cdot (p \sqcap b)$.

1. $p \leq \Box q \Leftrightarrow p \leq q$.
2. $p \leq \Box p$.
3. $\mathsf{ran}\, p = p$.
4. $p \leq \overline{\overline{1} \cdot \overline{1}}$.

Proof. 1. $p \leq \Box q$

$$\begin{aligned} &\Leftrightarrow \quad \{\!\mid \text{definition and shunting} \mid\!\} \\ &\quad p \sqcap \mathsf{F} \cdot \neg q \cdot \top \leq 0 \\ &\Leftrightarrow \quad \{\!\mid \text{assumption twice} \mid\!\} \\ &\quad (p \sqcap \mathsf{F}) \cdot (p \sqcap \neg q) \cdot (p \sqcap \top) \leq 0 \\ &\Leftrightarrow \quad \{\!\mid p \leq \mathsf{F} \text{ and meet on tests} \mid\!\} \\ &\quad p \cdot p \cdot \neg q \cdot p \leq 0 \\ &\Leftrightarrow \quad \{\!\mid \text{commutativity and idempotence of tests} \mid\!\} \\ &\quad p \cdot \neg q \leq 0 \\ &\Leftrightarrow \quad \{\!\mid \text{test shunting} \mid\!\} \\ &\quad p \leq q\ . \end{aligned}$$

2. Set $q = p$ in 1.
3. Using the Galois connection (1) and 1., we have

$$\mathsf{ran}\, p \leq q \Leftrightarrow p \leq \Box q \Leftrightarrow p \leq q\ .$$

   Now the claim follows by indirect equality.
4. We have $p \sqcap \overline{1} \cdot \overline{1} = (p \sqcap \overline{1}) \cdot (p \sqcap \overline{1}) = 0 \cdot 0 = 0$. □

By Lemma 3.1.3 properties 1. to 4. hold in PRO. In REL, however, subidentities can be decomposed into non-subidentities (unless the underlying base set is a singleton); so these properties do not hold there. The element $\overline{\overline{1} \cdot \overline{1}}$ has been called step in von Karger's work; it represents the elements that cannot be decomposed into non-subidentities. Note that in arbitrary Boolean semirings property 4. is equivalent to $\overline{1} \cdot \overline{1} \leq \overline{1}$, which roughly says that progress in time cannot be undone.

5 A Sufficient Criterion for Safety-Closedness

For the technical developments of this section we need additional operators. In any left quantale, the *left residual* a/b exists and is characterised by the Galois connection

$$x \leq a/b \overset{\text{def}}{\Leftrightarrow} x \cdot b \leq a\ .$$

In PRO, this operation is characterised pointwise by $t \in V/U \Leftrightarrow \forall\, u \in U : t \cdot u \in V$ (provided $t \cdot u$ is defined). Based on the left residual, in a Boolean quantale the *right detachment* $a\lfloor b$ can be defined as

$$a\lfloor b \stackrel{\text{def}}{=} \overline{\overline{a}/b}\ .$$

The pointwise characterisation in PRO reads $t \in V\lfloor U \Leftrightarrow \exists\, u \in U : t \cdot u \in V$. By de Morgan's laws, the Galois connection for $/$ transforms into the exchange law $a\lfloor b \leq x \Leftrightarrow \overline{x} \cdot b \leq \overline{a}$ for $\lfloor$ that generalises the Schröder rule of relational calculus. A straightforward consequence is $(\Box p)\lfloor a \leq \Box p$ (box detachment). Now we can prove

Lemma 5.1 *If S is locally linear [11], i.e., $(a \cdot b)\lfloor a = a \cdot (b\lfloor c) + a\lfloor(c\lfloor b)$, and right-distributive then S is safety-closed.*

Proof. First, by the definition of diamond, local linearity and box detachment,

$$\begin{aligned}(\Diamond\neg p)\lfloor(\Box p) &= \mathsf{F}\cdot\neg p\cdot(\top\lfloor(\Box p)) + (\mathsf{F}\cdot\neg p)\lfloor((\Box p)\lfloor\top)\\ &\leq \mathsf{F}\cdot\neg p\cdot(\top\lfloor(\Box p)) + (\mathsf{F}\cdot\neg p)\lfloor(\Box p) \qquad (*)\\ &\leq \Diamond\neg p + (\mathsf{F}\cdot\neg p)\lfloor(\Box p)\ .\end{aligned}$$

Hence

$$\begin{aligned}
&(\Box p)\cdot(\Box p) \leq \Box p\\
\Leftrightarrow\quad &\{\!|\ \text{exchange law}\ |\!\}\\
&(\Diamond\neg p)\lfloor(\Box p) \leq \Diamond\neg p\\
\Leftarrow\quad &\{\!|\ \text{by } (*)\ |\!\}\\
&\Diamond\neg p + (\mathsf{F}\cdot\neg p)\lfloor(\Box p) \leq \Diamond\neg p\\
\Leftrightarrow\quad &\{\!|\ \text{lattice algebra}\ |\!\}\\
&(\mathsf{F}\cdot\neg p)\lfloor(\Box p) \leq \Diamond\neg p\\
\Leftrightarrow\quad &\{\!|\ \text{exchange law}\ |\!\}\\
&(\Box p)\cdot(\Box p) \leq \overline{\mathsf{F}\cdot\neg p}\\
\Leftrightarrow\quad &\{\!|\ \text{Boolean algebra}\ |\!\}\\
&(\Box p)\cdot(\Box p) \leq \mathsf{N} + \mathsf{F}\cdot p\\
\Leftrightarrow\quad &\{\!|\ \text{by Lemma 4.2.1}\ |\!\}\\
&(\Box p)\cdot(\Box p)\cdot p \leq \mathsf{N} + \mathsf{F}\cdot p\\
\Leftrightarrow\quad &\{\!|\ \Box p = f_p + i_p \text{ (p. 6), distributivity and } \mathsf{fin}/\mathsf{inf} \text{ laws}\ |\!\}\\
&f_p\cdot f_p\cdot p \leq \mathsf{F}\cdot p\\
\Leftrightarrow\quad &\{\!|\ f_p \text{ finite and } \mathsf{F} \text{ closed under } \cdot\ |\!\}\\
&\mathrm{TRUE}\ .
\end{aligned}$$

□

Local linearity of PRO can be proved as in the case of the semiring of formal languages, as done in [8]; hence PRO is safety-closed. Next, we have

Lemma 5.2 *Assume a right-distributive and safety-closed left test quantale S.*

1. $a \cdot b \sqcap f_p \cdot \Box q \;=\; (a \sqcap f_p) \cdot (b \sqcap f_p \cdot \Box q) + (a \sqcap f_p \cdot \Box q) \cdot (b \sqcap \Box q)$.
2. $a \cdot b \sqcap i_p \;=\; (a \sqcap f_p) \cdot (b \sqcap i_p) + (a \sqcap i_p)$.
3. $a \cdot b \sqcap \Box p \cdot \Box q \;=\; (a \sqcap f_p) \cdot (b \sqcap \Box p \cdot \Box q) + (a \sqcap \Box p \cdot \Box q) \cdot (b \sqcap \Box q)$.

The proofs are straightforward and omitted for lack of space. An application of Lemma 5.2.1 is to combine safety requirements like $R_{[l,u]}$. Since $f_p \cdot \Box q = \Box p \frown \Box q$, a safety requirement of this form guarantees that the process $\Box q$ is actually entered.

6 Temporal Operators

Specifications are particular processes that express desired patterns. Following Sintzoff [16], we define the following quantifier-like operators relating a specification W with a process B supposed to implement it. If one considers the values in V as states then the set $\{t(0) \,|\, t \in B \cap W\}$ gives all starting states of the trajectories in B admitted by W as well. However, it is more convenient to represent this set as a test in the left test semiring of processes, viz. as $\{\overline{t(0)} \,|\, t \in B \cap W\}$. But this compacts simply into $\ulcorner(B \cap W)$. Therefore, a first definition of Sintzoff's quantifiers reads as follows (the primes indicate that we will use a different definition later on):

$$\mathsf{E}'B\,.\,W \stackrel{\text{def}}{=} \ulcorner(B \cap W)\;, \qquad \mathsf{A}'B\,.\,W \stackrel{\text{def}}{=} \neg\mathsf{E}'B\,.\,\overline{W} = \neg\ulcorner(B \cap \overline{W})\;,$$
$$\mathsf{AE}'B\,.\,W \stackrel{\text{def}}{=} \mathsf{A}'B\,.\,W \cap \mathsf{E}'B\,.\,W\,.$$

This definition works in general Boolean left domain semirings. However, as the resulting quantifiers are operators of type PRO $\to$ (PRO $\to$ test(PRO)), they cannot easily be composed. Therfore, Sintzoff gives a different semantics to combinations of these quantifiers. We want to avoid this by introducing new quantifiers that omit the final projection into test(PRO). Doing this, we also allow a look into the "future" of trajectories and not only at the starting states. In other words, our new quantifiers in PRO should model formulas like

$$t \in \mathsf{E}B\,.\,W \stackrel{\text{def}}{\Leftrightarrow} \exists\, u \in B : t \cdot u \in W,$$
$$t \in \mathsf{A}B\,.\,W \stackrel{\text{def}}{\Leftrightarrow} \forall\, u \in B : t \cdot u \in W.$$

These quantifiers are operators of type PRO $\to$ PRO and their sequential composition simply is function composition. If a projection into test(PRO) is desired it can be added at the outermost level by finally applying one of the three quantifiers above. For their algebraic characterisation we use again the detachment operator.

Lemma 6.1 *In a Boolean test quantale, one has*

$$\ulcorner(b \sqcap w) \;=\; w \lfloor b \sqcap 1 \;=\; b \lfloor w \sqcap 1\;.$$

In the detachment formulas of this lemma, forming the meet with 1 performs the projection into the test algebra, and we obtain our revised operators by omitting this meet. There is a choice in which of these two formulas to use. We take the first one, since it results in a more direct translation of the universal quantifier A'. Assume a Boolean quantale S and $a, b \in S$. Then

$$\mathsf{E}b\,.\,w \stackrel{\text{def}}{=} w\lfloor b\ , \qquad \mathsf{A}b\,.\,w \stackrel{\text{def}}{=} \overline{\mathsf{E}b\,.\,\overline{w}} = w/b\ ,$$
$$\mathsf{AE}b\,.\,w \stackrel{\text{def}}{=} (\mathsf{A}b\,.\,w) \sqcap (\mathsf{E}b\,.\,w)\ .$$

In PRO the process $\mathsf{E}B\,.\,W$ consists of all trajectories that can be completed by a B-trajectory to yield a trajectory in W. Thus, $\mathsf{E}B\,.\,W$ is the inverse image of W under the operation $\cdot B$, while $\mathsf{A}B\,.\,W$ is the largest process whose image under $\cdot B$ is contained in W. This suggests the following modal view of these quantifiers: E is a kind of diamond, whereas A forms a box operator. Correspondingly, we have the following properties that are typical for modal operators.

Lemma 6.2

1. *$\mathsf{E}a$ is universally disjunctive and $\mathsf{A}a$ is universally conjunctive.*
2. *$\mathsf{E}(a \cdot b)\,.\,c = \mathsf{E}a\,.\,(\mathsf{E}b\,.\,c)$ and $\mathsf{A}(a \cdot b)\,.\,c = \mathsf{A}a\,.\,(\mathsf{A}b\,.\,c)$.*
3. *If $\cdot$ is positively disjunctive in its right argument then E is positively disjunctive and A is positively antidisjunctive.*

7 Linking with Game Theory

As Sintzoff [16] has shown, the theory of games helps in understanding control systems as well as hybrid and reactive ones, since it deals with interaction between dynamics. For example, a control system can be presented as a game where the *controlling* and the *controlled* components are, respectively, the proponent and the opponent [10]. As the controller has to counteract all possible failures induced by "moves" of the controlled system, it has to force the opponent into a "losing" position where nothing can go wrong anymore. In PRO, moves correspond to process transformers of the shapes $\mathsf{E}B$ and $\mathsf{A}B$. They describe the possible and guaranteed reachabilities from a game position using B-trajectories.

Abstractly, a *game* consists of one or more *players* who interact with each other. A *move* is an action of one player. Obviously, there are various kinds of games, like games with finite or infinite duration. In the second case, one can distinguish games with finite and infinite move duration. Another possibility of classifying games are the categories of *cooperate*, *non-cooperate* and *semi-cooperate* games, depending on the methods by which the players will interact. Further, we can split all games into *disjoint* and *non-disjoint* ones. Non-disjoint games allow several moves at the same time, while in a disjoint game there is one move at a time.

In the remainder, we restrict ourselves to disjoint games with finite move duration. In a *game round*, each player, one by one, makes a move. Hence, if S_i is defined as the a move of player i, a game round is represented by $(S_1 \cdot S_2 \cdots S_n)$.

In that case, we can use the * and ω operators; $(S_1 \cdot S_2 \cdot \dots \cdot S_n)^*$ describes a finite game and $(S_1 \cdot S_2 \cdot \dots \cdot S_n)^\omega$ a game with infinitely many game rounds. In the latter case, the game has infinite duration if the S_i have positive durations.

In a game with player X and opponent Y, represented by their respective moves $\mathsf{E}a$ and $\mathsf{A}b$, we can interpret a game round in which X has the possibility of "winning" as the product $\mathsf{E}a \circ \mathsf{A}b$ (cf. [6]), where $\circ$ is composition of process transformers. Finite or infinite games can then be described as $(\mathsf{E}a \circ \mathsf{A}b)^*$ or $(\mathsf{E}a \circ \mathsf{A}b)^\omega$ from which winning and losing "positions" can be calculated by fixpoint iteration (e.g according to *Kleene's theorem*); for details see e.g. [1, 5]. Since we have now established the connection to the modal view of games started in [1] and treated abstractly in [5], we can re-use the analysis of winning and losing positions provided in these papers. This allows us to unify several results (e.g. [16]). A more thorough analysis of the game-theoretic connection will be the subject of further papers.

8 Conclusion and Outlook

This paper provides a starting point for developing an algebraic theory of hybrid systems. The theory of *Lazy Kleene algebras* [15] finds a useful further application here, generalising some similar results for the strict setting in [11]. Although one has to take some care with the modified laws relative to standard (modal) Kleene algebra, things work out reasonably well and many results come for free.

The aim of further work in this area is to develop a suitable specialisation of the general results to form new, more convenient algebraic calculi, both for safety and liveness proofs, and to provide a connection with the algebraic view of the duration calculus started in [11, 9]. Another aim is to use the game-theoretic approach to obtain improved controllers for hybrid systems. Finally, it has to be checked in how far hybrid (I/O) automata can be treated in this style to make the theory even more useful. It seems that the semantic models used in [4, 13] can be made into left domain quantales, too, so that our results would carry over to these frameworks.

Acknowledgements. We are grateful to M. Sintzoff for preparing the ground so well and to J. Desharnais, G. Struth and the anonymous referees for helpful discussions and remarks.

References

1. R. Backhouse, D. Michaelis: Fixed-Point Characterisation of Winning Strategies in Impartial Games. In R. Berghammer, B. Möller, G. Struth (eds.): Relational and Kleene-Algebraic Methods in Computer Science. LNCS 3051. Springer 2004, 34–47
2. E. Cohen: Separation and Reduction. In R. Backhouse, J. N. Oliveira (eds.): Mathematics of Program Construction. LNCS 1837. Springer 2000, 45–59
3. J. H. Conway: Regular Algebra and Finite Machines. Chapman & Hall, 1971

4. J. M. Davoren, A. Nerode: Logics for Hybrid Systems. Proc. IEEE 88, 985–1010 (2000)
5. J. Desharnais, B. Möller, G. Struth: Kleene Algebra with Domain. ACM Trans. Computational Logic (to appear 2006). Preliminary version: Universität Augsburg, Institut für Informatik, Report No. 2003-07, June 2003
6. J. Desharnais, B. Möller, G. Struth: Modal Kleene Algebra and Applications – A Survey. J. Relational Methods in Computer Science 1, 93–131 (2004) `http://www.cosc.brocku.ca/Faculty/Winter/JoRMiCS/`
7. T. Henzinger: The Theory of Hybrid Automata. Proc. 11th Annual IEEE Symposium on Logic in Computer Science, New Brunswick, New Jersey, 1996, 278–292
8. P. Höfner: From Sequential Algebra to Kleene Algebra: Interval Modalities and Duration Calculus. Technical Report 2005-5, Institut für Informatik, Universität Augsburg, 2005
9. P. Höfner: An Algebraic Semantics for Duration Calculus. 17th European Summer School in Logic, Language and Information (ESSLLI), Proc. 10th ESSLLI Student Session, Heriot-Watt University Edinburgh, Scotland, August 2005, 99–111
10. R. Isaacs: Differential Games. Wiley, 1965. Republished: Dover, 1999
11. B. von Karger: Temporal Algebra. Habilitation thesis, University of Kiel 1997
12. D. Kozen: Kleene Algebra with Tests. ACM Trans. Programming Languages and Systems 19, 427–443 (1997)
13. N. A. Lynch, R. Segala, F. W. Vaandrager: Hybrid I/O Automata. Information and Computation 185, 105–157 (2003)
14. B. Möller: Complete Tests do not Guarantee Domain. Technical Report 2005-6, Institut für Informatik, Universität Augsburg, 2005
15. B. Möller: Lazy Kleene Algebra. In D. Kozen (ed.): Mathematics of Program Construction. LNCS 3125. Springer 2004, 252–273
16. M. Sintzoff: Iterative Synthesis of Control Guards Ensuring Invariance and Inevitability in Discrete-Decision Games. In O. Owe, S. Krogdahl, T. Lyche (eds.): From Object-Orientation to Formal Methods — Essays in Memory of Ole-Johan Dahl. LNCS 2635. Springer 2004, 272–301

Relational Correspondences for Lattices with Operators*

Jouni Järvinen[1] and Ewa Orłowska[2]

[1] Turku Centre for Computer Science, Lemminkäisenkatu 14 A, 20520 Turku, Finland
Jouni.Jarvinen@it.utu.fi

[2] National Institute of Telecommunications, Szachowa 1, 04-894 Warszawa, Poland
E.Orlowska@itl.waw.pl

Abstract. In this paper we present some examples of relational correspondences for not necessarily distributive lattices with modal-like operators of possibility (normal and additive operators) and sufficiency (co-normal and co-additive operators). Each of the algebras $(P, \vee, \wedge, 0, 1, f)$, where $(P, \vee, \wedge, 0, 1)$ is a bounded lattice and f is a unary operator on P, determines a relational system (frame) $(X(P), \lesssim_1, \lesssim_2, R_f, S_f)$ with binary relations $\lesssim_1$, $\lesssim_2$, R_f, S_f, appropriately defined from P and f. Similarly, any frame of the form $(X, \lesssim_1, \lesssim_2, R, S)$ with two quasi-orders $\lesssim_1$ and $\lesssim_2$, and two binary relations R and S induces an algebra $(L(X), \vee, \wedge, 0, 1, f_{R,S})$, where the operations $\vee$, $\wedge$, and $f_{R,S}$ and constants 0 and 1 are defined from the resources of the frame. We investigate, on the one hand, how properties of an operator f in an algebra P correspond to the properties of relations R_f and S_f in the induced frame and, on the other hand, how properties of relations in a frame relate to the properties of the operator $f_{R,S}$ of an induced algebra. The general observations and the examples of correspondences presented in this paper are a first step towards development of a correspondence theory for lattices with operators.

1 Introduction

Let Lan be a formal language the semantics of which is determined by a class Frm of frames (relational systems). A correspondence theory aims at finding relationships between the truth of formulas in a frame and properties of the relations of that frame. Typically, a correspondence has the following form:

(1) A formula α is true in a Kripke frame (X, R) if and only if the relation R has a certain property.

The well known examples of such correspondences are provided by modal correspondence theory [12]. For example, a modal formula $p \to \Diamond p$ is true in a Kripke frame (X, R) if and only if R is reflexive.

In this paper we deal with correspondences which arise in connection with both frame semantics and algebraic semantics of formal languages. We assume that these

* This research was supported by COST Action 274 "Theory and Applications of Relational Structures as Knowledge Instruments" (http://www.tarski.org).

W. MacCaull et al. (Eds.): RelMiCS 2005, LNCS 3929, pp. 134–146, 2006.

two classes of semantic structures are related according to a duality via truth [7, 8]. Given a class Lan of formal languages, a class of frames Frm which determines a frame semantics for Lan and a class Alg of algebras (being signature and/or axiomatic extensions of the class of lattices) which determines its algebraic semantics, a duality via truth theorem says that these two kinds of semantics are equivalent in the following sense.

(DvT) A formula $\alpha \in$ Lan is true in every algebra of Alg if and only if α is true in every frame of Frm.

In order to prove such a theorem, we define some special algebras and frames, and we prove some lemmas about them. To each algebra $P \in$ Alg we assign a canonical frame $X(P)$, to each frame $X \in$ Frm we associate a complex algebra $L(X)$, and we prove that $X(P) \in$ Frm and $L(X) \in$ Alg. The following, what is called a complex algebra theorem, is essential:

(CA) For every frame $X \in$ Frm, a formula $\alpha \in$ Lan is true in X if and only if α is true in $L(X)$.

Furthermore, we need a representation theorem:

(R) Every algebra $P \in$ Alg is isomorphically embeddable into the complex algebra of its canonical frame $L(X(P))$.

With these results, we can prove the duality via truth theorem (DvT). The part (DvT $\rightarrow$) follows from (CA $\leftarrow$), and the part (DvT $\leftarrow$) follows from (CA $\rightarrow$) and (R).

Example 1. Let Alg be the class of the classical Boolean algebras with operators, $(B, \vee, \wedge, \neg, 0, 1, \Diamond)$, where $(B, \vee, \wedge, \neg, 0, 1)$ is a Boolean algebra and $\Diamond\colon B \to B$ is an additive and normal operator. The class Frm of the corresponding frames consists of relational systems (X, R), where X is a nonempty set and R is a binary relation on X. The *complex algebra* of the frame (X, R) is the algebra $(\wp(X), \cup, \cap, -, \emptyset, X, \langle R\rangle)$, where $\langle R\rangle\colon \wp(X) \to \wp(X)$ is the usual possibility operator determined by R. For a possibility algebra $(B, \vee, \wedge, \neg, 0, 1, \Diamond)$, its *canonical frame* is a pair $(X(B), R_\Diamond)$, where $X(B)$ is the set of ultrafilters of B and the relation $R_\Diamond$ on $X(B)$ is defined by $x\, R_\Diamond\, y$ whenever $\Diamond a \in x$ for all $a \in y$. It is well known that B can be embedded into the complex algebra of its canonical frame $X(B)$.

Let *Var* be a set of propositional variables. A *model* based on a frame (X, R) is a triple (X, R, m), where $m\colon$ *Var* $\to \wp(X)$ is a meaning function, which extends homomorphically to all propositional formulas. In particular, $m(\Diamond\alpha) = \{x \in X \mid (\exists y)\, x\, R\, y \wedge y \in m(\alpha)\}$. A formula α is true in a model (X, R, m), if $m(\alpha) = X$, and α is true in (X, R) if it is true for all possible choices of m.

Algebraic semantics may be defined as usual. Let $(B, \vee, \wedge, \neg, 0, 1, \Diamond)$ be a Boolean possibility algebra. An *assignment* for *Var* is a function $v\colon$ *Var* $\to B$ that can be extended to all propositional formulas. Now a formula α is true in B if and only if $v(\alpha) = 1$ for all assignments v. The complex algebra theorem states that a formula α is true in a frame (X, R) if and only if α is true in the complex algebra $(\wp(X), \cup, \cap, -, \emptyset, X, \langle R\rangle)$. Further, by the duality via truth theorem, α is true in every Boolean possibility algebra if and only if it is true in every Kripke frame [7, 8].

This paper is devoted to study algebras with operators and the corresponding frames in a more general setting than that of Boolean lattices considered in Example 1. Given the classes Alg and Frm satisfying the duality via truth theorem with respect to a language Lan, we consider the following correspondences:

(Csp1) Relations $R_1, \ldots, R_n$ of a frame $X \in \mathsf{Frm}$ have a certain property if and only if a formula $\alpha \in \mathsf{Lan}$ is true in the complex algebra $L(X)$ of X.
(Csp2) A formula $\alpha \in \mathsf{Lan}$ is true in an algebra $P \in \mathsf{Alg}$ if and only if the relations $R_1, \ldots, R_n$ of the canonical frame $X(P)$ have a certain property.

Observe that the following relationships between these correspondences and the previously discussed theorems hold.

(i) (Csp1 $\rightarrow$) and (CA $\leftarrow$) imply the part (1 $\leftarrow$) of the classical correspondence (1).
(ii) (Csp1 $\leftarrow$) and (CA $\rightarrow$) imply the part (1 $\rightarrow$) of the classical correspondence (1).

In this paper we deal with correspondences (Csp1) and (Csp2) for the lattice-based logics with possibility and sufficiency operators. Notice that in proving correspondences of type (Csp1), we employ the Ackermann Theorem [1, 5, 10]. In order to make the paper self-contained we briefly recall the construction leading to the Urquhart representation theorem for not necessarily distributive lattices, the notions of lattice-based possibility algebras and frames, and lattice-based sufficiency algebras and frames, together with the corresponding constructions of complex algebras and canonical frames. A detailed presentation of their representation theorems and duality via truth theorems can be found in [4, 6].

2 Preliminary Definitions and Results

2.1 Doubly Ordered Sets

Here we consider doubly ordered sets and present some essential results from [11].

A binary relation $\lesssim$ is a *quasi-order* if $\lesssim$ is reflexive and transitive. By a *doubly ordered set* we mean a relational system $(X, \lesssim_1, \lesssim_2)$, where $\lesssim_1$ and $\lesssim_2$ are quasi-orders on a non-empty set X such that for all $x, y \in X$, $x \lesssim_1 y$ and $x \lesssim_2 y$ imply $x = y$. A set A is $\lesssim_i$*-increasing* ($i = 1, 2$) when for all $x, y \in X$, $x \in A$ and $x \lesssim_i y$ imply $y \in A$. Let us denote by $\mathcal{O}_i(X)$ the set of all $\lesssim_i$-increasing subsets of X. The sets $\mathcal{O}_1(X)$ and $\mathcal{O}_2(X)$ are complete lattices of sets.

We may define two mappings $l \colon \wp(X) \to \wp(X)$ and $r \colon \wp(X) \to \wp(X)$ as follows. For any $A \subseteq X$, let

$$lA = \{x \in X \mid (\forall y \in X)\, x \lesssim_1 y \Rightarrow y \notin A\},$$
$$rA = \{x \in X \mid (\forall y \in X)\, x \lesssim_2 y \Rightarrow y \notin A\}.$$

The pair (r, l) forms a Galois connection between $(\mathcal{O}_1(X), \subseteq)$ and $(\mathcal{O}_2(X), \subseteq)$.

A set $A \subseteq X$ is said to be *l-stable* (resp. *r-stable*) whenever $lrA = A$ (resp. $rlA = A$). Let us denote by $L(X)$ (resp. by $R(X)$) the family of all l-stable sets (resp. r-stable sets). Now, $(L(X), \subseteq)$ is a lattice such that

$$A \wedge B = A \cap B \quad \text{and} \quad A \vee B = l(rA \cap rB).$$

Furthermore, $0 = \emptyset$ and $1 = X$. Obviously, also $(R(X), \subseteq)$ is a lattice. The system $(L(X), \vee, \wedge, 0, 1)$ is called the *complex algebra of the doubly ordered set* $(X, \lesssim_1, \lesssim_2)$.

Example 2. Let us consider the doubly-ordered set $(X, \lesssim_1, \lesssim_2)$, where $X = \{a, b, c\}$ and the quasi-orders $\lesssim_1$ and $\lesssim_2$ are defined by $a \lesssim_1 b$ and $a \lesssim_2 c$. Now

$$\mathcal{O}_1(X) = \{\emptyset, \{b\}, \{c\}, \{a, b\}, \{b, c\}, X\},$$
$$\mathcal{O}_2(X) = \{\emptyset, \{b\}, \{c\}, \{a, c\}, \{b, c\}, X\}.$$

It is easy to see that $L(X) = \{\emptyset, \{b\}, \{c\}, \{a, b\}, X\}$. Since $(L(X), \subseteq)$ is isomorphic to $\mathbf{N_5}$, it is not distributive.

2.2 Urquhart Representation of Lattices

Here we shortly consider some results by Urquhart [11].

Let $(P, \vee, \wedge, 0, 1)$ be a bounded lattice. By a *filter-ideal pair* of P we mean a pair (x_1, x_2) such that x_1 is a filter of P, x_2 is an ideal of P, and $x_1 \cap x_2 = \emptyset$. We may define an order $\leq$ on the family of filter-ideal pairs of P by setting

$$(x_1, x_2) \leq (y_1, y_2) \overset{\text{def}}{\iff} x_1 \subseteq y_1 \text{ and } x_2 \subseteq y_2.$$

A filter-ideal pair (x_1, x_2) is said to be *maximal* if it is maximal with respect to the order $\leq$. Let $X(P)$ be the family of the maximal filter-ideal pairs of a bounded lattice $(P, \vee, \wedge, 0, 1)$. Let us define relations $\subseteq_1$ and $\subseteq_2$ on $X(P)$ as follows. Let $x = (x_1, x_2)$ and $y = (y_1, y_2)$ be elements of $X(P)$. Then

$$x \subseteq_1 y \overset{\text{def}}{\iff} x_1 \subseteq y_1,$$
$$x \subseteq_2 y \overset{\text{def}}{\iff} x_2 \subseteq y_2.$$

For any bounded lattice $(P, \vee, \wedge, 0, 1)$, the system $(X(P), \subseteq_1, \subseteq_2)$ is a doubly ordered set. The frame $(X(P), \subseteq_1, \subseteq_2)$ is referred to as the *canonical frame of the lattice* $(P, \vee, \wedge, 0, 1)$.

Theorem 3 (Representation Theorem). *Every bounded lattice is isomorphic to a subalgebra of the complex algebra of its canonical frame.*

3 Lattice-Based Possibility Algebras

In this section we present the basic properties of lattice-based possibility algebras [6].

By a *possibility algebra* we mean an algebra $(P, \vee, \wedge, 0, 1, \Diamond)$, where $(P, \vee, \wedge, 0, 1)$ is a bounded lattice and $\Diamond$ is a unary operation on P satisfying for all $a, b \in P$ the conditions

(P1) $\Diamond(a \vee b) = \Diamond a \vee \Diamond b$ (additive)
(P2) $\Diamond 0 = 0$ (normal)

Here $\Diamond$ is referred to as a *possibility operator*. It is easy to see that each possibility operator is order-preserving, that is, $a \leq b$ implies $\Diamond a \leq \Diamond b$.

A *possibility frame* is a relational system $(X, \lesssim_1, \lesssim_2, R, S)$ such that $(X, \lesssim_1, \lesssim_2)$ is a doubly ordered set, and R and S are binary relations on X satisfying the following conditions for all $x, y, x', y' \in X$:

(Mono $R_\Diamond$) $(x \lesssim_1 x') \wedge (x \, R \, y) \wedge (y' \lesssim_1 y) \to x' \, R \, y'$
(Mono $S_\Diamond$) $(x' \lesssim_2 x) \wedge (x \, S \, y) \wedge (y \lesssim_2 y') \to x' \, S \, y'$
(SC $R_\Diamond S_\Diamond$) $x \, R \, y \to (\exists y' \in X) \, (y \lesssim_1 y') \wedge (x \, S \, y')$
(SC $S_\Diamond R_\Diamond$) $x \, S \, y \to (\exists x' \in X) \, (x \lesssim_2 x') \wedge (x' \, R \, y)$

The conditions (Mono $R_\Diamond$) and (Mono $S_\Diamond$) are called *possibility monotonicity conditions*, and the conditions (SC $R_\Diamond S_\Diamond$) and (SC $S_\Diamond R_\Diamond$) are referred to as *possibility stability conditions*.

We may define unary operators $[S] \colon \wp(X) \to \wp(X)$ and $\langle R \rangle \colon \wp(X) \to \wp(X)$ as usual. For any $A \subseteq X$,

$$[S]A = \{x \in X \mid \text{ for all } y \in X, x \, S \, y \text{ implies } y \in A\},$$
$$\langle R \rangle A = \{x \in X \mid \text{ there is } y \in X \text{ such that } x \, R \, y \text{ and } y \in A\}.$$

It is known that for all $A \subseteq X$, the set $[S]A$ is $\lesssim_2$-increasing and the set $\langle R \rangle A$ is $\lesssim_1$-increasing. Further, if $A \in R(X)$, then

$$(\star) \qquad [S]A = r\langle R \rangle lA \in R(X).$$

Given a possibility frame, we define its complex algebra as follows. A *complex algebra of a possibility frame* $(X, \lesssim_1, \lesssim_2, R, S)$ is an algebra $(L(X), \vee, \wedge, 0, 1, \Diamond)$, where $(L(X), \vee, \wedge, 0, 1)$ is the complex algebra of the doubly ordered set $(X, \lesssim_1, \lesssim_2)$ and $\Diamond$ is a unary operator defined for any $A \in L(X)$ by

$$\Diamond A = l[S]rA.$$

The operator $\Diamond \colon L(X) \to L(X)$ can be expressed also by means of R, namely,

$$\Diamond A = lr\langle R \rangle A.$$

Now, for every possibility frame $(X, \lesssim_1, \lesssim_2, R, S)$, its corresponding complex algebra $(L(X), \vee, \wedge, 0, 1, \Diamond)$ is a possibility algebra.

Next we define canonical frames of possibility algebras. By a filter-ideal pair of a possibility algebra $(P, \vee, \wedge, 0, 1, \Diamond)$, we mean a filter-ideal pair of the lattice $(P, \vee, \wedge, 0, 1)$. Let $X(P)$ be the family of all maximal filter-ideal pairs of P and let $(X(P), \subseteq_1, \subseteq_2)$ be the canonical frame of the lattice $(P, \vee, \wedge, 0, 1)$.

For every $A \subseteq P$, we set $\Diamond A = \{a \in P \mid \Diamond a \in A\}$. Trivially, for any $A, B \subseteq P$, $A \subseteq B$ implies $\Diamond A \subseteq \Diamond B$. We can now define binary relations $S_\Diamond$ and $R_\Diamond$ on $X(P)$ as follows. Let $x = (x_1, x_2)$ and $y = (y_1, y_2)$ be elements of $X(P)$. Then

$$(x, y) \in R_\Diamond \overset{\text{def}}{\iff} y_1 \subseteq \Diamond x_1,$$
$$(x, y) \in S_\Diamond \overset{\text{def}}{\iff} \Diamond x_2 \subseteq y_2.$$

The frame $(X(P), \subseteq_1, \subseteq_2, R_\Diamond, S_\Diamond)$ satisfies the possibility monotonicity and stability conditions, and it is called the *canonical frame of a possibility algebra P*.

The representation theorem can be extended to possibility algebras. Furthermore, the complex algebra theorem holds [6].

4 Correspondence Theorems for Possibility Algebras

We begin by recalling the following lemma which is essential for our considerations [1, 5, 10]. Let us write $A(\sigma := \delta)$ for the formula obtained from A by replacing every occurrence of expression σ by expression δ.

Lemma 4. *Let P be a predicate and let $A(x_1, \ldots, x_n)$ and $B(P)$ be classical first-order formulas without second-order quantification. Let P occur in B only positively and let A contain no occurrences of P at all.*

(a) *The formula*

$$(\exists P)(\forall x_1, \ldots, x_n)\ (P(x_1, \ldots, x_n) \vee A(x_1, \ldots, x_n)) \wedge B(P := \neg P)$$

is equivalent to

$$B(P := A(x_1, \ldots, x_n)),$$

where, in the second formula, the arguments $x_1, \ldots, x_n$ of A are substituted by the actual arguments of P (with renaming the bound variables whenever necessary).

(b) *The formula*

$$(\exists P)(\forall x_1, \ldots, x_n)\ (\neg P(x_1, \ldots, x_n) \vee A(x_1, \ldots, x_n)) \wedge B(P)$$

is equivalent to

$$B(P := A(x_1, \ldots, x_n)),$$

where, in the second formula, the arguments $x_1, \ldots, x_n$ of A are substituted by the actual arguments of P (with renaming the bound variables whenever necessary).

4.1 Reflexivity vs. Extensivity

First we consider correspondences of type (Csp1).

Proposition 5. *Let $(X, \lesssim_1, \lesssim_2, R, S)$ be a possibility frame. Then the following conditions are equivalent:*

(C I) $(\forall A \subseteq X)\ A \subseteq lr\langle R\rangle A$;
(C II) $(\forall x, y \in X)\ x \lesssim_1 y \rightarrow (\exists z)\ y \lesssim_2 z \wedge z\, R\, x$.

Proof. Condition (C I) can be presented in the form

$$(\forall P)(\forall x)\ P(x) \rightarrow (\forall y)\ x \lesssim_1 y \rightarrow (\exists z)\ y \lesssim_2 z \wedge (\exists w)\ z\, R\, w \wedge P(w).$$

We negate the above formula and obtain

$$(\exists x, y)(\exists P)(\forall w)\ (\neg P(w) \vee (\forall z)\ y \not\lesssim_2 z \vee (z, w) \notin R) \wedge P(x) \wedge x \lesssim_1 y.$$

By Lemma 4(b), this is equivalent to

$$(\exists x, y)\ x \lesssim_1 y \wedge (\forall z)\ y \not\lesssim_2 z \vee (z, x) \notin R.$$

Since we negated the initial formula, we have to negate the above formula in order to obtain the equivalent first order formula for (C I). We have the formula

$$(\forall x, y)\, x \lesssim_1 y \to (\exists z)\, y \lesssim_2 z \wedge z\, R\, x.$$

Notice that if $\lesssim_1$ and $\lesssim_2$ are identity relations on X, then l and r are the regular complement operators of $\wp(X)$, $L(X) = \wp(X)$, and the equivalence of (C I) and (C II) is the ordinary correspondence result "R is reflexive if and only if $A \subseteq \langle R \rangle A$ for all $A \subseteq X$".

We may now write the following observation.

Lemma 6. *Let* $(X, \lesssim_1, \lesssim_2, R, S)$ *be a possibility frame. If* R *or* S *are reflexive, then condition* (C II) *holds.*

Proof. Suppose that R is reflexive and $x \lesssim_1 y$. Then $y \lesssim_1 y$, $y\, R\, y$, and $x \lesssim_1 y$ imply $y\, R\, x$ by (Mono $R_\Diamond$). Now, $y \lesssim_2 y$ and $y\, R\, x$, which means that condition (C II) holds.

Assume that S is reflexive and $x \lesssim_1 y$. Then $y\, S\, y$ implies that there exists a $z \in X$ such that $y \lesssim_2 z$ and $z\, R\, y$ by (SC $S_\Diamond R_\Diamond$). Now, $z \lesssim_1 z$, $z\, R\, y$, and $x \lesssim_1 y$ imply $z\, R\, x$ by (Mono $R_\Diamond$). Hence, (C II) holds.

By Proposition 5, conditions (C I) and (C II) are equivalent. It is also obvious that (C I) implies $A \subseteq \Diamond A$ for all $A \in L(X)$. Further, by Lemma 6, reflexivity of R or S implies (C II). Thus, if R or S are reflexive, then $A \subseteq \Diamond A$ for all $A \in L(X)$. This is illustrated in Fig. 1.

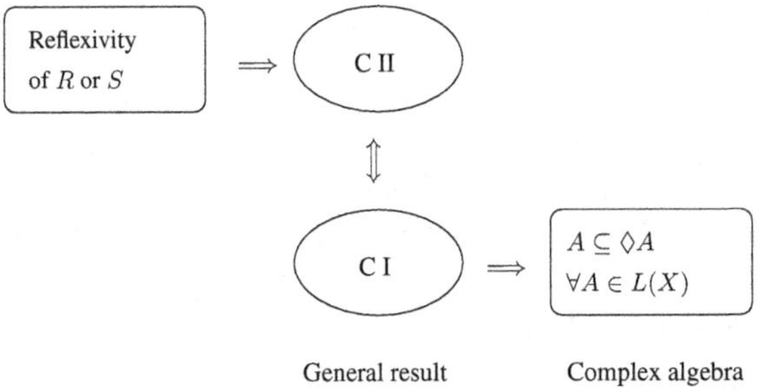

Fig. 1.

Note that (C II) is equivalent to the condition that $A \subseteq lr\langle R \rangle A$ for *all* $A \in \wp(X)$, not just for all $A \in L(X)$. This means that Proposition 5 is not a real correspondence between a possibility frame $(X, \lesssim_1, \lesssim_2, R, S)$ and its complex algebra $(L(X), \vee, \wedge, 0, 1, \Diamond)$, and (C II) seems to be too "strong". Therefore, we should find a first-order formula corresponding the formula

$$(\forall A \subseteq X)\, A \in L(X) \to A \subseteq lr\langle R \rangle A.$$

However, we omit such considerations here.

Because there are two ways to express the operator $\Diamond \colon L(X) \to L(X)$, we may also write the following proposition. Notice that the proof can be obtained "automatically", for example, by applying the DLS* algorithm[1].

[1] `http://www.ida.liu.se/labs/kplab/projects/dlsstar/`

Proposition 7. *Let* $(X, \lesssim_1, \lesssim_2, R, S)$ *be a possibility frame. Then the following conditions are equivalent:*

(C I)° $(\forall A \subseteq X)\, A \subseteq l[S]rA$;
(C II)° $(\forall x, y)\, x \lesssim_1 y \to (\exists z)\, z \lesssim_2 x \wedge y\, S\, z$.

As above, it is now obvious that (C I)° implies $A \subseteq \Diamond A$ for all $A \in L(X)$. It is also interesting to notice the following implication.

Lemma 8. *Let* $(X, \lesssim_1, \lesssim_2, R, S)$ *be a possibility frame. Then,* (C II)° *implies reflexivity of* S.

Proof. Let $x \in X$. Because $x \lesssim_1 x$, there exists a $z \lesssim_2 x$ such that $x\, S\, z$ by (C II)°. Now $x \lesssim_2 x$, $x\, S\, z$, and $z \lesssim_2 x$, which imply $x\, S\, x$ by (Mono $S_\Diamond$).

Next we consider correspondences of type (Csp2).

Lemma 9. *Let* $(P, \vee, \wedge, 0, 1, \Diamond)$ *be a possibility algebra. If* $\Diamond$ *is extensive, then the following conditions hold.*

(a) *If* A *is a filter of* P*, then* $A \subseteq \Diamond A$.
(b) *If* B *is an ideal of* P*, then* $\Diamond B \subseteq B$.

Proof. (a) Let A be a filter and $a \in A$. Then $a \leq \Diamond a$ implies $\Diamond a \in A$, that is, $a \in \Diamond A$.

(b) Suppose B is an ideal and $a \notin B$. Then $a \leq \Diamond a$ implies $\Diamond a \notin B$, which is equivalent to $a \notin \Diamond B$.

Proposition 10. *Let* $(P, \vee, \wedge, 0, 1, \Diamond)$ *be a possibility algebra. If* $\Diamond$ *is extensive, then in the corresponding canonical possibility frame* $(X(P), \subseteq_1, \subseteq_2, R_\Diamond, S_\Diamond)$ *the relations* $R_\Diamond$ *and* $S_\Diamond$ *are reflexive.*

Proof. Let $x = (x_1, x_2) \in X(P)$. Since x_1 is a filter and x_2 is an ideal, $(x, x) \in R_\Diamond$ and $(x, x) \in S_\Diamond$ follow from the previous lemma.

For the other direction, we have a weaker result. Let $0 \neq a \in P$, and denote

$$F_a = \bigcap \{x_1 \mid (x_1, x_2) \in X(P) \text{ and } a \in x_1\}.$$

Obviously, F_a is a filter of P containing the element a. This gives $\uparrow a \subseteq F_a$. Let us introduce the condition

(PF) $\quad (\forall a \in P)\, a \neq 0 \to \uparrow a = F_a$

The above condition means that for all $a \neq 0$, the set F_a is the principal filter of a.

Proposition 11. *Let* $(P, \vee, \wedge, 0, 1, \Diamond)$ *be a possibility algebra satisfying* (PF). *If* $R_\Diamond$ *is reflexive, then* $\Diamond \colon P \to P$ *is extensive.*

Proof. Assume that $R_\Diamond$ is reflexive and $a \in P$. If $a = 0$, then obviously $a \leq \Diamond a$. Let $a \neq 0$ and assume that $x = (x_1, x_2)$ is an element of $X(P)$ such that $a \in x_1$. Then $a \in \Diamond x_1$ by reflexivity of $R_\Diamond$. This means that $\Diamond a \in x_1$, and we obtain $\Diamond a \in F_a$. Since $F_a \subseteq \uparrow a$ by assumption, we get $a \leq \Diamond a$, that is, $\Diamond$ is extensive.

For possibility algebras satisfying (PF), we may now state the correspondence: $\Diamond$ is extensive if and only if $R_\Diamond$ is reflexive.

4.2 Transitivity vs. Closedness

Here we are able to present only correspondences of type (Csp1 $\rightarrow$) and (Csp2 $\rightarrow$).

Proposition 12. *Let* $(X, \lesssim_1, \lesssim_2, R, S)$ *be a possibility frame. If* S *is transitive, then in the corresponding complex possibility algebra* $(L(X), \vee, \wedge, 0, 1, \Diamond)$ *the operator* $\Diamond\colon L(X) \to L(X)$ *is closed, that is, for all* $A \subseteq X$, $\Diamond\Diamond A \subseteq \Diamond A$.

Proof. Let $A \in L(X)$. Because $rA \in R(X)$, $[S]rA \in R(X)$ by ($\star$). Hence, $rl[S]rA = [S]rA$. This implies

$$\Diamond\Diamond A = l[S]rl[S]rA = l[S][S]rA \subseteq l[S]rA = \Diamond A.$$

Corollary 13. *Let* $(X, \lesssim_1, \lesssim_2, R, S)$ *be a possibility frame. If* S *is reflexive and transitive, then in the corresponding complex possibility algebra* $(L(X), \vee, \wedge, 0, 1, \Diamond)$ *the operator* $\Diamond\colon L(X) \to L(X)$ *is a closure operator.*

Lemma 14. *Let* $(P, \vee, \wedge, 0, 1, \Diamond)$ *be a possibility algebra. If* $\Diamond$ *is closed, then the following conditions hold.*

(a) *If* A *is a filter of* P*, then* $\Diamond\Diamond A \subseteq \Diamond A$.
(b) *If* B *is an ideal of* P*, then* $\Diamond B \subseteq \Diamond\Diamond B$.

Proof. (a) Let A be a filter and $a \in \Diamond\Diamond A$, that is, $\Diamond\Diamond a \in A$. Because $\Diamond\Diamond a \leq \Diamond a$, we have $\Diamond a \in A$, that is, $a \in \Diamond A$. Claim (b) can be proved in an analogous manner.

Proposition 15. *Let* $(P, \vee, \wedge, 0, 1, \Diamond)$ *be a possibility algebra. If* $\Diamond$ *is closed, then in the corresponding canonical possibility frame* $(X(P), \subseteq_1, \subseteq_2, R_\Diamond, S_\Diamond)$ *the relations* $R_\Diamond$ *and* $S_\Diamond$ *are transitive.*

Proof. Let $x = (x_1, x_2)$, $y = (y_1, y_2)$, and $z = (z_1, z_2)$ be elements of $X(P)$. Assume that $(x, y) \in R_\Diamond$ and $(y, z) \in R_\Diamond$, which is equivalent to $y_1 \subseteq \Diamond x_1$ and $z_1 \subseteq \Diamond y_1$. This implies $z_1 \subseteq \Diamond y_1 \subseteq \Diamond\Diamond x_1 \subseteq \Diamond x_1$. Thus, $x\, R_\Diamond\, z$. The other part can be proved analogously.

5 Lattice-Based Sufficiency Algebras

A *sufficiency algebra* is an algebra of the form $(P, \vee, \wedge, 0, 1, [\![\Box]\!])$, where $(P, \vee, \wedge, 0, 1)$ is a bounded lattice and $[\![\Box]\!]\colon P \to P$ satisfies for all $a, b \in P$ the following conditions

(S1) $[\![\Box]\!](a \vee b) = [\![\Box]\!]a \wedge [\![\Box]\!]b$ (co-additive)
(S2) $[\![\Box]\!]0 = 1$ (co-normal)

The operator $[\![\Box]\!]$ is clearly order-reversing, that is, $a \leq b$ implies $[\![\Box]\!]a \geq [\![\Box]\!]b$.

A *sufficiency frame* is a relational system $(X, \lesssim_1, \lesssim_2, R, S)$ such that $(X, \lesssim_1, \lesssim_2)$ is a doubly ordered set, and R and S are binary relations on X satisfying the following conditions for all $x, y, x', y' \in X$:

(Mono $R_{|\Box|}$) $(x' \lesssim_1 x) \wedge (x\,R\,y) \wedge (y \lesssim_2 y') \to x'\,R\,y'$
(Mono $S_{|\Box|}$) $(x \lesssim_2 x') \wedge (x\,S\,y) \wedge (y' \lesssim_1 y) \to x'\,S\,y'$
(SC $R_{|\Box|}S_{|\Box|}$) $x\,R\,y \to (\exists x' \in X)\,(x \lesssim_1 x') \wedge (x'\,S\,y)$
(SC $S_{|\Box|}R_{|\Box|}$) $x\,S\,y \to (\exists y' \in X)\,(y \lesssim_1 y') \wedge (x\,R\,y')$

Given a sufficiency frame, we define its complex algebra as follows. A *complex algebra of a sufficiency frame* $(X, \lesssim_1, \lesssim_2, R, S)$ is an algebra $(L(X), \vee, \wedge, 0, 1, |\Box|)$, where $(L(X), \vee, \wedge, 0, 1)$ is the complex algebra of the doubly ordered set $(X, \lesssim_1, \lesssim_2)$ and $|\Box|$ is a unary operator defined for any $A \in L(X)$ by

$$|\Box|A = [R]rA.$$

The operator $|\Box|\colon L(X) \to L(X)$ can also be expressed by means of the relation S:

$$|\Box|A = l\langle S\rangle A.$$

A *canonical frame of the necessity algebra* $(P, \vee, \wedge, 0, 1, |\Box|)$ is the system $(X(P), \subseteq_1, \subseteq_2, R_{|\Box|}, S_{|\Box|})$ such that $(X(P), \subseteq_1, \subseteq_2)$ is the canonical frame of $(P, \vee, \wedge, 0, 1)$ and the relations $R_{|\Box|}$ and $S_{|\Box|}$ on $X(P)$ are defined as follows:

$$(x, y) \in R_{|\Box|} \overset{\text{def}}{\iff} |\Box|x_1 \subseteq y_2,$$
$$(x, y) \in S_{|\Box|} \overset{\text{def}}{\iff} y_1 \subseteq |\Box|x_2.$$

As before, for every $A \subseteq P$ it is defined $|\Box|A = \{a \in P \mid |\Box|a \in A\}$. Obviously, for all $A, B \subseteq P$, $A \subseteq B$ implies $|\Box|A \subseteq |\Box|B$.

It is known that the canonical frame of a sufficiency algebra is a sufficiency frame. Furthermore, the representation and complex algebra theorems hold for sufficiency algebras also [6].

6 Correspondence Theorems for Sufficiency Algebras

We begin this section with the following observation.

Proposition 16. *Let $(X, \lesssim_1, \lesssim_2, R, S)$ be a sufficiency frame. Then the following conditions are equivalent.*

(D I) $(\forall A \subseteq X)\; A \subseteq l\langle S\rangle l\langle S\rangle A$;
(D II) $(\forall x, y, z \in x)\; x \lesssim_1 y \wedge y\,S\,z \to (\exists t)\; z \lesssim_1 t \wedge t\,S\,x$.

Proof. Condition (D I) may be written in form

$$(\forall P)(\forall x)\,P(x) \to (\forall y)\,x \lesssim_1 y \to (\forall z)\,y\,S\,z \to (\exists t)\,z \lesssim_1 t \wedge (\exists w)\,t\,S\,w \wedge P(w).$$

We negate the above formula and get the formula

$$(\exists x, y, z)(\exists P)(\forall w)\;(\neg P(w) \vee (\forall t)\,z \not\lesssim_1 t \vee (t, w) \notin S) \wedge P(x) \wedge x \lesssim_1 y \wedge y\,S\,z.$$

By Lemma 4(b), this is equivalent to

$$(\exists x, y, z)\,x \lesssim_1 y \wedge y\,S\,z \wedge (\forall t)\,z \not\lesssim_1 t \vee (t, x) \notin S.$$

By negating this, we obtain the desired formula.

If $\lesssim_1$ and $\lesssim_2$ are identity relations, then the equivalence of (D I) and (D II) means that "S is symmetric if and only if $A \subseteq [S]\langle S \rangle A$ for all $A \subseteq X$".

Lemma 17. *Let* $(X, \lesssim_1, \lesssim_2, R, S)$ *be a sufficiency frame. If* S *is symmetric or* $R = S^{-1}$, *then* (D II) *holds.*

Proof. Suppose that $x \lesssim_1 y$ and $y \, S \, z$. Because S is symmetric, also $z \, S \, y$ holds. Now $z \lesssim_2 z$, $z \, S \, y$, and $x \lesssim_1 y$ imply $z \, S \, x$ by (Mono $S_{|\Box|}$), that is, (D II) holds.

Let $R = S^{-1}$. If $x \lesssim_1 y$ and $y \, S \, z$, then $z \, R \, y$, and by (SC $R_{|\Box|} S_{|\Box|}$), there exists a t such that $z \lesssim_1 t$ and $t \, S \, y$. Now $t \lesssim_2 t$, $t \, S \, y$, and $x \lesssim_1 y$ imply $t \, S \, x$ by (Mono $S_{|\Box|}$).

Obviously, $A \subseteq |\Box||\Box|A$ for all $A \in L(X)$, whenever S is symmetric or $R = S^{-1}$. The previous discussion is illustrated in Fig. 2.

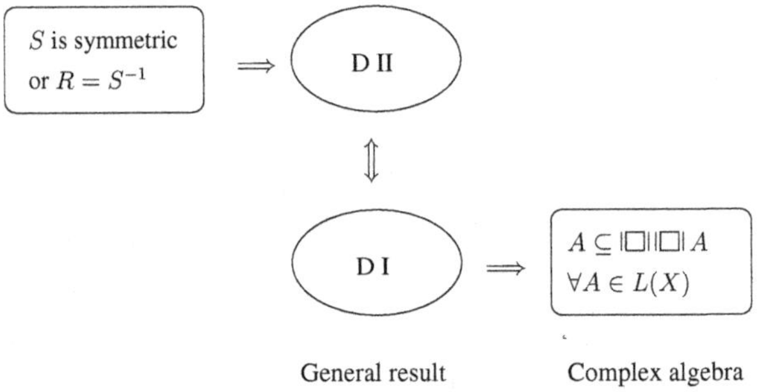

Fig. 2.

We may also present the following correspondence.

Proposition 18. *Let* $(X, \lesssim_1, \lesssim_2, R, S)$ *be a sufficiency frame. Then the following conditions are equivalent.*

(D I)° $(\forall A \subseteq X)\ A \subseteq [R]r[R]r(A)$;
(D II)° $(\forall x, y, z \in x)\ x \, R \, y \wedge y \lesssim_2 z \to (\exists t)\ t \lesssim_2 x \wedge z \, R \, t$.

Also the following lemma holds.

Lemma 19. *Let* $(X, \lesssim_1, \lesssim_2, R, S)$ *be a sufficiency frame. If* (D II)° *holds, then* R *is symmetric.*

Proof. Suppose that $x \, R \, y$. Since $y \lesssim_2 y$, we obtain that there exists a $t \in X$ such that $t \lesssim_2 x$ and $y \, R \, t$ by (D II)°. Now $y \lesssim_1 y$, $y \, R \, t$, and $t \lesssim_2 x$ imply $y \, R \, x$ by (Mono $R_{|\Box|}$).

Notice also that since there are two ways to define the operator $|\Box| \colon L(X) \to L(X)$, we could also find correspondences for conditions $(\forall A \subseteq X)\ A \subseteq [R]rl\langle S \rangle A$ and $(\forall A \subseteq X)\ A \subseteq l\langle S \rangle [R]r(A)$.

We end this work by considering sufficiency algebras and their canonical frames.

Lemma 20. *Let* $(P, \vee, \wedge, 0, 1, |\Box|)$ *be a sufficiency algebra such that* $a \leq |\Box||\Box|a$ *for all* $a \in P$.

(a) *If A is a filter of P, then $A \subseteq |\square||\square|A$.*
(b) *If B is an ideal of P, then $|\square||\square|B \subseteq B$.*

The proof is similar to that of Lemma 9.

Proposition 21. *Let $(P, \vee, \wedge, 0, 1, |\square|)$ be a sufficiency algebra. If $|\square| : P \to P$ satisfies $a \leq |\square||\square|a$, then in the canonical sufficiency frame $(X, \lesssim_1, \lesssim_2, R_{|\square|}, S_{|\square|})$, $R_{|\square|} = S_{|\square|}^{-1}$ holds.*

Proof. Suppose that $x\, R_{|\square|}\, y$, that is, $|\square|x_1 \subseteq y_2$. This implies $x_1 \subseteq |\square||\square|x_1 \subseteq |\square|y_2$, which means that $y\, S_{|\square|}\, x$. On the other hand, if $x\, S_{|\square|}\, y$, that is, $y_1 \subseteq |\square|x_2$, then $|\square|y_1 \subseteq |\square||\square|x_2 \subseteq x_2$. Hence, $y\, R_{|\square|}\, x$.

7 Conclusions

In this paper a first step is made towards an extension of the modal correspondence theory to the general lattice-based logics of possibility and sufficiency. We presented few examples of correspondences for these logics which show that the classical correspondences do not always carry over to the case of lattice-based logics. Further work is needed on general theorems characterising the classes of elementary frames for lattice-based logics or on the Sahlqvist-style theorems [9]. Some basic facts leading to a correspondence theory for the classical sufficiency logics based on Boolean algebras can be found in [2, 3].

References

1. W. Ackermann, Untersuchungen uber des Eliminationsproblem der mathematischen Logik, *Mathematische Annalen* **110**, 390–413 (1935).
2. S. Demri and E. Orłowska, *Incomplete Information: Structure, Inference, Complexity*, EATCS Monographs in Theoretical Computer Science (Springer-Verlag, Berlin Heidelberg, 2002).
3. I. Düntsch and E. Orłowska, Beyond modalities: sufficiency and mixed algebras, in: E. Orłowska and A. Szałas (eds.), *Relational Methods for Computer Science Applications. Studies in Fuzziness and Soft Computing* **65** (Physica-Verlag, Heidelberg, 2001) 263–285.
4. I. Düntsch, E. Orłowska, A. Radzikowska, D. Vakarelov, Relational representation theorems for some lattice-based structures, *Journal of Relational Methods in Computer Science* **1**, 132–160 (2005).
5. A. Nonnengart, A. Szałas, A fixpoint approach to second-order quantifier elimination with applications to correspondence theory, in: E. Orłowska (ed.), *Logic at Work. Essays dedicated to the memory of Helena Rasiowa* (Physica-Verlag, Heidelberg, 1999) 89–108.
6. E. Orłowska, D. Vakarelov, Lattice-based modal algebras and modal logics, in: P. Hájek, L. Valdes-Villanueva, D. Westerstahl (eds.), *Logic, Methodology and Philosophy of Science. Proceedings of the 12th International Congress* (KCL Publications, London, 2005) 147–170.
7. E. Orłowska, I. Rewitzky, Duality via truth: semantic frameworks for lattice-based logics, *Logic Journal of the IGPL* **13**, 467–490 (2005).
8. E. Orłowska, Relational semantics through duality, in: I. Düntsch, M. Winter (eds.), *Proceedings of the Eighth International Conference on Relational Methods in Computer Science (RelMiCS 8)* (Department of Computer Science, Brock University, 2005) p. 187.

9. H. Sahlqvist, Completeness and correspondence in the first and second order semantics for modal logics, in: S. Kanger (ed.), *Third Scandinavian Logic Symposium, Uppsala, Sweden, 1973* (North Holland, Amsterdam, 1975) 110–143.
10. A. Szałas, On the correspondence between modal and classical logic: an automated approach, *Journal of Logic and Computation* **3**, 605–620 (1993).
11. A. Urquhart, A topological representation theory for lattices, *Algebra Universalis* **8** (1978) 45–58.
12. J. van Benthem, Correspondence theory, in: D. Gabbay and F. Guenthner (eds.), *Handbook of Philosophical Logic* **II** (Reidel, Dordrecht, 1984) 167–247.

Control-Flow Semantics for Assembly-Level Data-Flow Graphs

Wolfram Kahl*, Christopher K. Anand, and Jacques Carette

SQRL, McMaster University, Hamilton,
Ontario, Canada

Abstract. As part of a larger project, we have built a declarative assembly language that enables us to specify multiple code paths to compute particular quantities, giving the instruction scheduler more flexibility in balancing execution resources for superscalar execution.

Since the key design points for this language are to only describe data flow, have built-in facilities for redundancies, and still have code that *looks like* assembler, by virtue of consisting mainly of assembly instructions, we are basing the theoretical foundations on data-flow graph theory, and have to accommodate also relational aspects.

Using functorial semantics into a Kleene category of "hyper-paths", we formally capture the *data-flow-with-choice* aspects of this language and its implementation, providing also the framework for the necessary correctness proofs.

1 Introduction

Magnetic resonance imaging (MRI) relies on highly efficient signal processing software — for example in medical applications, higher efficiency can make quite an important qualitative difference. The state of the art in the development of such software is that a scientist starts from a mathematical model and produces an appropriate signal processing algorithm; as first step towards an implementation this algorithm is directly translated into a prototype program, with reasonable confidence in its correctness. This is then turned over to a "digital signal processing guru" who will apply — manually! — different kinds of code transformations and optimisations, up to manual rearrangement of assembly code.

It is obvious that it is not easy to be fully confident in the correctness of software coming out of such a process. However, particularly in medical applications, correctness can be crucial, and defects in MRI signal processing software could manifest themselves as visible artifacts in the generated images that could introduce problems in the medical uses of these images.

The Coconut project prepares to produce a system that provides a coherent and consistent path from a mathematical specification of signal processing problems to verified *and* highly optimised machine code [2]. As part of this project,

* This research has been supported by an NSERC Discovery Grant.

W. MacCaull et al. (Eds.): RelMiCS 2005, LNCS 3929, pp. 147–160, 2006.

we encountered a need for a language of a quite peculiar nature: we needed to specify choices amongst different "equivalent" computation paths made up of low-level assembler. An intelligent instruction scheduler will choose the best path, using built-in knowledge of the intricacies of a modern, vectorised and pipelined CPU architecture. Our collective experience told us that we should be specifying our problem in a declarative manner, to give maximal freedom to the scheduler. We decided to see if we could get these rather different paradigms (declarative and assembly) to coexist, and serve as the main language for our compiler's back end.

The central idea is that this approach should allow us a separation of concerns in the code generation part of our special-purpose compiler:

- The *(assembly) code generator* will use knowledge of the *mathematical semantics* of assembly instructions to generate *correct* assembly code, but it will leave control flow decisions open as far as possible.
- The *scheduler* (or *assembler*) uses knowledge about the *resource consumption* of assembly instructions to generate *fast* machine code; correctness is guaranteed by the fact that the scheduler essentially performs only a selection of one path among those proposed by the code generator.

Our intermediate *declarative assembly language* therefore represents the semantics of its programs mostly as data flow; it allows to express some control flow constraints, but essentially leaves all efficiency-related instruction selection and scheduling decisions open.

Our targets are vectorised and pipelined CPUs, currently PowerPC 745X and 970, that are commonly used in signal processing applications. We design the scheduler (with appropriate support from the code generation component) to be able to automate a number of "tricks" used in manual optimisation. For example, it will be able to take advantage of limited precision requirements, and more generally, choose between "equivalent" machine code computations, which includes choosing instructions that produce the same results with different resource consumption, and choosing between computations that produce different intermediate values which can be used interchangeably. Other tricks avoid register spill for example via recomputation of previously available values, or via the use of renaming registers (used internally by some PowerPC versions) as non-addressable intermediate storage.

These requirements motivated our decision to base our declarative assembly language essentially on data flow graphs, and to add choices of computation paths as a new feature.

Therefore, branches are eliminated from our declarative assembly language, and we express all control flow which cannot be eliminated by use of permutation and selection in special-purpose 'combinators' (not considered in this paper). Only the non-branching instructions (of the PowerPC 745X and PowerPC 970) can be used as labels in our *code graphs*, which we understand to be the abstract syntax of our assembly language.

2 Code Graphs

Term graphs are usually represented by graphs where nodes are labelled with function symbols and edges connect function calls with their arguments [12]. An alternative representation was introduced with the name of *jungle* by Hoffmann and Plump [7] for the purpose of efficient implementation of term rewriting systems (it is called "term graph" in [11]).

A *jungle* is a directed hypergraph where nodes are only labelled with type information (if applicable), function names are hyperedge labels, each hyperedge has a sequence of input tentacles and exactly one output tentacle, and for each node, there is at most one hyperedge that has its output tentacle incident with that node.

For representing our declarative assembly code fragments, we use a generalisation of the jungle concept, corresponding to Ştefănescu's "flow graphs" [14]:

Definition 2.1. A *code graph* $G = (\mathcal{N}, \mathcal{E}, \mathsf{In}, \mathsf{Out}, \mathsf{src}, \mathsf{trg}, \mathsf{eLab})$ over an edge label set ELab consists of

- a set $\mathcal{N}$ of *nodes* and a set $\mathcal{E}$ of *hyperedges* (or *edges*),
- two node sequences $\mathsf{In}, \mathsf{Out} : \mathcal{N}^*$ containing the *input nodes* and *output nodes* of the code graph,
- two functions $\mathsf{src}, \mathsf{trg} : \mathcal{E} \to \mathcal{N}^*$ assigning each hyperedge the sequence of its *source nodes* and *target nodes* respectively, and
- a function $\mathsf{eLab} : \mathcal{E} \to \mathsf{ELab}$ assigning each hyperedge its *edge label*, where the label has to be compatible with the numbers of source and target nodes of the edge. □

In COCONUT, nodes are actually labelled with *types*, but this is not relevant for the current paper. Edge labels are either opcodes or constants, as can be seen in the example code graph above, where output tentacles are arrows from hyper-edges to nodes, and input tentacles are arrows from nodes to hyperedges — the ordering relation between in- resp. output tentacles incident with the same hyperedge is not made explicit in the drawing, but is part of the graph structure.

Acyclic code graphs where all edges have exactly one target node and no node is the target of several edges correspond to the jungles of [7] (called "term graphs" in [11]), which are essentially a hypergraph version of conventional term graphs. Since some operations produce more than one result, our hyperedges can have multiple output tentacles just as the primitives of Ştefănescu's flow graphs [14]; this also corresponds to the use of "hypersignatures" in [3]. In the application to PowerPC, the second result is always a condition code, i.e. carry or overflow, but we think it will be better to treat all results uniformly.

The more radical departure from conventional term-graph formalisms is that we allow *several* output tentacles to be incident with any one node — such *joining* tentacles are used for results that can be obtained in different ways, and also for situations where different intermediate values could be used interchangeably. In Ştefănescu's flownomials there are joins, too — Ştefănescu proposes two interpretations for the "branch/join" pair of operations in data flow networks, one as "copy/equality test" and one as "split/merge" [13]. Our use is closer to a control-flow interpretation of join; the following list of typical applications of joins shows how this single feature opens up a large bag of "tricks" for code generation from code graphs:

- **Multiple entry points:** Many common mathematical functions are implemented (for the sake of efficiency) via algorithms with extra preconditions, and initial code ensures that those preconditions are satisfied. For example,
  - trigonometric functions are only calculated on a fundamental domain, and modulo calculations are first performed to put arguments into the fundamental domain;
  - some functions have a standard interface (e.g. choice of units), but an alternative interface is much more efficient, so the initial statements perform the necessary conversions.

  In both cases, we can eliminate the respective initial instructions if we can verify the stricter preconditions, or if we can rewrite the upstream calculation to produce results matched to the more efficient interface.
- **Instruction selection:** For example merging disjoint bit-fields by either logical `or` or arithmetic `add` instructions uses different functional units on some processors. In such cases, conventional optimising compilers switch instructions to get better schedules; in our approach, we emit a join in the assembly code and let the code graph scheduler select the better branch in each case.
- **Multiple code paths** (beyond single-instruction alternatives): It is possible to do some computations in different units, e.g., evaluating polynomials in the scalar floating point unit or the vector floating point unit. Such alternative code paths can be used in two ways:
  - `map f`, where `f` is a simple function, can be unrolled and performed simultaneously on different data in different execution units;
  - the code can be in-lined in different contexts where relative demand on execution units, register pressure, etc., vary enough to make one code path more efficient than another.

As an example, the following code graph arises in an implementation of a non-uniform Fourier transform — we always show code graphs with the sequences of input and output nodes indicated by arrows from, respectively to, numbered triangles; here, there is only one input and one output. Rectangles are instruction hyperedges, and ellipses are (zero-input) constant hyperedges. The light grey instructions require the vector integer unit; the dark grey instructions require the vector permutation unit, and all other instructions the vector floating point unit. The small solid circles are nodes; arrows are drawn from source nodes to hyperedges and from hyperedges to target nodes. In this example, the three nodes that serve as inputs for more than one source tentacle are enlarged; the central grey node is in addition a join node since it is the target of more than one hyperedge.

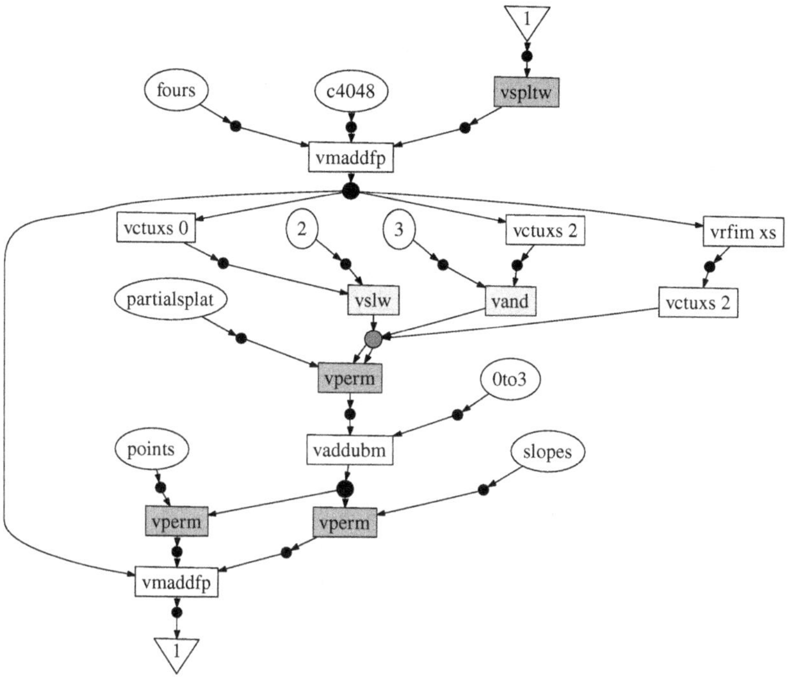

This particular join can be used by the scheduler to take some pressure off the floating point unit. If in a larger context, sharing of one of the constants used by the integer instructions becomes possible, the affected branch becomes preferable over the other branch since this reduces register pressure.

The flavour of this kind of joins is very similar to the instruction sequence alternatives produced by superoptimisers [10, 6], but we also use joins for alternatives that produce *different* results that still meet, for example, appropriate precision requirements. This approach is justified by the relational semantics we present in Sect. 5.

Further discussion of examples and the speed-ups we achieved using our approach can be found in [2].

Reachability in code graphs is defined via the node successor relation, where n' is a successor of n if there is an edge for which n is a source node and n' a target node. We use this to define two basic node and edge properties:

Definition 2.2. A node in a code graph is called *used* iff an output node is reachable from it, and *supported* iff it is either an input node or a target node of a supported edge.

An edge in a code graph is called *used* iff *at least one* of its target nodes is used, and *supported* iff *all* its source nodes are supported. □

The following code graphs provide some illustration of these concepts.

In the left graph, the A edge and its output node are unused, and the C edge and its input node are unsupported. In the middle graph, the left output node of the Q edge is unused, but the Q edge itself is still used, since its right output node is an output node of the graph. In the right graph (which has one output and zero inputs), nothing is supported:

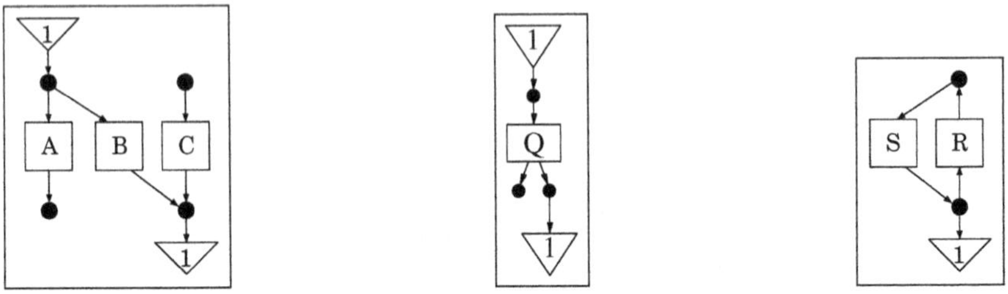

Building on top of these node and edge properties, we define a number of important graph properties:

Definition 2.3. A code graph is called:
- *acyclic* iff the node successor relation is acyclic,
- *join-free* iff each node occurs at most once in the concatenation of the target node lists of all edges with the input node list of the graph,
- *forward-garbage-free* iff all edges and non-output nodes are supported (no computations that cannot be performed because of lack of input),
- *backward-garbage-free* iff all *edges* are used (no computations for which no result is used),
- *garbage-free* iff it is both forward- and backward-garbage-free,
- *lean* iff it is garbage-free and join-free (and therefore acyclic),
- *coherent* iff all output nodes are supported,
- *solid* iff it is garbage-free and coherent,
- *executable* iff it is solid and lean. □

For example, the A edge above is backward-garbage (it will be collected in a backward direction) and the C edge is forward-garbage; the middle graph above is garbage-free, and furthermore lean and solid, and therefore executable, and the right graph has only forward-garbage edges and is not coherent.

The (forward-, resp. backward-) garbage-collected version of a code graph is obtained by iteratively deleting all nodes and edges that violate the respective condition — it is easy to see that the result is uniquely determined, always defined, and has the same input and output nodes as the original graph.

A first, simplified understanding of the use of code graphs in Coconut is the following:

- The code generator produces a coherent code graph G from a library of code graph fragments.
- G is garbage collected into a solid code graph S.
- The scheduler selects an executable subgraph E of S (or, more precisely, an executable "hyper-path" through S, see Sect. 5).
- The scheduler (already during selection) sequentialises the instructions in E in a way that maximises instruction-level parallelism in the target CPU.

Joins enable cycles, and since this may be particularly surprising in a data flow context, we discuss an (artificial) example here:

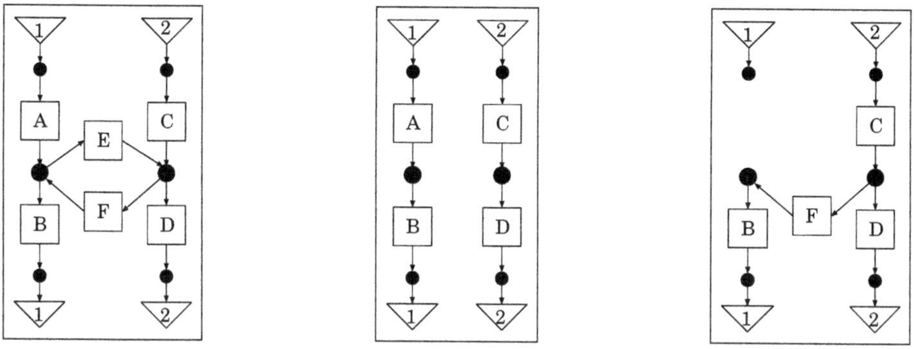

The graph on the left should be understood as describing a computation where the operations B and D require (at the hollow nodes) intermediate results that

- can be obtained from inputs (via A and C), and alternatively
- can be obtained from each other (via E resp. F).

The two graphs drawn beside it are both executable subgraphs that could be selected by the scheduler when processing the cyclic graph as input, and which of these will be more efficient may well depend on the context in which they are used.

3 Data-Flow Categories of Code Graphs

We now summarise the theory of our code graphs, which is essentially a reformulation of Ştefănescu's data-flow network algebra, in the language of category theory. In particular, we use the gs-monoidal categories proposed by Corradini and Gadducci for modelling acyclic term graphs [4].

The following definition serves mainly to introduce our notation:

Definition 3.1. A *category* $\mathbf{C}$ is a tuple $(\mathsf{Obj}, \mathsf{Mor}, \mathsf{src}, \mathsf{trg}, \mathbb{I}, ;)$ with the following constituents:

- Obj is a collection of *objects*.
- Mor is a collection of *arrows* or *morphisms*.
- src (resp. trg) maps each morphism to its source (resp. target) object. We write "$f : \mathcal{A} \to \mathcal{B}$" for "$f \in \mathsf{Mor} \wedge \mathsf{src}(f) = \mathcal{A} \wedge \mathsf{trg}(f) = \mathcal{B}$". The collection of all morphisms f of category $\mathbf{C}$ with $f : \mathcal{A} \to \mathcal{B}$ is denoted as $\mathsf{Mor}_{\mathbf{C}}[\mathcal{A}, \mathcal{B}]$ and also called a *homset*.
- "$;$" is the binary *composition* operator, and composition of two morphisms $f : \mathcal{A} \to \mathcal{B}$ and $g : \mathcal{B}' \to \mathcal{C}$ is defined iff $\mathcal{B} = \mathcal{B}'$, and then $(f;g) : \mathcal{A} \to \mathcal{C}$; composition is associative.
- $\mathbb{I}$ associates with every object $\mathcal{A}$ a morphism $\mathbb{I}_{\mathcal{A}}$ which is both a right and left unit for composition. □

The objects of the untyped code graph category over a set of edge labels ELab are natural numbers; in the typed case we would have sequences of types. A morphism from m to n is a code graph with m input nodes and n output nodes (more precisely, it is an isomorphism class of code graphs, since node and edge identities do not matter). Composition $F{;}G$ "glues" together the output nodes of F with the respective input nodes of G. The identity on n consists only of n input nodes which are also, in the same sequence, output nodes, and no edges.

A *primitive* code graph is a code graph that corresponds to a single operation, i.e., a code graph with a single edge where each node is the target of exactly one tentacle, and the target node sequence of the edge coincides with the output node sequence of the graph, and the source sequence with the input sequence.

Definition 3.2. A *symmetric strict monoidal category* $\mathbf{C} = (\mathbf{C}_0, \otimes, \mathbb{1}, \mathbb{X})$ consists of a category $\mathbf{C}_0$, a strictly associative monoidal bifunctor $\otimes$ with $\mathbb{1}$ as its strict unit, and a transformation $\mathbb{X}$ that associates with every two objects $\mathcal{A}$ and $\mathcal{B}$ an arrow $\mathbb{X}_{\mathcal{A},\mathcal{B}} : \mathcal{A} \otimes \mathcal{B} \to \mathcal{B} \otimes \mathcal{A}$ with:

$$(F \otimes G){;}\mathbb{X}_{\mathcal{C},\mathcal{D}} = \mathbb{X}_{\mathcal{A},\mathcal{B}}{;}(G \otimes F), \qquad \mathbb{X}_{\mathcal{A},\mathcal{B}}{;}\mathbb{X}_{\mathcal{B},\mathcal{A}} = \mathbb{I}_{\mathcal{A}} \otimes \mathbb{I}_{\mathcal{B}},$$
$$\mathbb{X}_{\mathcal{A}\otimes\mathcal{B},\mathcal{C}} = (\mathbb{I}_{\mathcal{A}} \otimes \mathbb{X}_{\mathcal{B},\mathcal{C}}){;}(\mathbb{X}_{\mathcal{A},\mathcal{C}} \otimes \mathbb{I}_{\mathcal{B}}), \qquad \mathbb{X}_{\mathbb{1},\mathbb{1}} = \mathbb{I}_{\mathbb{1}}. \qquad \square$$

For code graphs, $\mathbb{1}$ is the number 0 and $\otimes$ on objects is addition. On morphisms, $\otimes$ forms the disjoint union of code graphs, concatenating the input and output node sequences. $\mathbb{X}_{m,n}$ differs from $\mathbb{I}_{m+n}$ only in the fact that the two parts of the output node sequence are swapped.

Definition 3.3. $\mathbf{C} = (\mathbf{C}_0, \otimes, \mathbb{1}, \mathbb{X}, !)$ is a *strict g-monoidal category* iff

- $(\mathbf{C}_0, \otimes, \mathbb{1}, \mathbb{X})$ is a symmetric strict monoidal category, and
- $!$ associates with every object $\mathcal{A}$ of $\mathbf{C}_0$ an arrow $!_{\mathcal{A}} : \mathcal{A} \to \mathbb{1}$,

such that $\mathbb{I}_{\mathbb{1}} = !_{\mathbb{1}}$, and *monoidality of termination* holds: $!_{\mathcal{A}\otimes\mathcal{B}} = !_{\mathcal{A}} \otimes !_{\mathcal{B}}$ $\square$

For code graphs, $!_n$ differs from $\mathbb{I}_n$ only in the fact that the output node sequence is empty. The "g" of "g-monoidal" stands for "garbage": all edges of code graph $G : m \to n$ are backward-garbage in $G{;}!_n$.

Note that $!_n$ itself is garbage free, coherent, and lean, and therefore solid and even executable.

Definition 3.4. $\mathbf{C} = (\mathbf{C}_0, \otimes, \mathbb{1}, \mathbb{X}, \nabla)$ is a *strict s-monoidal category* $\mathbf{C}$ iff

- $(\mathbf{C}_0, \otimes, \mathbb{1}, \mathbb{X})$ is a symmetric strict monoidal category, and
- ∇ associates with every object $\mathcal{A}$ of $\mathbf{C}_0$ an arrow $\nabla_{\mathcal{A}} : \mathcal{A} \to \mathcal{A} \otimes \mathcal{A}$,

such that $\mathbb{I}_{\mathbb{1}} = \nabla_{\mathbb{1}}$, and the *coherence* axioms

- *associativity of duplication*: $\nabla_{\mathcal{A}}{;}(\mathbb{I}_{\mathcal{A}} \otimes \nabla_{\mathcal{A}}) = \nabla_{\mathcal{A}}{;}(\nabla_{\mathcal{A}} \otimes \mathbb{I}_{\mathcal{A}})$,
- *commutativity of duplication*: $\nabla_{\mathcal{A}}{;}\mathbb{X}_{\mathcal{A},\mathcal{A}} = \nabla_{\mathcal{A}}$

and the *monoidality* axiom

- *monoidality of duplication*: $\nabla_{\mathcal{A}\otimes\mathcal{B}}{;}(\mathbb{I}_{\mathcal{A}} \otimes \mathbb{X}_{\mathcal{B},\mathcal{A}} \otimes \mathbb{I}_{\mathcal{B}}) = \nabla_{\mathcal{A}} \otimes \nabla_{\mathcal{B}}$

are satisfied. $\square$

For code graphs, ∇_n differs from $\mathbb{I}_n$ only in the fact that the output node sequence is the concatenation of the input node sequence with itself. The "s" of "s-monoidal" stands for "sharing: every input of $\nabla_k ; (F \otimes G)$ is shared by $F : k \to m$ and $G : k \to n$.

Definition 3.5. $\mathbf{C} = (\mathbf{C}_0, \otimes, \mathbb{1}, \mathbb{X}, \nabla, !)$ is a *strict gs-monoidal category* iff
- $(\mathbf{C}_0, \otimes, \mathbb{1}, \mathbb{X}, !)$ is a strict g-monoidal category, and
- $(\mathbf{C}_0, \otimes, \mathbb{1}, \mathbb{X}, \nabla)$ is a strict s-monoidal category,

such that the *coherence* axiom
- *right-inverse of duplication* holds: $\nabla_{\mathcal{A}} ; (\mathbb{I}_{\mathcal{A}} \otimes !_{\mathcal{A}}) = \mathbb{I}_{\mathcal{A}}$ □

Code graphs (and term graphs) over a fixed edge label set form a gs-monoidal category, but not a *cartesian* category, where in addition ! and ∇ are *natural* transformations, i.e., for all $F : \mathcal{A} \to \mathcal{B}$ we have $F ; !_{\mathcal{B}} = !_{\mathcal{A}}$ and $F ; \nabla_{\mathcal{B}} = \nabla_{\mathcal{A}} ; (F \otimes F)$. To see how these naturality conditions are violated, the first five code graphs in the following drawing can be obtained as, in this sequence, $F : 1 \to 1$, $!_1$, $F ; !_1$, $F ; \nabla_1$, and $\nabla_1 ; (F \otimes F)$:

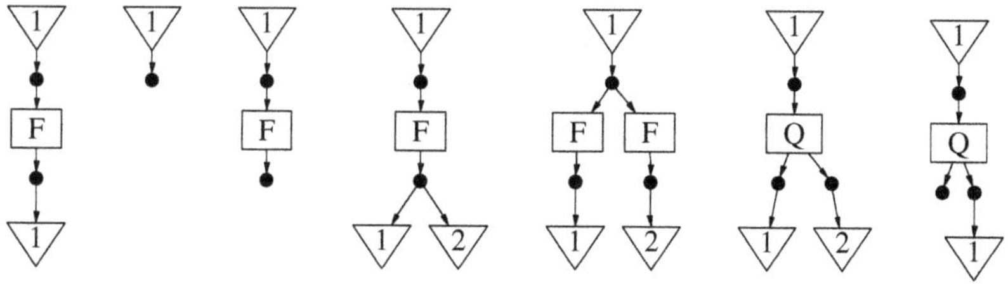

It is easy to see that we obtain naturality of termination if we consider equivalence classes of code graphs up to backward-garbage collection. Therefore, we introduce a special "garbage-collecting" variant:

Definition 3.6. A *gc-s-monoidal category* is a gs-monoidal category with natural termination, i.e., with $G ; !_{\mathcal{B}} = !_{\mathcal{B}}$ for all $G : \mathcal{A} \to \mathcal{B}$. □

From the last two code graphs drawn above, namely $Q : 1 \to 2$ and $Q ; (!_1 \otimes \mathbb{I}_1)$, we can also see why backward-garbage collection had to be defined so carefully: In the last graph, the Q edge is not backward-garbage since one of its results is still needed as output.

The code graph definition itself is completely symmetric with respect to inputs and outputs, so the duals of the termination and duplication are defined, too, and also satisfy all the corresponding laws, thus turning the category of code graphs over a set of primitives (with arities) into the free bi-gs-monoidal category over these primitives, or, equivalently, the free data-flow network algebra [14].

The dual to duplication is *join* $\Delta_{\mathcal{A}} : \mathcal{A} \otimes \mathcal{A} \to \mathcal{A}$, which, as a code graph, differs from $\mathbb{I}_{\mathcal{A}}$ only in the fact that the input node sequence is the concatenation of the output node sequence with itself.

The dual to termination is *co-termination* $¡_{\mathcal{A}} : \mathbb{1} \to \mathcal{A}$, which introduces forward-garbage and differs from $\mathbb{I}_{\mathcal{A}}$ only in the fact that the input node sequence is empty, so the corresponding code graph is not coherent.

Using primitives $F, H : 1 \to 1$ and $A : 2 \to 2$, we show in the following drawing the code graphs obtained as A alone, then $(\text{¡}_1 \otimes \mathbb{I}_1) ; A$, and finally $(\text{¡}_1 ; F) \otimes H$, which could also be obtained as $(\text{¡}_1 \otimes \mathbb{I}_1) ; (F \otimes H)$ because of functoriality of $\otimes$.

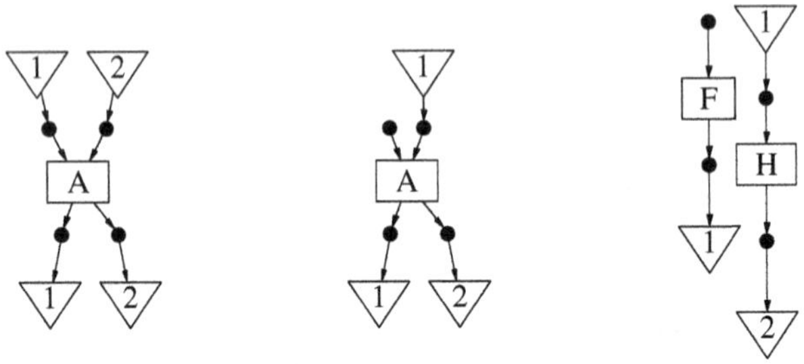

In the category of "code graphs up to forward garbage collection", co-termination is also a left-zero for composition, i.e., $\text{¡}_{\mathcal{A}} ; F = \text{¡}_{\mathcal{B}}$ for every $F : \mathcal{A} \to \mathcal{B}$, and also satisfies for each primitive $P : \mathcal{A} \to \mathcal{B}$ and each decomposition $\mathcal{A} = \mathcal{A}_1 \otimes \mathcal{A}_2 \otimes \mathcal{A}_3$ the following equation which corresponds in more detail to forward garbage collection:

$$(\mathbb{I}_{\mathcal{A}_1} \otimes \text{¡}_{\mathcal{A}_2} \otimes \mathbb{I}_{\mathcal{A}_3}) ; P = \text{¡}_{\mathcal{B}}$$

However, not even there we have $F \otimes \text{¡}_{\mathcal{C}} = \text{¡}_{\mathcal{B} \otimes \mathcal{C}}$ in general for $F : \mathcal{A} \to \mathcal{B}$.

All important kinds of code graphs as morphisms form at least gs-monoidal categories (co-s-monoidal categories also have joins and the corresponding laws, and co-g-monoidal categories correspondingly have co-termination):

Definition 3.7. With operations as defined above, natural numbers as objects, and given primitives (i.e., edge labels with input and output arities) we define the following categories:

- CG with code graphs as morphisms,
- CCG with coherent code graphs as morphisms.

Replacing direct code graph composition with composition that "performs automatic garbage collection", we further define:

- LCG with lean code graphs as morphisms,
- SCG with solid code graphs as morphisms,
- ECG with executable code graphs as morphisms. □

Proposition 3.8. The categories in Def. 3.7 are well defined, and, with primitive code graphs as generators, we have:

- CG is the free bi-gs-monoidal category;
- CCG is the free gs-monoidal and co-s-monoidal category;
- LCG is the free gc-s-monoidal and co-g-monoidal category;
- SCG is the free gc-s-monoidal and co-s-monoidal category;
- ECG is the free gc-s-monoidal category. □

4 Control Flow Aspects of Code Graphs

As already mentioned in Sect. 2, the task of the scheduler is to find (efficient) executable "hyper-paths" through a solid code graph.

If both graphs and edges were restricted to be one-input and one-output, then branching could occur only for the purpose of later joins, and a code graph would become a finite automaton, where the input node is the start state, the output node is the (only) accepting state, and edges are labelled with machine instructions. For such an automaton, we obtain its *operational semantics* in the *free Kleene algebra* of sets of instruction execution sequences.

Since each machine instruction induces a state transition relation, and this in turn induces another state transition relation for each set of instruction execution sequences, we obtain the *denotational semantics* of such an automaton in the *Kleene algebra* of relations.

To prepare the generalisation to arbitrary code graphs, we note that a more graph-theoretic view of the operational semantics is to consider it as the set of all paths from input to output, or, equivalently (and easier to generalise), as the set of all *line graphs* for which there is an input- and output-preserving homomorphism into the original code graph.

In comparison with finite automata, the additional code graph features are term graph features, namely

- n-ary operations enabled by parallel composition $\otimes$,
- multiple use of results, enabled by duplication ∇, and
- unused (additional) results, enabled by termination $!$.

Therefore, we need to enrich the Kleene algebra semantics with gs-monoidal features; we only need the technically simple complete variant of Kleene categories (see also [8]):

Definition 4.1. A *locally ordered category* is a category $\mathbf{C}$ such that

- for each two objects $\mathcal{A}$ and $\mathcal{B}$, the relation $\sqsubseteq_{\mathcal{A},\mathcal{B}}$ is a partial order on the homset $\mathsf{Mor}_{\mathbf{C}}[\mathcal{A},\mathcal{B}]$ (the indices will usually be omitted), and
- composition is monotonic with respect to $\sqsubseteq$ in both arguments.

A *complete Kleene category* is a locally ordered category where each homset is a complete upper semilattice and composition distributes over arbitrary joins from both sides. □

This implies the existence of zero morphisms $\perp\!\!\!\perp$, binary union, and Kleene star, all obeying the usual laws for typed Kleene algebras [9].

Now the denotational semantics of code graphs can use the gs-monoidal Kleene category of relations. For the operational semantics, we use the standard construction of set-based Kleene categories:

It is well-known that for any category $\mathbf{C} = (\mathsf{Obj}_{\mathbf{C}}, \mathsf{Mor}_{\mathbf{C}}, \mathsf{src}, \mathsf{trg}, \mathbb{I}_{\mathbf{C}}, ;_{\mathbf{C}})$, a complete Kleene category $\mathbf{C}^{\mathbb{P}}$ can be obtained by defining its components as follows:

- The objects are the same: $\mathsf{Obj}_{\mathbf{C}^{\mathbb{P}}} = \mathsf{Obj}_{\mathbf{C}}$
- Morphisms are subsets of the corresponding **C**-homsets:
$$\mathsf{Mor}_{\mathbf{C}^{\mathbb{P}}}[\mathcal{A},\mathcal{B}] = \mathbb{P}\ \mathsf{Mor}_{\mathbf{C}}[\mathcal{A},\mathcal{B}]$$
- Identities are singletons: $\mathbb{I}_{\mathbf{C}^{\mathbb{P}},\mathcal{A}} = \{\mathbb{I}_{\mathbf{C},\mathcal{A}}\}$
- Composition is set composition: $F ;_{\mathbf{C}^{\mathbb{P}}} G = \{f ;_{\mathbf{C}} g \mid f \in F \wedge g \in G\}$
- The ordering is set inclusion: $F \sqsubseteq G \Leftrightarrow F \subseteq G$. This ordering is complete, and composition distributes over arbitrary joins, so we have:
  - Least elements: $\perp\!\!\!\perp_{\mathcal{A},\mathcal{B}} = \{\}$
  - Binary joins: $F \sqcup G = F \cup G$
  - If $F : \mathcal{A} \to \mathcal{A}$, then Kleene star: $F^* = \bigcup\{F^n \mid n \in \mathbb{N}\}$, with the understanding that $\mathbb{N} = \{0,1,2,\ldots\}$, and $F^0 = \mathbb{I}_{\mathcal{A}}$ and $F^{n+1} = F;F^n$.

This construction preserves gs-monoidality:

Theorem 4.2. If $\mathbf{C} = (\mathbf{C}_0, \otimes, \mathbb{1}, \mathbb{X}, \nabla, !)$ is a gs-monoidal category, then we obtain a gs-monoidal category, again, by extending $\mathbf{C}^{\mathbb{P}}$ with the following constants:

- $\otimes$ and $\mathbb{1}$ on objects are the same as in **C**.
- monoidal composition is defined as monoidal set composition:
$$F \otimes_{\mathbf{C}^{\mathbb{P}}} G = \{f \otimes g \mid f \in F \wedge g \in G\}$$
- the constants are singleton sets:
  - $\mathbb{X}_{\mathbf{C}^{\mathbb{P}},\mathcal{A},\mathcal{B}} = \{\mathbb{X}_{\mathcal{A},\mathcal{B}}\}$
  - $!_{\mathbf{C}^{\mathbb{P}},\mathcal{A}} = \{!_{\mathcal{A}}\}$
  - $\nabla_{\mathbf{C}^{\mathbb{P}},\mathcal{A}} = \{\nabla_{\mathcal{A}}\}$

PROOF: This gs-monoidal category is well defined since all constants are defined as singleton sets, and in each axiom of gs-monoidal categories, all variables occur exactly once on *both* sides of the equality in such a way that the resulting sets are always isomorphic via axiom instances on the elements. □

5 Code Graph Semantics and Scheduling

We now are in a position to provide the details of the semantical aspects of the use of code graphs in COCONUT as sketched in Sect. 2.

The principle we use is that of *functorial semantics* [5]: Since all our gs-monoidal code graph categories are, according to Proposition 3.8, freely generated from the primitives modulo some additional constants and/or laws, any gs-monoidal category **C** providing these constants and laws immediately provides a semantics $[\![G]\!]^{\mathbf{C}}$ for each code graph G from the respective code graph category via the unique gs-monoidal functor from the free (i.e., initial) category into the chosen semantical category.

- We start with a (total) relational specification R.
- The code generator produces a coherent code graph G from a library of code graph fragments — G is a morphism of CCG. The denotational semantics

of CCG is considered in the gs-monoidal and co-s-monoidal category of *total relations*, where all primitives are interpreted as *total functions* (we do not consider halting or interrupting machine instructions).

The task of the code generator therefore is to ensure that $[\![G]\!]^{TotRel} \sqsubseteq R$.

- G is garbage collected into a solid code graph S, a morphism of SCG, still interpreted as the same total relation $[\![S]\!]^{TotRel} = [\![G]\!]^{TotRel}$.
- The scheduler selects an executable code graph E, i.e., a morphism of ECG, from the set of executable code graphs that form the functorial semantics $[\![S]\!]^{\mathsf{ECG}^{\mathbb{P}}}$ of S in the gs-monoidal Kleene category $\mathsf{ECG}^{\mathbb{P}}$.

  This selection is of course not arbitrary, but attempts to select a code graph that is minimal to some resource consumption metric, normally execution time, or, more precisely, total throughput through the resulting program, occasionally influenced by register consumption. On a pipelined architecture, these metrics are not fully compositional, so the gs-monoidal Kleene category structure is only of limited use for designing appropriate optimisation strategies for the scheduler.

  Executable code graphs are interpreted in the cartesian category of total functions. Since $\{E\} \sqsubseteq [\![S]\!]^{\mathsf{ECG}^{\mathbb{P}}}$ in $\mathsf{ECG}^{\mathbb{P}}$, we also have $[\![E]\!]^{TotRel} \sqsubseteq [\![S]\!]^{TotRel}$ in the category of total relations, which establishes that E with its functional semantics $[\![E]\!]^{Set} = [\![E]\!]^{TotRel}$ satisfies its relational specification $[\![S]\!]^{TotRel}$ and therefore R.
- By construction, there is a code graph homomorphism from the executable (and therefore join-free) code graph E to S, so E can also be considered as a generalised path ("hyper-path") through S — the scheduler is therefore implemented as a kind of shortest path search.

6 Conclusion and Outlook

The code graphs introduced in Sect. 2 serve as concrete syntax for computation fragments at the assembly level in the special-purpose compiler suite of the CoConut project. We have shown that through the device of functorial semantics, these code graphs, essentially considered as data-flow graphs, can be equipped with a relational denotational semantics, which is used for reasoning about their correctness.

The joins in these data-flow graphs have been assigned a novel interpretation which amounts to giving control-flow semantics to this aspect of data-flow graphs — this is realised by defining a functorial operational semantics in a Kleene category of "hyper-paths".

While joins give us obviously the opportunity to integrate a superoptimiser into the code generator similar to [6], the fact that we use a relational semantics also allows us to be more flexible and make use of the fact that, for example, many results in image processing only need to be correct up to a relatively low precision, so mathematically radically different algorithms can be explored.

The coherent semantical treatment of code graphs as presented in Sect. 5 has already proven beneficial in guiding the design of the scheduler for computation fragments without branches and loops.

The most important next step is to integrate this with proper control flow structure, in particular for wrapping loop structures around computation fragments serving as loop bodies, and for the separation of loop bodies into several stages for the purpose of modulo scheduling [1] which is another useful optimisation trick in the context of heavily pipelined architecture.

We would like to thank the anonymous referees for their useful comments.

References

[1] V. H. Allan, R. B. Jones, R. M. Lee, S. J. Allan. *Software pipelining.* ACM Comput. Surv. **27**(3) 367–432, 1995.

[2] C. K. Anand, J. Carette, W. Kahl, C. Gibbard, R. Lortie. *Declarative Assembler.* SQRL Report 20, Software Quality Research Laboratory, McMaster University, 2004. available from http://sqrl.mcmaster.ca/sqrl_reports.html.

[3] M. Coccia, F. Gadducci, A. Corradini. *GS-Λ Theories: A Syntax for Higher-Order Graphs.* Electronic Notes in Computer Science **69** 18, 2002.

[4] A. Corradini, F. Gadducci. *An Algebraic Presentation of Term Graphs, via GS-Monoidal Categories.* Applied Categorical Structures **7**(4) 299–331, 1999.

[5] A. Corradini, F. Gadducci. *Functorial Semantics for Multi-Algebras and Partial Algebras, with Applications to Syntax.* Theoretical Computer Science **286**(2) 293–322, 2002.

[6] T. Granlund, R. Kenner. *Eliminating Branches Using a Superoptimizer and the GNU C Compiler.* In: Programming Language Design and Implementation, PLDI '92, pp. 341–352. acm, 1992.

[7] B. Hoffmann, D. Plump. *Jungle Evaluation for Efficient Term Rewriting.* In J. Gabrowski, P. Lescanne, W. Wechler, eds., Algebraic and Logic Programming, ALP '88, Mathematical Research **49**, pp. 191–203. Akademie-Verlag, 1988.

[8] W. Kahl. *Refactoring Heterogeneous Relation Algebras around Ordered Categories and Converse.* J. Relational Methods in Comp. Sci. **1** 277–313, 2004.

[9] D. Kozen. *Typed Kleene Algebra.* Technical Report 98-1669, Computer Science Department, Cornell University, 1998.

[10] H. Massalin. *Superoptimizer: A Look at the Smallest Program.* In: ASPLOS-II: Proceedings of the Second International Conference on Architectual Support for Programming Languages and Operating Systems, pp. 122–126, Los Alamitos, CA, USA, 1987. IEEE Computer Society Press.

[11] D. Plump. *Term Graph Rewriting.* In H. Ehrig, G. Engels, H.-J. Kreowski, G. Rozenberg, eds., Handbook of Graph Grammars and Computing by Graph Transformation, Vol. 2: Applications, Languages and Tools, Chapt. 1, pp. 3–61. World Scientific, Singapore, 1999.

[12] M. Sleep, M. Plasmeijer, M. van Eekelen, eds. *Term Graph Rewriting: Theory and Practice.* Wiley, 1993.

[13] Gheorghe Ştefănescu. *Algebra of Flownomials — Part 1: Binary Flownomials; Basic Theory.* Technical Report TUM-I9437, Technische Universität München, Institut für Informatik, 1994.

[14] Gheorghe Ştefănescu. *Network Algebra.* Springer, London, 2000.

Relational Implementation of Simple Parallel Evolutionary Algorithms

Britta Kehden, Frank Neumann, and Rudolf Berghammer

Institut für Informatik und Praktische Mathematik,
Christian-Albrechts-Univ. zu Kiel,
Olshausenstraße 40, 24098 Kiel, Germany

Abstract. Randomized search heuristics, among them evolutionary algorithms, are applied to problems whose structure is not well understood, as well as to hard problems in combinatorial optimization to get near-optimal solutions. We present a new approach implementing simple parallel evolutionary algorithms by relational methods. Populations are represented as relations which are implicitly encoded by (reduced, ordered) binary decision diagrams. Thereby, the creation and evaluation is done in parallel, which increases efficiency considerably.

1 Introduction

In the past, randomized search heuristics (see [7] for an overview), among them *evolutionary algorithms* (abbreviated as EAs), became quite popular. Initially developed in the 1960s and 1970s, they have found many applications as simple, robust, and efficient solvers in wide areas of real world problems whose structure is not well understood. But they also have been used in the case of many hard problems in combinatorial optimization to get solutions, which do not differ too much from optimal solutions, in reasonable (for instance, polynomial) time.

Also since some time, relational algebra [10] has been successfully used for algorithm development, especially in the case of discrete structures. Relational programs are implementations of algorithms that are mainly based on a datatype for binary relations. This approach has some benefits. Such programs can be specified and developed in a rigid way, which helps to prove their correctness very formally. Since relational algebra has a fixed and surprisingly small set of base constants and operations, it also opens the possibility of computer-support. Especially, it offers to execute programs with the aid of a relation-algebraic manipulation and programming system like the Kiel RelView tool [1,3]. This does not only allow to increase their trustworthiness. It can also be used to get a feeling of the programs' behaviour, for instance, concerning runtimes or the quality of the computed solutions if they implement search heuristics or approximation algorithms. Furthermore, since relational programs frequently are quite simple, it enables to play and experiment with them. In case of evolutionary algorithms this may help to adjust free parameters in the right way.

The aim of this paper is to show how to develop relational versions of randomized search heuristics, how to implement them as RelView-programs, and

W. MacCaull et al. (Eds.): RelMiCS 2005, LNCS 3929, pp. 161–172, 2006.

how RELVIEW's specific representation of relations positively influences the runtimes. We concentrate on covering problems and consider minimum vertex covers on undirected graphs and minimum set covers on hypergraphs.

For the first problem in [2] a heuristics is presented. It combines a well-known approximation algorithm (in [5] attributed to Gavril and Yannakakis) with a generic program for computing a minimal subset fullfilling a certain property. Such an approach does not have the chance to escape from local optima. However, this can be different in the case of evolutionary algorithms since they, even when only based on mutation, sample a much larger neighbourhood. Of course, one has to pay for such an approach, normally with a higher runtime.

One can reduce the runtime of evolutionary algorithms if the individuals are created and evaluated in parallel. In this case one speaks of *parallel evolutionary algorithms*, or PEAs for short. Doing parallel random search, it is important how to create and evaluate individuals. Often this is done by a distributed system. The disadvantage of such an approach is that the system can get rather huge and a lot of resources will be necessary. To avoid this drawback, we will use an approach based on one single system. The key idea is to describe the process of creating and evaluating individuals by relation-algebraic expressions, which immediately can be formulated in RELVIEW and then executed using this tool. And here is the place where the specific representation of relations in RELVIEW comes into the play. The system represents them as (reduced, ordered) binary decision diagrams; see [4, 8, 9]. This implicitely means that the creation and evaluation of individuals is performed in parallel.

After having motivated this line of research, we present in Section 2 the algorithm to be implemented using relational methods. In Section 3, we introduce the basics of relational algebra and describe the RELVIEW system. Section 4 presents an implementation of the parallel evolutionary algorithm in RELVIEW. Results of our numerous tests with the tool are reported in Section 5. In Section 6, we finish with some conluding remarks.

2 The (1+λ)-EA and Covering Problems

We start with the minimum vertex cover problem. This classical optimization task has the following description. Given an undirected graph $G = (V, E)$ with vertex set $V = \{v_1, \ldots, v_n\}$ and edge set $E = \{e_1, \ldots, e_m\}$, the task is to find a subset of $V' \subseteq V$ with a smallest number of vertices such that each edge of G is incident to at least one vertex of V'. Computing minimum vertex covers is NP-hard and can be approximated within a worst case ratio of $2 - \frac{2 \ln \ln n}{\ln n}(1 - o(1))$ as shown in [6]. This is up to now the best known approximability result achievable in polynomial time. Nevertheless, on most instances heuristics like evolutionary algorithms have a good chance to achieve better solutions than approximation algorithms. To garantee the best possible performance ratio for an evolutionary algorithm, too, one can integrate such a solution into the initial population. In addition, such a starting point ensures that one always has a vertex cover as the current solution.

In the following, we consider a variant of the well-known $(1+\lambda)$ evolutionary algorithm. It is based on one single individual x and produces independently λ children from x. The selection procedure selects an individual from the children that is not inferior to x. If there is not such an individual in the offspring population, x is chosen for the next generation. Here is our base algorithm:

Algorithm ($(1+\lambda)$-EA)

1. Choose $x \in \{0,1\}^k$
2. Create a population P of size λ from x. Each individual y of P is created by flipping each bit of x with probability $1/k$.
3. Let Q be the set of individuals of P that are not inferior to x. If Q is not emtpy, then replace x by one individual of Q. Otherwise, do nothing.
4. Repeat Steps 2 and 3.

In the rest of this paper, we assume the loop to be stopped if the individual x has not been improved by an individual of P in t iterations. Creating the λ children in parallel leads to a variant, which is the perhaps simplest parallel evolutionary algorithm that can be considered.

To approximate a minimum vertex cover of $G = (V, E)$ with the aid of $(1+\lambda)$-EA, we take $k := n$ and represent subsets V' of V as bitstrings $x \in \{0,1\}^k$ such that $x_i = 1$ if and only if $v_i \in V'$. The algorithm also can be applied to approximate a solution of the NP-hard minimum hyperedge cover problem for a hypergraph $H = (V, E)$ with vertex set $V = \{v_1, \ldots, v_n\}$ and hyperedge set $E = \{h_1, \ldots, h_m\}$, where hyperedges are non-empty subsets of V. The problem asks for finding a subset $E' \subseteq E$ with a smallest number of hyperedges such that each vertex from V is contained in at least one hyperedge of E'. As we are searching for a subset of the hyperedges in the hypergraph, we use the above algorithm working on bitstrings of length $k := m$ and, consequently, flip each bit with probability $1/m$ in the mutation step.

3 Relational Preliminaries

In this section, we introduce some basics of relational algebra and consider vectors and points, which form specific classes of relations and subsequently are used for representing individuals. We also have a short look at RelView.

3.1 Relational Algebra

We write $R : X \leftrightarrow Y$ if R is a relation with domain X and range Y, i.e., a subset of $X \times Y$. If the sets X and Y of R's *type* $X \leftrightarrow Y$ are finite and of cardinality m and n, respectively, we may consider R as a Boolean matrix with m rows and n columns. Since this Boolean matrix interpretation is well suited for many purposes, in the following we often use matrix terminology and matrix notation. Especially, we speak of the rows, columns and entries of R and write R_{ij} instead of $(i, j) \in R$. We assume the reader to be familiar with the basic operations on relations, viz. R^T (*transposition*), $\overline{R}$ (*negation*), $R \cup S$ (*union*),

$R \cap S$ (*intersection*), $R \cdot S$ (*composition*), $R \subseteq S$ (*inclusion*), and the special relations O (*empty relation*), L (*universal relation*), and I (*identity relation*).

If X and Y are finite, then $|R|$ denotes the cardinality of $R : X \leftrightarrow Y$. In Boolean matrix terminology, $|R|$ equals the number of true-entries.

For modeling individuals, in this paper we will use (column) *vectors*, which are relations x with $x = x \cdot \mathsf{L}$. For $x : X \leftrightarrow Y$ this condition means: whatever set Z and universal relation $\mathsf{L} : Y \leftrightarrow Z$ we choose, an element $i \in X$ is either in relationship $(x \cdot \mathsf{L})_{ij}$ to no element $j \in Z$ or to all elements $j \in Z$. As for a vector, therefore, the range is irrelevant, we consider in the following mostly vectors $x : X \leftrightarrow \mathbf{1}$ with a specific singleton set $\mathbf{1} = \{\perp\}$ as range and omit in such cases the second subscript, i.e., write x_i instead of $x_{i\perp}$. Such a vector can be considered as a bitstring if X equals an intervall $[1..k] := \{i \in \mathbb{N} : 1 \leq i \leq k\}$. Then x_i corresponds to the fact that the i-th component of the bitstring is 1. A non-empty vector x is said to be a *point* if it is injective, which relation-algebraically can be specified by $x \cdot x^{\mathsf{T}} \subseteq \mathsf{I}$. In the bitstring model a point $x : [1..k] \leftrightarrow \mathbf{1}$ is a bitstring in which exactly one component is 1.

3.2 The RelView System

RELVIEW is a tool for calculating with relations and relational programming. It has been developed at Kiel University since 1993, in particular in the course of the Ph.D. theses [8, 9] and a series of Diploma theses. As already mentioned, the tool uses a representation of relations based on binary decision diagrams (BDDs). The latter have been shown to be very compact representations of Boolean functions (see e.g., [11]). Exactly this property is used to implement a relation $R : X \leftrightarrow Y$ via a Boolean function f_R such that R_{ij} if and only if $f_R(x, y) = 1$, where x and y are the binary representation of i and j, respectively. In doing so, the relational constants and operations can be implemented using the constant BDDs TRUE and FALSE and standard operations on BDDs.

The main purpose of RELVIEW is the evaluation of relation-algebraic expressions. These are constructed from the relations of the system's workspace using pre-defined operations and tests, user-defined relational functions, and user-defined relational programs.

Relational functions are of the form $F(R_1, \ldots, R_n) = exp$, where F is the function name, the R_i, $1 \leq i \leq n$, are the formal parameters (standing for relations), and *exp* is a relation-algebraic expression over the relations of the system's workspace that can additionally contain the formal parameters. A relational program is much like a function procedure in the programming languages Pascal or Modula 2, except that it only uses relations as data type. It starts with the headline, i.e., the name of the program and the list of formal parameters. Then the declaration part follows, which consists of the declarations of local relational domains (direct products and sums), local relational functions, and local variables. The third part is the body, a sequence of statements, which are separated by semicolons and terminated by the `RETURN`-clause.

The choice of a point contained in a non-empty vector is fundamental in relational programming. Therefore, RELVIEW possesses a corresponding pre-defined operations `point`. It is deterministic. The implementation uses that the system only deals with relations on finite carrier sets, which are linearely ordered by an internal enumeration. But RELVIEW allows to generate random relations, too, using the pre-defined operation `random`. If it is applied to two relations $R : X \leftrightarrow Y$ and $S : U \leftrightarrow V$, then a relation $T : X \leftrightarrow Y$ is generated uniformly at random with the additional property that for all $i \in X$ and $j \in Y$ the probability of T_{ij} being true is $|S|/(|U|\,|V|)$. Especially, for S as point $p : U \leftrightarrow \mathbf{1}$ we obtain a random relation of the same type as R and $1/|U|$ as probability of a pair to be contained in it due to $|p| = 1$.

4 Relational Implementation of PEAs

Now, we demonstrate how the instances of $(1+\lambda)$-EA for minimum vertex cover and minimum hyperedge cover, respectively, can be implemented using relational methods and RELVIEW. We start with the development of the basic modules, i.e., the parallel creation of the children of a given individual and the parallel test whether the new individuals are feasible solutions. Based on these modules and their RELVIEW-implementations, we obtain the final RELVIEW-programs as specific instantiations of a general program.

4.1 Relational Creation of Childrens

We assume that K denotes the intervall $[1..k]$ of the natural numbers as introduced in Section 3.1. Individuals of $(1+\lambda)$-EA are bitstrings from $\{0,1\}^k$, which we represent relation-algebraically by vectors of type $K \leftrightarrow \mathbf{1}$ as shown in Section 3.1, too. In the following, we treat the construction of a population of λ children from a given individual (current solution).

Let $x : K \leftrightarrow \mathbf{1}$ be the vector-representation of the current solution. We model the population of the λ children of x by means of a relation P such that each column of P, considered as a vector of type $K \leftrightarrow \mathbf{1}$, represents a single child of x. To create P, we first constitute a relation A consisting of λ columns equal to x. Defining L as intervall $[1..\lambda]$, this easily can be obtained as follows:

$$A : K \leftrightarrow L \qquad A = x \cdot \mathsf{L}, \text{ where } \mathsf{L} : \mathbf{1} \leftrightarrow L\,. \tag{1}$$

After that, we flip each entry in A with probability $1/k$ following the procedure of Section 3.2. Let $F : K \leftrightarrow L$ denote the relation where each entry has probability $1/k$ of being true. Then column l of P (i.e., the l-th child of x) is obtained by putting its i-th component, $1 \le i \le k$, to the entry A_{il} if F_{il} is false, and to the negation of the entry A_{il} in case of F_{il} being true. This precisely means that P equals the symmetric difference of A and F, in terms of relational algebra:

$$P : K \leftrightarrow L \qquad P = (A \cap \overline{F}) \cup (\overline{A} \cap F)\,. \tag{2}$$

Translated into the language of RELVIEW, the relation P of (2) is computed if the following relational program `children` is applied to the relation A of (1).

```
children(A)
  DECL p, F
  BEG  p = point(Ln1(A));
       F = random(A,p)
       RETURN (A & -F) | (-A & F)
  END.
```

In this code, Ln1 computes an universal vector the domain of which equals the domain of the argument and the range of which is **1**. The symbols -, |, and & denote in RelView negation, union, and intersection, respectively.

4.2 Testing Individuals to be Vertex Covers

Having solved the task of creating λ children of x in parallel in form of a relation $P : K \leftrightarrow L$, we now tackle the problem of testing feasibility. In this subsection, we concentrate on vertex covers. Hyperedge covers are considered in Section 4.5.

It is well-known how to test a single vector to represent a vertex cover of a graph; see e.g., [10]. However, essential for a fast relational version of $(1 + \lambda)$-EA is to decide in parallel which columns / individuals of $P : K \leftrightarrow L$ represent a vertex cover of a graph G. We solve this task by developing a vector of type $L \leftrightarrow \mathbf{1}$ such that its l-th entry is true if and only if the l-th column of P represents a vertex cover of G. For example, consider this graph with six vertices and its adjacency Relation R.

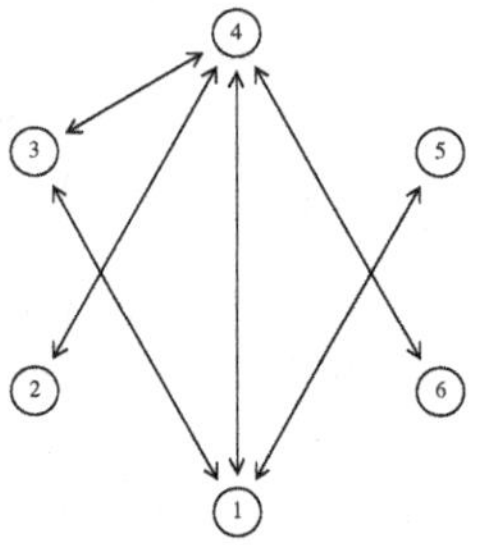

R =

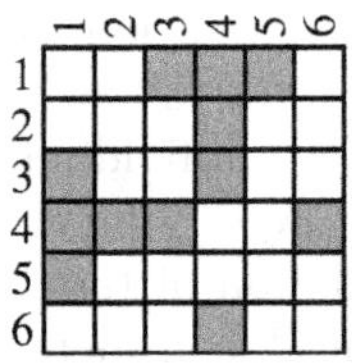

Given a population P consisting of 8 individuals, the task is to find out, which individuals are vertex covers and which are not. This means, we have to compute a vector (in this picture represented as a row vector), which specifies the columns of P that are vertex covers of the graph.

P =

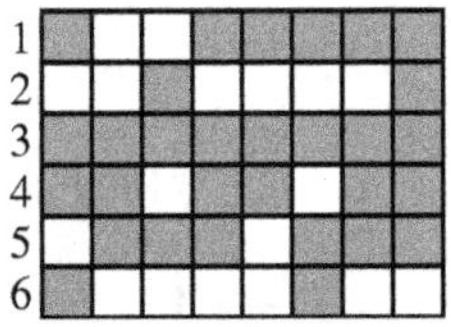

IsVertexCover(R,P) =

Using first-order logic, the l-th entry of the vector has to be true if and only if the following formula holds:

$$\forall i,j : R_{ij} \rightarrow (P_{il} \vee P_{jl}) . \tag{3}$$

Here i,j range over K and $R : K \leftrightarrow K$ is the *adjacency relation* of G, i.e., for all $i,j \in K$ the vertices v_i and v_j are connected via an edge if and only if R_{ij}.

Starting with (3), we can calculate as given below. In doing so, we only use simple logical and relation-algebraic laws in combination with well-known correspondences between logical and relation-algebraic constructions.

$$\begin{aligned}
\forall i,j : R_{ij} \rightarrow (P_{il} \vee P_{jl}) &\iff \neg\exists i,j : R_{ij} \wedge \overline{P}_{jl} \wedge \overline{P}_{il} \\
&\iff \neg\exists i : (R \cdot \overline{P})_{il} \wedge \overline{P}_{il} \\
&\iff \neg\exists i : (R \cdot \overline{P} \cap \overline{P})_{il} \\
&\iff \neg\exists i : \mathsf{L}_{\perp i} \wedge (R \cdot \overline{P} \cap \overline{P})_{il} \\
&\iff \neg(\mathsf{L} \cdot (R \cdot \overline{P} \cap \overline{P}))_{\perp l} \\
&\iff \overline{\mathsf{L} \cdot (R \cdot \overline{P} \cap \overline{P})}^{\mathsf{T}}_{l} .
\end{aligned}$$

From the last expression, we get $\overline{\mathsf{L} \cdot (R \cdot \overline{P} \cap \overline{P})}^{\mathsf{T}} : L \leftrightarrow \mathbf{1}$ as relation-algebraic specification of the vector we are interested in, where $\mathsf{L} : \mathbf{1} \leftrightarrow K$ is a "row vector". An immediate consequence is the following RelView-function, which computes this vector from R and P:

```
IsVertexCover(R,P) = -(L1n(R) * (R * -P & -P))^.
```

4.3 Subsets and Supersets

Using $\overline{\mathsf{L} \cdot (R \cdot \overline{P} \cap \overline{P})}^{\mathsf{T}}$, we can easily obtain a child of x that represents a vertex cover. But the algorithm has to select an individual of the new population, which is not inferior to x. Therefore, we introduce an additional vector of type $L \leftrightarrow \mathbf{1}$. This will allow us to get a smaller set of candidates for the next generation. First, we have to solve the following problem. Given two Relations M and N of the same type, we want to compute a vector that specifies each column of M, that is contained in the according column of N. More formally this means for each l in the range of M and N:

The l-th column of M is a subset of the l-th column of N

$$\begin{aligned}
&\iff \forall i : M_{il} \rightarrow N_{il} \\
&\iff \neg\exists i : M_{il} \wedge \neg N_{il} \\
&\iff \neg\exists i : \mathsf{L}_{\perp i} \wedge (M \cap \overline{N})_{il} \\
&\iff \neg(\mathsf{L} \cdot (M \cap \overline{N}))_{\perp l} \\
&\iff \overline{(\mathsf{L} \cdot (M \cap \overline{N}))}^{\mathsf{T}}_{l} .
\end{aligned}$$

We obtain the following RelView-function

```
IsSubset(M,N) = -(L1n(M) * (M & -N))^.
```

The function is used to test the individuals of a population P to be supersets or subsets of an present solution x. For example consider a graph with six vertices and this present solution x consisting of four vertices.

x =

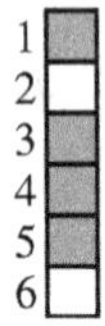

The relations A and P are computed like described in Section 4.1. Then the complement of `IsSubset(A,P)` specifies the columns of P, which aren't supersets of x and `IsSubset(P,A)` specifies the columns of P, which are subsets of x. By computing the intersection of both vectors we obtain a vector specifying the columns of P that are strictly contained in x. In this example, among the 8 individuals in P there are 2 real subsets of x.

A =

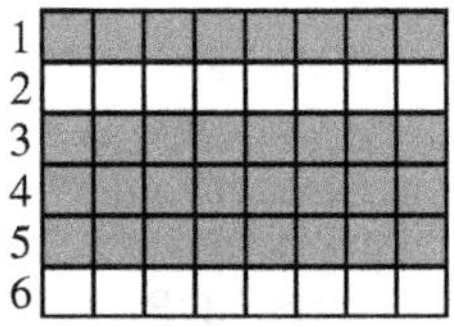

P =

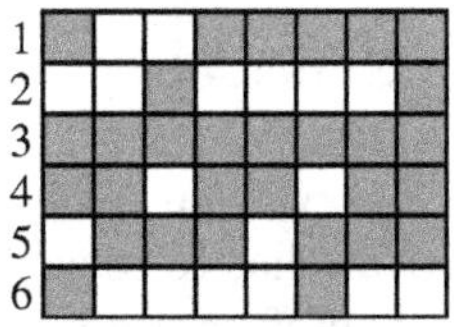

-IsSubset(A,P) =

IsSubset(P,A) =

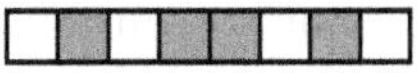

-IsSubset(A,P) & IsSubset(P,A) =

4.4 The Relational (1+λ)-EA for Minimum Vertex Covers

After these preparations, we are in a position to present a parallel implementation of $(1+\lambda)$-EA for the minimum vertex cover problem in RelView. The following program, called `PEA`, has four input relations: the adjacency relation $R : K \leftrightarrow K$ of G, a vector $s : K \leftrightarrow \mathbf{1}$ representing an initial vertex cover, a vector $lambda : L \leftrightarrow \mathbf{1}$ with $\lambda = |L|$ as size of the generated populations, and a vector $term$ the cardinality $|term|$ of which defines the maximal number t of non-improving iterations of the loop until the algorithm terminates.

```
PEA(R,s,lambda,term)
  DECL A, P, c, d, p, x, y, z
  BEG  x = s; z = term;
       WHILE -empty(z) DO
         z = z & -point(z);
         A = x * L(lambda)^;
         P = children(A);
         c = IsVertexCover(R,P) & -IsSubset(A,P);
         d = c & IsSubset(P,A);
         IF -empty(d) THEN
           p = point(d); x = P * p; z = term
         ELSE
           IF -empty(c) THEN
             p = point(c);
             y = P * p;
             c = c & -p;
             WHILE cardlt(x,y) & -empty(c) DO
               p = point(c);
               y = P * p;
               c = c & -p OD;
             IF cardlt(y,x) THEN x = y; z = term FI;
             IF cardeq(y,x) THEN x = y FI FI FI OD
       RETURN x
  END.
```

Using x as present solution, each iteration of the main loop of PEA works as follows. First, a new population is produced in form of a relation P using the specifications (1) and (2) and the relational program `children` of Section 4.1. After that, the vectors c and d are computed by using the RelView-functions `IsVertexCover` and `IsSubset` as described in Sections 4.2 and 4.3. If $d \neq \mathsf{O}$, i.e., there is a real subset of the present solution x in P that is a vertex cover, one of these individuals is selected with the aid of a point and the present solution is changed accordingly. As this signifies an improvement, the number of the longest sequence of non-improving iterations of the loop is set to t. The latter is modeled by the assignment $z = term$. If, however, no such individual exists in P, then the individuals of P that are vertex covers and no supersets of x are checked one after another by means of an inner loop. If an individual y is discovered that is smaller than x, then y is taken as new present solution and z is changed to *term* since an improvement has been achieved. An individual y is also taken as new present solution if $|x| = |y|$. Since this case, however, does not mean an improvement, z remains unchanged.

4.5 The Case of Minimum Hyperedge Covers

Now, we consider the hypergraph $H = (V, E)$ of Section 2. That is, we take $k := m$. Furthermore, we assume $R : N \leftrightarrow K$ to be the *incidence relation* of H.

This means that N equals the intervall $[1..n]$ and that for all $i \in N$ and $j \in K$ vertex v_i is contained in hyperedge h_j precisely if R_{ij}.

Based on the relations $P : K \leftrightarrow L$ of (2) and R, we can derive $\overline{\mathsf{L} \cdot \overline{R \cdot P}}^{\mathsf{T}} : L \leftrightarrow \mathbf{1}$ as vector the l-th component of which is true if and only if the l-th column of P is a hyperedge cover of H. A formulation as RelView-function is obvious and its use in `PEA` (instead of `IsVertexCover`) yields a parallel implementation of $(1+\lambda)$-EA for the minimum hyperedge cover problem in RelView.

5 Experiments

We have carried out a lot of experiments with random relations of various sizes and densities on a Sun-Fire 880 workstation running Solaris 9 at 750 MHz. In each test we set the product λt to ek^2, because in the case that the current solution x is not inclusion minimal, the probability that a single child of x improves x is at least $\frac{1}{k}(1-\frac{1}{k})^{k-1} \geq \frac{1}{ek}$. Hence, the expected number of individuals that have to be created to reach an improvement is at most ek. Using Markovs' inequality, the probability that in the run of `PEA` a inclusion minimal solution has not been produced is upper bounded by $1/k$. Figure 1 shows some experimental results for minimum vertex covers, graphs with $k := n = 100$ vertices and approx. 500 edges, and the set of all vertices (represented by $\mathsf{L} : V \leftrightarrow \mathbf{1}$) as initial vertex cover. In it the number of children λ is listed at the x-axis and the arithmetic mean of the execution times of 20 experiments (in seconds) at the y-axis. The curve has been obtained by varying λ from 25 to $\lceil en \rceil$ so that, for example, $\lambda := n/2$ implies $t = 544$ (after rounding).

Apart from the execution times the pictures of all other tests look very similar to the above one. We also have experimented with $\lambda \leq en$ being a variable but

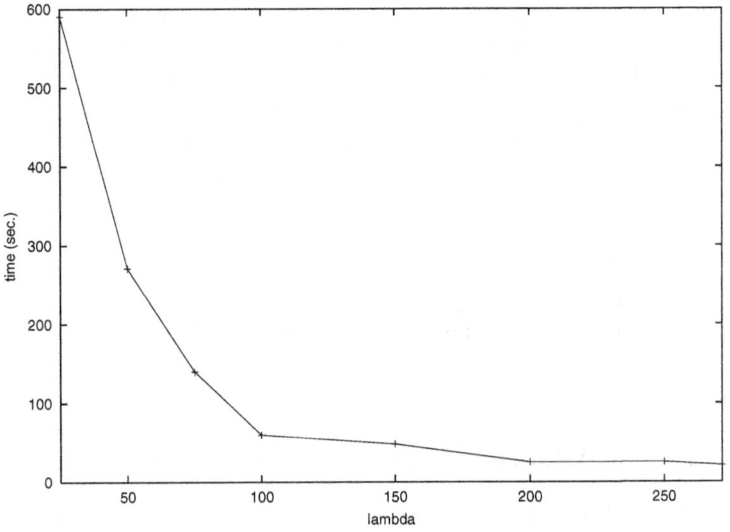

Fig. 1. Runtimes for different offspring population sizes

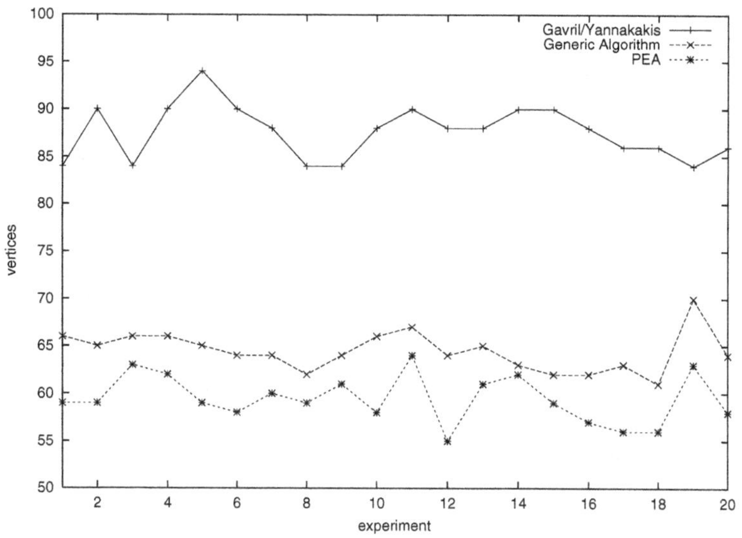

Fig. 2. Quality of solutions

t being a constant, for instance $t = n$. Especially for a small λ this leads to a much better runtime, however also to a worse result.

To get a feeling for the quality of the results, we have compared `PEA` with Gavril/Yannakakis and an instantiation of the program of [2] for vertex covers. `PEA` achieves better results than the other two algorithms in each experiment. Figure 2 shows the results of 20 tests on graphs with 100 vertices and again approx. 500 edges. The size of the computed result is listed at the y-axis. The upper curve belongs to Gavril/Yannakakis, that in the middle to the instantiation of the generic program, and the lower one to our algorithm.

6 Conclusions

We have presented a new approach to implement parallel evolutionary algorithms based on relational algebra and BDDs. The relational version for the vertex cover problem is derived by formal specification of the different modules of the PEA. Each population of children is represented as one relation which is implicitly encoded by one single BDD. Therefore, creation and evaluation of the children is done in parallel using relational operations. Our experiments for the vertex cover problem show that this approach reduces the runtime in comparison to a sequential implementation. We have also shown that the relational PEA can be adapted to other problems (e.g., the minimum set cover problem in hypergraphs) very fast.

Our future work will concentrate on applications of our approach to other problem domains. We also want to investigate parallel evolutionary algorithms based on larger parent populations as well as the implementation of other selection procedures and crossover operators.

References

1. Behnke R. et al.: RelView — A system for calculation with relations and relational programming. In: Astesiano E. (ed.): Proc. 1st Conf. *Fundamental Approaches to Software Engineering*, LNCS 1382, Springer, 318-321 (1998).
2. Berghammer R.: A generic program for minimal subsets with applications. In: Leuschel M., editor, Proc. 12th Int. Workshop *Logic-based Program Development and Transformation*, LNCS 2664, Springer, 144-157 (2003).
3. Berghammer R., Hoffmann T., Leoniuk B., Milanese U.: Prototyping and programming with relations. Electronic Notes in Theoretical Computer Science 44 (2003).
4. Berghammer R., Leoniuk B., Milanese U.: Implementation of relational algebra using binary decision diagrams. In: de Swart H. (ed.): Proc. 6th Int. Workshop *Relational Methods in Computer Science*, LNCS 2561, Springer, 241-257 (2002).
5. Cormen T.T., Leiserson C.E., Rivest R.L.: Introduction to algorithms. The MIT Press (1990).
6. Halperin, E.: Improved approximation algorithms for the vertex cover problem in graphs and hypergraphs, Proc. 11th Ann. ACM-SIAM Symp. on Discrete Algorithms, ACM-SIAM (2000).
7. Hromkovic J.: Algorithms for hard problems. Introduction to combinatorial optimization, randomization, approximation, and heuristics. Springer (2001).
8. Leoniuk B.: ROBDD-based implementation of relational algebra with applications (in German). Ph.D. thesis, Inst. für Inf. und Prak. Math., Univ. Kiel (2001).
9. Milanese U.: On the implementation of a ROBDD-based tool for the manipulation and visualization of relations (in German). Ph.D. thesis, Inst. für Inf. und Prak. Math., Univ. Kiel (2003).
10. Schmidt G., Ströhlein T.: Relations and graphs. Springer (1993).
11. Wegener, I.: Branching programs and binary decision diagrams – theory and applications. SIAM Monographs on Discr. Math. and Appl. (2000).

Lattice-Based Paraconsistent Logic*

Wendy MacCaull[1,**] and Dimiter Vakarelov[2,***]

[1] Department of Mathematics, Statistics and Computer Science, St.Francis Xavier University, Antigonish, Canada
wmaccaul@stfx.ca

[2] Department of Mathematical Logic, Sofia University, Sofia, Bulgaria
dvak@fmi.uni-sofia.bg

Abstract. In this paper we describe a procedure for developing models and associated proof systems for two styles of paraconsistent logic. We first give an Urquhart-style representation of bounded not necessarily discrete lattices using (grill, cogrill) pairs. From this we develop Kripke semantics for a logic permitting 3 truth values: true, false and both true and false. We then enrich the lattice by adding a unary operation of negation that is involutive and antimonotone and show that the representation may be extended to these lattices. This yields Kripke semantics for a nonexplosive 3-valued logic with negation.

Keywords: paraconsistent logic, lattice representation, Kripke semantics, negation, graded information, multi-valued logic.

1 Introduction

If $\models$ is the relation of logical consequence, defined either semantically or proof theoretically, $\models$ is said to be explosive iff for any formulas F and G, $\{F, \neg F\} \models G$. Standard logics are explosive. A logic is said to be *paraconsistent* if its relation of logical consequence is not explosive. Paraconsistent logics are motivated not only by philosophical considerations, stemming from the reality of inconsistent but non-trivial theories (for example, in physics) and the existence of "true contradictions", but also by numerous applications in automated reasoning, artificial intelligence and mathematics. A number of candidates for paraconsistent logic have been suggested in the literature: see, for example, Arieli and Avron

* This work was performed within the framework of the COST Action 274, entitled: "Theory and Applications of Relational Structures as Knowledge Instruments" (www.tarski.org). Both authors were supported by a NATO Science and Technology Collaborative Linkages Grant.

** The author was supported by a grant from Science and Engineering Research Canada (NSERC).

*** The author was partially supported by the Bulgarian Ministry of Science and Education, project No 123.

W. MacCaull et al. (Eds.): RelMiCS 2005, LNCS 3929, pp. 173–187, 2006.

[3], Belnap [4], da-Costa [5], Fitting [6] and Ginsberg [7]. As computer and information systems become more highly distributed, the problem of finding non-explosive frameworks to deal with inconsistent information becomes more and more critical.

The above notion of paraconsistency depends on the operation of negation. A more liberal view of paraconsistency is simply to allow the semantics of the logic to permit sentences that are true or false or both true and false. Examples of such logics may be found in [16]. All such examples, however are distributive: that is, they have algebraic semantics based on distributive lattices. In this paper we show that from bounded, not necessarily distributive lattices, we can develop Kripke semantics for a logic called LAT with the property that truth at a world may be true or false or both. We then introduce *bounded lattices with negation* enriching the lattice structure by adding a unary operator, which is involutive and antimonotone and show that this gives us Kripke semantics for a 3-valued logic with negation with a nonexplosive consequence relation, called NLAT.

We begin the paper by reviewing the Urquhart Representation Theorem for bounded, not necessarily distributive lattices [13]. This representation permits the development of Kripke semantics for a logic LAT introduced in [11]. Using the ideas from Allwein and Dunn [1] and Allwein and MacCaull [2], we describe a 3-valued Kripke semantics for the logic LAT arising from the Urquhart representation, with the truth values – *true, false* and *neither true nor false.* Next, we prove an analog of (which is, in a sense, dual to) the Urquhart Representation Theorem for bounded lattices. This new representation theorem yields a new Kripke semantics for LAT with truth values *true, false* and *both true and false,* which is paraconsistent. These results were announced in [10]. Then, we extend the representation to bounded lattices with negation, which yields Kripke semantics for a nonexplosive logic with negation, called NLAT. We end by discussing some possible extensions of NLAT.

2 Urquhart-Style Representation

In this section we review the Urquhart representation of lattices; our interest is in the algebra and the logic so we do not assume any topological structure. We denote a bounded lattice by the 5-tuple $\mathcal{W} = (W, \vee, \wedge, 1, 0)$ (or, more simply, by $\mathcal{W}$) where the lattice partial order is $a \leq b$ iff $a \vee b = b$.

Definition 2.1. *Let $\mathcal{W} = (W, \vee, \wedge, 1, 0)$ be a bounded lattice and let $\mathcal{A} \subseteq W$; then:*

(1) $\mathcal{A}$ is a filter iff: (i) $1 \in \mathcal{A}$; (ii) $(x \in \mathcal{A} \wedge x \leq y) \Rightarrow y \in \mathcal{A}$; and (iii) $x, y \in \mathcal{A} \Rightarrow x \wedge y \in \mathcal{A}$.
(2) $\mathcal{A}$ is an ideal iff: (i) $0 \in \mathcal{A}$; (ii) $(y \in \mathcal{A} \wedge x \leq y) \Rightarrow x \in \mathcal{A}$; and (iii) $x, y \in \mathcal{A} \Rightarrow x \vee y \in \mathcal{A}$.

Lemma 2.1. *Let $\mathcal{W} = (W, \vee, \wedge, 1, 0)$ be a bounded lattice and let $a \in W$; then:*
(a) $\uparrow a = \{b | a \leq b\}$ is a filter, the smallest filter in $\mathcal{W}$ containing a.
(b) $\downarrow a = \{b | b \leq a\}$ is an ideal, the smallest ideal in $\mathcal{W}$ containing a.

Let $\mathcal{W} = (W, \vee, \wedge, 1, 0)$ be a bounded lattice. A *filter-ideal pair in* $\mathcal{W}$ is a two-tuple (x_1, x_2) such that $x_1, x_2 \subseteq W$, x_1 is a filter, x_2 is an ideal and $x_1 \cap x_2 = \emptyset$. We denote such pairs by x, y, z, etc.; the filter part of the pair x is x_1, while the ideal part of x is x_2. Define an ordering $\leq$ on the family of filter-ideal pairs of $\mathcal{W}$ as follows:

$$x \leq y \text{ iff } x_1 \subseteq y_1 \text{ and } x_2 \subseteq y_2.$$

A filter-ideal pair is said to be *maximal* whenever it is maximal with respect to the partial order $\leq$. It is easy to show that every increasing chain of filter-ideal pairs starting with x has an upper bound, so by Zorn's Lemma, the collection $\{y : x \leq y\}$ has a maximal element.

Let $X^M(W)$ be the family of maximal filter-ideal pairs of the bounded lattice $\mathcal{W}$. Define relations $\leq_1, \leq_2$ on $X^M(W)$ as follows:

$$\text{for all } x, y \in X^M(W),\ x \leq_1 y \text{ iff } x_1 \subseteq y_1 \text{ and } x \leq_2 y \text{ iff } x_2 \subseteq y_2.$$

The system $(X^M(W), \leq_1, \leq_2)$ is referred to as the *canonical* ^{M}*frame* of the bounded lattice $\mathcal{W}$.

Definition 2.2. *A doubly ordered set is a system $(X, \leq_1, \leq_2)$ where X is a nonempty set and $\leq_1, \leq_2$ are partial orders on X such that for all $x, y \in X$, if $x \leq_1 y$ and $x \leq_2 y$ then $x = y$.*

Lemma 2.2. *For every bounded lattice $(W, \vee, \wedge, 1, 0)$, its canonical ^{M}frame $(X^M(W), \leq_1, \leq_2)$ is a doubly ordered set.*

Given a doubly ordered set $(X, \leq_1, \leq_2)$, maps $l : 2^X \to 2^X$ and $r : 2^X \to 2^X$ are defined as follows: for $A \subseteq X$,

$$l(A) = \{x \in X : \forall y \in X\ x \leq_1 y \Rightarrow y \notin A\};$$
$$r(A) = \{x \in X : \forall y \in X\ x \leq_2 y \Rightarrow y \notin A\}.$$

A set $A \subseteq X$ is *l-stable* (respectively, *r-stable*) whenever $lr(A) = A$ (respectively, $rl(A) = A$); A is $\leq_i$*-increasing* ($i = 1, 2$) whenever for all $x, y \in X$, if $x \in A$ and $x \leq_i y$ then $y \in A$.

Having in mind the Kripke semantics of intuitionistic negation by means of some ordering allows us to treat $l(A)$ and $r(A)$ as two intuitionistic negations.

Given a doubly ordered set $(X, \leq_1, \leq_2)$, let $L^M(X)$ denote the family of l-stable subsets of X. Define the following operations and constants on $L^M(X)$: $A \wedge B = A \cap B$; $A \vee B = l(r(A) \cap r(B))$, $0 = \emptyset$, $1 = X$. It then follows that:

Lemma 2.3. *For every doubly ordered set $(X, \leq_1, \leq_2)$ and for all $A, B \in L^M(X)$, $A \wedge B$, $A \vee B$, 0 and 1 are l-stable sets. Further, the tuple $(L^M(X), \vee, \wedge, 1, 0)$ is a bounded lattice.*

We refer to the lattice $(L^M(X), \vee, \wedge, 1, 0)$ as the *complex* ^{M}*algebra* of the doubly ordered set $(X, \leq_1, \leq_2)$. It is not necessarily distributive.

Let $\mathcal{W}$ be a bounded lattice. Define a map $h : W \to 2^{X^M(W)}$ by $h(a) = \{x \in X^M(W) : a \in x_1\}$.

Theorem 2.1. *Urquhart [13]. For every bounded lattice $\mathcal{W}$ and for every $a \in W$:*
(a) $h(a)$ is an l-stable set.
(b) h is a lattice embedding of $\mathcal{W}$ into $L^M(X^M(W))$.

A consequence of the above theorem is the Urquhart representation theorem.

Theorem 2.2. *(Representation theorem) Every bounded lattice is isomorphic to a subalgebra of the complex ^{M}algebra of its canonical ^{M}frame.*

Let LAT denote a propositional logic whose algebraic semantics is the class of lattices. The formulas of LAT are built from a countably infinite set $\mathcal{P}$ of propositional variables and the constants **T** (true) and **F** (false) using the propositional connectives $\vee$ (disjunction) and $\wedge$ (conjunction). Sequents are expressions in the form $a \vdash b$ where a, b are formulas. Let $\mathcal{W} = (W, 0, 1, \vee, \wedge)$ be a lattice. Any mapping v from the set of propositional variables to W is called a valuation. The pair $\mathcal{M} = (\mathcal{W}, v)$ is called a model. The valuation v is extended to all formulas inductively in a standard way: $v(\mathbf{T}) = 1$, $v(\mathbf{F}) = 0$, $v(a \vee b) = v(a) \vee v(b)$, $v(a \wedge b) = v(a) \wedge v(b)$. A sequent $a \vdash b$ is true in a model $\mathcal{M} = (\mathcal{W}, v)$ if $v(a) \leq v(b)$. A sequent $a \vdash b$ is true in the lattice $\mathcal{W}$ if it is true in all models over $\mathcal{W}$.

Following the presentation in [1] and [2], we present a 3-valued Kripke semantics arising from the Urquhart representation, where the truth values are: **t** – *true*, **f** – *false* and **n** – *neither true nor false.*

Definition 2.3. *A ^{M}Kripke model for LAT is a system $\mathcal{M} = (X, \leq_1, \leq_2, m)$, where $(X, \leq_1, \leq_2)$ is a doubly ordered set and m is a valuation which assigns to each proposition variable p an l-stable set $m(p) \subseteq X$. Then by a parallel induction on the construction of formulas we define three satisfaction relations $\models_i$, $i \in \{t, f, n\}$ between elements of X and formulas with the following intuitive meaning: $x \models_{\mathbf{t}} a$: "a is true at x", $x \models_{\mathbf{f}} a$: "a is false at x", $x \models_{\mathbf{n}} a$: "a is neither true nor false at x".*

- $x \models_{\mathbf{t}} p$ *iff* $x \in m(p)$; $x \models_{\mathbf{f}} p$ *iff* $x \in rm(p)$.
- $x \models_{\mathbf{t}} \mathbf{T}$, $x \not\models_{\mathbf{f}} \mathbf{T}$.
- $x \not\models_{\mathbf{t}} \mathbf{F}$, $x \models_{\mathbf{f}} \mathbf{F}$.
- $x \models_{\mathbf{t}} a \wedge b$ *iff* $x \models_{\mathbf{t}} a$ *and* $x \models_{\mathbf{t}} b$.
- $x \models_{\mathbf{f}} a \wedge b$ *iff* $\forall y($ *if* $x \leq_2 y$ *then* $y \not\models_{\mathbf{t}} a$ *or* $y \not\models_{\mathbf{t}} b)$.
- $x \models_{\mathbf{t}} a \vee b$ *iff* $\forall y($ *if* $x \leq_1 y$ *then* $(y \not\models_{\mathbf{f}} a$ *or* $y \not\models_{\mathbf{f}} b))$.
- $x \models_{\mathbf{f}} a \vee b$ *iff* $\forall y($ *if* $x \leq_2 y$ *then* $\exists z(y \leq_1 z$ *and* $z \models_{\mathbf{f}} a$ *and* $z \models_{\mathbf{f}} b))$.

For all formulas a we put
- $x \models_{\mathbf{n}} a$ *iff* $x \not\models_{\mathbf{t}} a$ *and* $x \not\models_{\mathbf{f}} a$.

A sequent $a \vdash b$ is true in the model $\mathcal{M}$ if $\forall x \in X(x \models_{\mathbf{t}} a$ implies $x \models_{\mathbf{t}} b)$.

The above definition ensures that if we let $m_{\mathbf{t}}(a) = \{x : x \models_{\mathbf{t}} a\}$ and $m_{\mathbf{f}}(a) = \{x : x \models_{\mathbf{f}} a\}$, then for all formulas a, $m_{\mathbf{t}}(a)$ is l-stable, $m_{\mathbf{t}}(a) = lm_{\mathbf{f}}(a)$, $m_{\mathbf{f}}(a) = rm_{\mathbf{t}}(a)$ and $m_{\mathbf{t}}(a) \cap m_{\mathbf{f}}(a) = \emptyset$. l-stable (r-stable) sets are $\leq_1$ (respectively, $\leq_2$)-increasing; hence:

Proposition 2.1. *For all formulas a:*

(a) if $x \models_{\mathbf{t}} a$ and $x \leq_1 x'$ then $x' \models_{\mathbf{t}} a$ and if $x \models_{\mathbf{f}} a$ and $x \leq_2 x'$ then $x' \models_{\mathbf{f}} a$.
(b) if $x \models_{\mathbf{t}} a$ then $x \not\models_{\mathbf{f}} a$ and if $x \models_{\mathbf{f}} a$ then $x \not\models_{\mathbf{t}} a$.

Note that from (b) we may conclude that in the above semantics it is not possible for one formula a to be both true and false at a state x, so this semantics is not paraconsistent. In the next section we will develop another representation theorem for lattices which yields a paraconsistent semantics for LAT.

Proposition 2.2. *The following conditions are true for all sequents $a \vdash b$ in LAT:*

(i) $a \vdash b$ is true in all lattices.
(ii) $a \vdash b$ is true in all lattices of the form $L^M(X)$ over doubly ordered sets.
(iii) $a \vdash b$ is true in all ^{M}Kripke models for LAT.

Proof. The equivalence of (i) and (ii) follows from the Urquhart Representation Theorem. The equivalence of (ii) and (iii) is obvious from the definition of validity of sequents in Kripke models. ■

3 Lattice-Based Models for Paraconsistent Logic

We develop a representation theorem for arbitrary not necessarily distributive lattices which will give a model where information is graded by two partial orders, bounded and inconsistent. The treatment using grills and cogrills was motivated by Vakarelov [15].

Definition 3.1. *Let $\mathcal{W} = (W, \vee, \wedge, 1, 0)$ be a bounded lattice and let $\mathcal{A} \subseteq W$; then:*

(1) $\mathcal{A}$ is a grill iff: (i) $0 \notin \mathcal{A}$; (ii) $(x \in \mathcal{A} \wedge x \leq y) \Rightarrow y \in \mathcal{A}$; (iii) $x \vee y \in \mathcal{A} \Rightarrow x \in \mathcal{A}$ or $y \in \mathcal{A}$.
(2) $\mathcal{A}$ is a cogrill iff: (i) $1 \notin \mathcal{A}$; (ii) $(y \in \mathcal{A} \wedge x \leq y) \Rightarrow x \in \mathcal{A}$; (iii) $x \wedge y \in \mathcal{A} \Rightarrow x \in \mathcal{A}$ or $y \in \mathcal{A}$.

Lemma 3.1. *Let $\mathcal{W} = (W, \vee, \wedge, 1, 0)$ be a bounded lattice, let $a \in W$ and let $\mathcal{A}_1, \mathcal{A}_2 \subseteq W$; then:*
(a) $\not\downarrow a = \{b | b \not\leq a\}$ is a grill, the largest grill not containing a.
(b) $\not\uparrow a = \{b | a \not\leq b\}$ is a cogrill, the largest cogrill not containing a.
(c) If $\mathcal{A}_1$ and $\mathcal{A}_2$ are grills, then $\mathcal{A}_1 \cup \mathcal{A}_2$ is a grill.
(d) If $\mathcal{A}_1$ and $\mathcal{A}_2$ are cogrills, then $\mathcal{A}_1 \cup \mathcal{A}_2$ is a cogrill.

Proof. For (a) observe: (i) $0 \leq a$ so $0 \notin \not\downarrow a$. (ii) Let $x \in \not\downarrow a$ and let $y \in W$ such that $x \leq y$. Assume $y \notin \not\downarrow a$. Then $y \leq a$; therefore $x \leq a$. But $x \in \not\downarrow a$ tells us $x \not\leq a$, a contradiction; we conclude that $y \in \not\downarrow a$. (iii) Let $x \vee y \in \not\downarrow a$. Then $x \vee y \not\leq a$. Assume $x \notin \not\downarrow a$ and $y \notin \not\downarrow a$. Thus $x \leq a$ and $y \leq a$; therefore $x \vee y \leq a$, which is a contradiction. We conclude $x \in \not\downarrow a$ or $y \in \not\downarrow a$. It is easy to show that $\not\downarrow a$ is the largest grill not containing a.

The proof of (b) is similar to that of (a) and the proofs of (c) and (d) are routine. ■

Let $\mathcal{W} = (W, \vee, \wedge, 1, 0)$ be a bounded lattice; a *grill-cogrill pair* in $\mathcal{W}$ is a two tuple (w_1, w_2), such that $w_1, w_2 \subseteq W$, w_1 is a grill, w_2 is a cogrill and $w_1 \cup w_2 = W$. For the remainder of this paper, the word *pair* denotes a grill-cogrill pair. This "exhaustive feature" (i.e., $w_1 \cup w_2 = W$) gives rise to 3 truth values: true, false and both true and false. Pairs in $\mathcal{W}$ are denoted by w, x, y, z, etc.; unless otherwise stated, the grill part of the pair x is x_1, while the cogrill part of x is x_2.

Definition 3.2. *Let $\mathcal{W}$ be a bounded lattice and let x, y be pairs in $\mathcal{W}$. Write $x \leq_1 y$ iff $x_1 \subseteq y_1$ and $x \leq_2 y$ iff $x_2 \subseteq y_2$.*

Definition 3.3. *Let x, y be pairs in a bounded lattice $\mathcal{W}$. Write $x \sqsubseteq y$ iff $y \leq_1 x$ and $y \leq_2 x$. The pair y is said to be a minimal pair iff it is maximal with respect to the $\sqsubseteq$-ordering.*

Lemma 3.2. *For every pair x in a bounded lattice $\mathcal{W}$ there exists a minimal pair y such that $y \leq_1 x$ and $y \leq_2 x$.*

Proof. Given a chain of pairs $x \sqsubseteq y^1 \sqsubseteq y^2 \sqsubseteq \ldots$, define $y = (\bigcap y_1^i, \bigcap y_2^i)$. It is straightforward to show that y is a pair, and an upper bound of this chain with respect to the $\sqsubseteq$-ordering. Then the result is a simple consequence of Zorn's Lemma. ■

For any bounded lattice $\mathcal{W}$, let $X_m(W)$ denote the family of minimal pairs of $\mathcal{W}$. It has the two orders $\leq_1$ and $\leq_2$. The system $(X_m(W), \leq_1, \leq_2)$ is referred to as the *canonical* $_m$*frame* of the bounded lattice $\mathcal{W}$.

Lemma 3.3. *For any bounded lattice $\mathcal{W}$, $(X_m(W), \leq_1, \leq_2)$ is a doubly ordered set.*

Proof. Let $x, y \in X_m(W)$ and suppose $x \leq_1 y$ and $x \leq_2 y$. Then $y \sqsubseteq x$, which implies that $y = x$ since both x and y are maximal with respect to the $\sqsubseteq$-ordering. ■

Now we develop another representation theorem for lattices, dual in a sense to the Urquhart representation. In the Urquhart theory, the operations $l(A)$ and $r(A)$, which have the properties of the intuitionistic negation, play a crucial role in the construction of a lattice from doubly ordered sets. In our construction, we consider the dual operations $\ulcorner_1 A$ and $\ulcorner_2 A$, introduced by Rauszer [12] under the name *dual intuitionistic negation.* The next definition is based on the Kripke semantics of this negation (see [14]).

Definition 3.4. *Let $(X, \leq_1, \leq_2)$ be a doubly ordered set and let $A \subseteq X$. Then:*
(1) $\ulcorner_1 A = \{x \in X : \exists y \in X(y \leq_1 x \wedge y \notin A)\}$.
(2) $\ulcorner_2 A = \{x \in X : \exists y \in X(y \leq_2 x \wedge y \notin A)\}$.

Lemma 3.4. *Let $(X, \leq_1, \leq_2)$ be a doubly ordered set, let $A, B \subseteq X$ and let $i \in \{1, 2\}$. Then:*

(a) $\ulcorner_i A$ is $\leq_i$-increasing.
(b) $A \cup \ulcorner_i A = X$.
(c) If A is $\leq_i$-increasing, then $\ulcorner_i \ulcorner_i A \subseteq A$.
(d) If $A \subseteq B$ then $\ulcorner_i B \subseteq \ulcorner_i A$.
(e) $\ulcorner_i X = \varnothing$, $\ulcorner_i \varnothing = X$.
(f) If A is $\leq_1$-increasing, then $\ulcorner_1 \ulcorner_2 A \subseteq A$.
(g) If A is $\leq_2$-increasing, then $\ulcorner_2 \ulcorner_1 A \subseteq A$.
(h) $\ulcorner_i(A \cap B) = \ulcorner_i A \cup \ulcorner_i B$.
(i) If A and B are $\leq_i$-increasing, then both $A \cap B$ and $A \cup B$ are $\leq_i$-increasing.

Proof. (a) Suppose $x \in \ulcorner_i A$ and let $x \leq_i x'$; by assumption there exists y such that $y \leq_i x$ and $y \notin A$. Thus $y \leq_i x \leq_i x'$ and hence $x' \in \ulcorner_i A$.

For (b), suppose $x \notin A$ and $x \notin \ulcorner_i A$. From $x \leq_i x$ and $x \notin A$ we obtain $x \in \ulcorner_i A$ – a contradiction.

For (c), let A be $\leq_i$-increasing, and suppose $x \in \ulcorner_i \ulcorner_i A$. Thus $\exists y(y \leq_i x$ and $y \notin \ulcorner_i A)$. (b) implies $y \in A$; since A is $\leq_i$-increasing, we conclude $x \in A$.

For (d) Suppose $A \subseteq B$ and let $x \in \ulcorner_i B$. Then $\exists y(y \leq_i x$ and $y \notin B)$. Consequently $y \notin A$ so $x \in \ulcorner_i A$.

(e) is straightforward.

For (f), suppose A is $\leq_1$-increasing and let $x \in \ulcorner_1 \ulcorner_2 A$. Then $\exists y(y \leq_1 x$ and $\forall z(z \leq_2 y$ implies $z \in A))$. $y \leq_2 y$ so $y \in A$; A is $\leq_1$-increasing so $x \in A$.

The proof of (g) is as (f). (h) and (i) follow easily from the definitions. ■

Remark. We see that the mapping $\ulcorner_1$ from the lattice of $\leq_2$-increasing sets to the lattice of $\leq_1$-increasing sets, and the mapping $\ulcorner_2$ from the lattice of $\leq_1$-increasing sets to the lattice of $\leq_2$-increasing sets, form a dual Galois connection (as introduced by Dunn [6]). The dual Galois properties allow us to show that $\ulcorner_2 \ulcorner_1$ and $\ulcorner_1 \ulcorner_2$ are closure operators.

Definition 3.5. *Let $(X, \leq_1, \leq_2)$ be a doubly ordered set, and let $A \subseteq X$. A is said to be $\ulcorner_1$-stable (respectively, $\ulcorner_2$-stable) iff $\ulcorner_1 \ulcorner_2 A = A$ (respectively, $\ulcorner_2 \ulcorner_1 A = A$).*

Lemma 3.5. *Let $(X, \leq_1, \leq_2)$ be a doubly ordered set, and let $A, B \subseteq X$. Then:*

(a) If A is a $\leq_2$-increasing set then $\ulcorner_1 A$ is a $\ulcorner_1$-stable set; if A is a $\leq_1$-increasing set then $\ulcorner_2 A$ is a $\ulcorner_2$-stable set.
(b) $\ulcorner_1 \ulcorner_2 A$ is $\ulcorner_1$-stable; $\ulcorner_2 \ulcorner_1 A$ is $\ulcorner_2$-stable.
(c) If A is $\ulcorner_1$-stable then $\ulcorner_2 A$ is $\ulcorner_2$-stable; if A is $\ulcorner_2$-stable then $\ulcorner_1 A$ is $\ulcorner_1$-stable.
(d) If A and B are $\ulcorner_1$-stable, then both $A \cup B$ and $\ulcorner_1(\ulcorner_2 A \cup \ulcorner_2 B)$ are $\ulcorner_1$-stable.
(e) Both $\emptyset$ and X are $\ulcorner_1$-stable.

Proof. (a) Suppose A is $\leq_2$-increasing; we show (i) $\ulcorner_1 A \subseteq \ulcorner_1 \ulcorner_2 \ulcorner_1 A$ and (ii) $\ulcorner_1 \ulcorner_2 \ulcorner_1 A \subseteq \ulcorner_1 A$. (i) follows directly from Lemma 3.4 (g) and (d) while (ii) follows from Lemma 3.4 (a) and (f). The proof of the other part of (a) is analogous.

(b) By Lemma 3.4(a), $\ulcorner_2 A$ is $\leq_2$-increasing. So by (a), $\ulcorner_1\ulcorner_2 A$ is $\ulcorner_1$-stable. The proof of the other part of (b) is similar.

(c) If A is $\ulcorner_1$-stable then $\ulcorner_1\ulcorner_2 A = A$; consequently $\ulcorner_2\ulcorner_1\ulcorner_2 A = \ulcorner_2 A$ which gives us the first result. The second result follows similarly.

(d) Suppose A and B are $\ulcorner_1$-stable. First we show that $\ulcorner_1\ulcorner_2(A \cup B) = A \cup B$. $A \subseteq A \cup B$ so (by Lemma 3.4(d)) $\ulcorner_2(A \cup B) \subseteq \ulcorner_2 A$; hence $A = \ulcorner_1\ulcorner_2 A \subseteq \ulcorner_1\ulcorner_2(A \cup B)$; similarly, $B \subseteq \ulcorner_1\ulcorner_2(A \cup B)$; thus (i) $A \cup B \subseteq \ulcorner_1\ulcorner_2(A \cup B)$. Now let $x \in \ulcorner_1\ulcorner_2(A \cup B)$. Then $\exists y(y \leq_1 x$ and $\forall z($if $z \leq_2 y$ then $z \in A \cup B)$. Hence $\exists y(y \leq_1 x$ and $y \in A \cup B)$; consequently, $\exists y(y \leq_1 x$ and $y \in A$ or $y \in B)$. $A = \ulcorner_1\ulcorner_2 A$ and $B = \ulcorner_1\ulcorner_2 B$ so by Lemma 3.4 (a) both A and B are $\leq_1$-increasing. Hence $x \in A$ or $x \in B$, i.e., $x \in A \cup B$; we conclude (ii) $\ulcorner_1\ulcorner_2(A \cup B) \subseteq A \cup B$. (i) and (ii) give us the first result. We now observe that $\ulcorner_1\ulcorner_2\ulcorner_1(\ulcorner_2 A \cup \ulcorner_2 B) =$ (by Lemma 3.4(h)) $\ulcorner_1\ulcorner_2\ulcorner_1\ulcorner_2(A \cap B) =$ (by (b)) $\ulcorner_1\ulcorner_2(A \cap B) =$ (by Lemma 3.4(h)) $\ulcorner_1(\ulcorner_2 A \cup \ulcorner_2 B))$.

(e) $x \in \ulcorner_1\ulcorner_2 \emptyset$ iff $\exists y(y \leq_1 x$ and $y \notin \ulcorner_2 \emptyset)$ iff $\exists y(y \leq_1 x$ and $\forall z($ if $z \leq_2 y$ then $z \in \emptyset))$. No such y exists with these properties, so $\ulcorner_1\ulcorner_2 \emptyset = \emptyset$. It remains to show that $X \subseteq \ulcorner_1\ulcorner_2 X$. Let $x \in X$ and suppose $x \notin \ulcorner_1\ulcorner_2 X$. Then $\forall y($if $y \leq_1 x$ then $y \in \ulcorner_2 X)$. Hence $\forall y($if $y \leq_1 x$ then $\exists z(z \leq_2 y$ and $z \notin X))$. This implies $\exists z(z \leq_2 x$ and $z \notin X)$, a contradiction; we conclude that $x \in \ulcorner_1\ulcorner_2 X$. ■

Proposition 3.1. *Let $(X, \leq_1, \leq_2)$ be a doubly ordered set, and let $L_m(X) = \{A \subseteq X : \ulcorner_1\ulcorner_2 A = A\}$, the $\ulcorner_1$-stable subsets of X. For $A, B \in L_m(X)$ let $A \vee B = A \cup B$, $A \wedge B = \ulcorner_1(\ulcorner_2 A \cup \ulcorner_2 B)$, $0 = \emptyset$ and $1 = X$. Then $(L_m(X), \vee, \wedge, 0, 1)$ is a bounded lattice, called the complex $_m$algebra of the doubly ordered set $(X, \leq_1, \leq_2)$.*

Proof. By Lemma 3.5(e) $L_m(X)$, the set of $\ulcorner_1$-stable sets, has a top element and a bottom element and by Lemma 3.5(d) it is closed under the operations $\vee$ and $\wedge$. $\vee$ is clearly a join; it remains to show that $\wedge$ is a meet. Let $A, B \in L_m(X)$; $\ulcorner_2 A \subseteq \ulcorner_2 A \cup \ulcorner_2 B$ so $\ulcorner_1(\ulcorner_2 A \cup \ulcorner_2 B) \subseteq \ulcorner_1\ulcorner_2 A = A$; the same holds for B. Suppose $C \in L_m(X)$ and $C \subseteq A$, $C \subseteq B$. Then $C \subseteq A \cap B$ so $C = \ulcorner_1\ulcorner_2 C \subseteq \ulcorner_1\ulcorner_2(A \cap B) =$ (by Lemma 3.4(h)) $\ulcorner_1(\ulcorner_2 A \cup \ulcorner_2 B)$. ■

Definition 3.6. *Let $\mathcal{W}$ be a bounded lattice. Define the map $h : W \to 2^{X_m(W)}$ by $h(a) = \{x \in X_m(W) : a \in x_1\}$.*

Theorem 3.1. *For every bounded lattice $\mathcal{W} = (W, \vee, \wedge, 1, 0)$ and for every $a \in W$:*

(a) Let $x = (x_1, x_2)$ be a pair:

(i) if $a \in x_1$ then there is a minimal pair y such that $y_1 \subseteq x_1$ and $a \notin y_2$;

(ii) if $a \in x_2$ then there is a minimal pair y such that $y_2 \subseteq x_2$ and $a \notin y_1$.

(b) $\ulcorner_2 h(a) = \{x \in X_m(W) : a \in x_2\}$.

(c) $h(a)$ is an $\ulcorner_1$-stable set.

(d) h is a lattice embedding of $\mathcal{W}$ into $L_m(X_m(W))$.

Proof. (a) (i) Suppose $a \in x_1$. Consider the cogrill $\not\triangledown a$ and assume there is $b \in W$ such that $b \notin \not\triangledown a$. This implies that $a \leq b$; $a \in x_1$ and x_1 is a grill so $b \in x_1$. Hence $x_1 \cup \not\triangledown a = W$; we conclude that $(x_1, \not\triangledown a)$ is a pair. By Lemma 3.2

there is a minimal pair y, such that $y_1 \subseteq x_1$ and $y_2 \subseteq \not\uparrow a$. Since $a \notin \not\uparrow a$ we conclude that $a \notin y_2$. (ii) Suppose $a \in x_2$. Consider the grill $\not\downarrow a$, and assume there is $b \in W$ such that $b \notin \not\downarrow a$. This tells us that $b \leq a$; we know $a \in x_2$, and since x_2 is a cogrill, $b \in x_2$. Hence $\not\downarrow a \cup x_2 = W$ and we conclude that $(\not\downarrow a, x_2)$ is a pair. By Lemma 3.2 we know there is a minimal pair y such that $y_1 \subseteq \not\downarrow a$ and $y_2 \subseteq x_2$. Since $y_1 \subseteq \not\downarrow a$, and $a \notin \not\downarrow a$, we conclude $a \notin y_1$.

(b) ($\subseteq$) $x \in \ulcorner_2 h(a)$ iff $\exists y \in X_m(W)(y \leq_2 x$ and $y \notin h(a))$ iff $\exists y \in X_m(W)(y_2 \subseteq x_2$ and $a \notin y_1)$. Since $y_1 \cup y_2 = W$, $a \in y_2$. Thus $a \in x_2$.

($\supseteq$) Let $x \in X_m(W)$, and $a \in x_2$. We need to find $y \in X_m(W)$ such that $y \leq_2 x$ and $y \notin h(a)$, i.e., such that $y_2 \subseteq x_2$ and $a \notin y_1$. This follows directly from (a)(ii).

(c) ($\subseteq$) $x \in \ulcorner_1 \ulcorner_2 h(a)$ iff $\exists y \in X_m(W)(y \leq_1 x$ and $y \notin \ulcorner_2 h(a))$ iff $\exists y \in X_m(W)(y \leq_1 x$ and $a \notin y_2)$. Since $y_1 \cup y_2 = W$, $a \in y_1$ which implies $a \in x_1$; hence $x \in h(a)$.

($\supseteq$) Let $x \in X_m(W)$ where $x \in h(a)$. Then $a \in x_1$ and we need to find $y \in X_m(W)$ such that $y \leq_1 x$ and $y \notin \ulcorner_2 h(a)$; that is, we need that $y_1 \subseteq x_1$ and $a \notin y_2$. This follows from (a)(i).

(d) First we show that h is injective. Assume $h(a) = h(b)$ and suppose $a \neq b$. Then either $a \not\leq b$ or $b \not\leq a$. Assume that $a \not\leq b$, and suppose there is $c \in W$ such that $c \notin \not\downarrow b \cup \not\uparrow a$. This tells us that $a \leq b$, a contradiction. We may conclude that $(\not\downarrow b, \not\uparrow a)$ is a pair. By Lemma 3.2 there is a minimal pair x such that $x_1 \subseteq \not\downarrow b$ and $x_2 \subseteq \not\uparrow a$. We will now show that (i) $x \in h(a)$ and (ii) $x \notin h(b)$, which will contradict the fact that $h(a) = h(b)$. For (i), $a \notin \not\uparrow a$, so $a \notin x_2$; since $x_1 \cup x_2 = W$, we conclude $a \in x_1$. Hence, $x \in h(a)$. For (ii), $x_1 \subseteq \not\downarrow b$ and $b \notin \not\downarrow b$; therefore $b \notin x_1$. Hence $x \notin h(b)$. The case $b \not\leq a$ leads into a contradiction in the same way. Thus $a = b$.

We show that $h(a \vee b) = h(a) \vee h(b) =$ (by definition) $h(a) \cup h(b)$. Let x_1 be a grill and $a \in x_1$; then $a \leq a \vee b$ implies $a \vee b \in x_1$. Similarly, $b \in x_1$ implies $a \vee b \in x_1$. Thus $a \vee b \in x_1$ iff $a \in x_1$ or $b \in x_1$. Hence: $x \in h(a \vee b)$ iff $a \vee b \in x_1$ iff $a \in x_1$ or $b \in x_1$ iff $x \in h(a)$ or $x \in h(b)$ iff $x \in h(a) \cup h(b)$ iff $x \in h(a) \vee h(b)$.

We show that $h(a \wedge b) = h(a) \wedge h(b) =$ (by definition) $\ulcorner_1(\ulcorner_2 h(a) \cup \ulcorner_2 h(b))$. For ($\subseteq$), let $x \in h(a \wedge b)$ i.e., $a \wedge b \in x_1$. We must find $y \in X_m(W)$ such that $y \leq_1 x$ and $y \notin (\ulcorner_2 h(a) \cup \ulcorner_2 h(b))$; i.e., such that $y \leq_1 x$ and (by (b)) $a \notin y_2$ and $b \notin y_2$; i.e., such that $y \leq_1 x$ and (since y_2 is a cogrill) $a \wedge b \notin y_2$. (a)(i) assures us such a minimal pair y exists. For ($\supseteq$), let $x \in \ulcorner_1(\ulcorner_2 h(a) \cup \ulcorner_2 h(b))$. Then $\exists y \in X_m(W)(y \leq_1 x$ and $y \notin (\ulcorner_2 h(a) \cup \ulcorner_2 h(b)))$. This implies that $y \notin \ulcorner_2 h(a)$ and $y \notin \ulcorner_2 h(b)$, so $a \notin y_2$ and $b \notin y_2$. y_2 is a grill so $a \wedge b \notin y_2$; hence $a \wedge b \in y_1$. Finally $y_1 \subseteq x_1$ so $a \wedge b \in x_1$ and therefore $x \in h(a \wedge b)$.

$h(0) = \emptyset$, since no grill contains 0. To prove that $h(1) = X_m(W)$ it is sufficient to show that $X_m(W) \subseteq h(1)$. Let $x \in X_m(W)$. Then $1 \notin x_2$, because x_2 is a co-grill. Consequently $1 \in x_1$, hence $x \in h(1)$. ■

We can immediately conclude:

Theorem 3.2. *(Representation theorem) Every bounded lattice is isomorphic to a subalgebra of the complex* $_m$*algebra of its canonical* $_m$*frame.*

Now we shall introduce a 3-valued Kripke semantics for LAT which is essentially paraconsistent. It is based on the following three truth values: **t** – *true*, **f** – *false*, and **b** – *both true and false*.

Definition 3.7. *A paraconsistent Kripke model for LAT is a system* $\mathcal{M} = (X, \leq_1, \leq_2, m)$*, where* $(X, \leq_1, \leq_2)$ *is a doubly ordered set and* m *is a valuation which assigns to each proposition variable* p *an* $\ulcorner_1$*-stable set* $m(p) \subseteq X$*. Then by a parallel induction on the construction of formulas we define three satisfaction relations* $\models_i$*,* $i \in \{\mathbf{t}, \mathbf{f}, \mathbf{n}\}$ *between elements of* X *and formulas with the following intuitive meaning:*
$x \models_{\mathbf{t}} a$*: "a is true at x",* $x \models_{\mathbf{f}} a$*: "a is false at x",* $x \models_{\mathbf{n}} a$*: "a is both true and false at x".*

- $x \models_{\mathbf{t}} p$ *iff* $x \in m(p)$*;* $x \models_{\mathbf{f}} p$ *iff* $x \in \ulcorner_2 m(p)$.
- $x \models_{\mathbf{t}} \mathbf{T}$, $x \not\models_{\mathbf{f}} \mathbf{T}$.
- $x \not\models_{\mathbf{t}} \mathbf{F}$, $x \models_{\mathbf{f}} \mathbf{F}$.
- $x \models_{\mathbf{t}} a \vee b$ *iff* $x \models_{\mathbf{t}} a$ *or* $x \models_{\mathbf{t}} b$.
- $x \models_{\mathbf{f}} a \vee b$ *iff* $\exists y(y \leq_2 x$ *and* $y \not\models_{\mathbf{t}} a$ *and* $y \not\models_{\mathbf{t}} b)$.
- $x \models_{\mathbf{t}} a \wedge b$ *iff* $\exists y(y \leq_1 x$ *and* $y \not\models_{\mathbf{f}} a$ *and* $y \not\models_{\mathbf{f}} b)$.
- $x \models_{\mathbf{f}} a \wedge b$ *iff* $\exists y(y \leq_2 x$ *and* $\forall z(z \leq_1 y$ *implies* $z \models_{\mathbf{f}} a$ *or* $z \models_{\mathbf{f}} b))$.

For all formulas a we put
- $x \models_{\mathbf{b}} a$ *iff* $x \models_{\mathbf{t}} a$ *and* $x \models_{\mathbf{f}} a$.

The sequent $a \vdash b$ *is true in the model* $\mathcal{M}$ *if* $\forall x \in X(x \models_{\mathbf{t}} a$ *implies* $x \models_{\mathbf{t}} b)$.

The above definition ensures that if we let $m_{\mathbf{t}}(a) = \{x : x \models_{\mathbf{t}} a\}$ and $m_{\mathbf{f}}(a) = \{x : x \models_{\mathbf{f}} a\}$, we can use Lemma 3.5 and structural induction on formulas to prove: for all formulas a, $m_{\mathbf{t}}(a)$ is $\ulcorner_1$-stable, $m_{\mathbf{f}}(a)$ is $\ulcorner_2$-stable, $m_{\mathbf{t}}(a) = \ulcorner_1 m_{\mathbf{f}}(a)$, $m_{\mathbf{f}}(a) = \ulcorner_2 m_{\mathbf{t}}(a)$ and $m_{\mathbf{t}}(a) \cup m_{\mathbf{f}}(a) = X$. This, together with Lemma 3.4(a), allows us to prove:

Lemma 3.6. *Let* $(X, \leq_1, \leq_2)$ *be a doubly ordered set and a be a LAT-formula; then:*

(a) If $\mathcal{M}$ *is a paraconsistent Kripke model of LAT then:*

(i) If $x \models_{\mathbf{t}} a$ *and* $x \leq_1 x'$ *then* $x' \models_{\mathbf{t}} a$ *and if* $x \models_{\mathbf{f}} a$ *and* $x \leq_2 x'$ *then* $x' \models_{\mathbf{f}} a$.

(ii) Either $x \models_t a$ *or* $x \models_f a$.

(b) For any $\ulcorner_1$*-stable set* $A \subseteq X$*, such that* $A \cap \ulcorner_2 A \neq \emptyset$*, if* $\mathcal{M}$ *is a model such that* $m_{\mathbf{t}}(p) = A$*,* $m_{\mathbf{f}}(p) = \ulcorner_2 A$*,* $m_{\mathbf{t}}(q) = \varnothing$*, then for any* $x \in A \cap \ulcorner_2 A$*,* $x \models_{\mathbf{b}} p$ *and for any* $y \in X$*,* $y \not\models_{\mathbf{t}} q$*.*

From (b) we see that p is both true and false at x, and that the sequent $p \vdash q$ is not true in $\mathcal{M}$. So even though p is "paradoxical" (it is both true and false) it does not imply everything. This shows that LAT under the above semantics has three truth values and the third value **b** is not "explosive", so LAT is indeed a paraconsistent logic.

The following proposition is a straightforward corollary of the representation theorem.

Proposition 3.2. *The following conditions are equivalent for all sequents $a \vdash b$ in LAT:*

(i) $a \vdash b$ is true in all lattices.
(ii) $a \vdash b$ is true in all lattices of the form $L_m(X)$ over doubly ordered sets X.
(iii) $a \vdash b$ is true in all paraconsistent models $\mathcal{M}$ of LAT.

4 Lattice-Based Models for Paraconsistent Logic with Negation

We now enrich the algebraic structure of bounded lattices with a unary operation $\sim$ satisfying:

(i) $\sim\sim a = a$ (involution) and (ii) $a \leq b$ implies $\sim b \leq \sim a$ (antimonotonicity). We call such a system a *bounded lattice with negation* and denote it by $\mathcal{W} = (W, \vee, \wedge, 1, 0, \sim)$. The following are well known and easily proved consequences of involution and antimonotonicity.

Lemma 4.1. *Let $\mathcal{W} = (W, \vee, \wedge, 1, 0, \sim)$ be a bounded lattice with negation:*
(a) $\sim 1 = 0$; $\sim 0 = 1$; (b) $\sim (a \vee b) = \sim a \wedge \sim b$; (c) $\sim (a \wedge b) = \sim a \vee \sim b$.

We augment the sequent style deduction system for LAT (see e.g., in [11]) by the following axioms and rules, and call the resulting system NLAT:

$$a \vdash \sim\sim a \qquad\qquad \sim\sim a \vdash a \qquad\qquad \frac{a \vdash b}{\sim b \vdash \sim a}.$$

It is easy to show that NLAT is sound and complete with respect to bounded lattices with negation.

Let $\mathcal{W}$ be a bounded lattice with negation. Using a definition inspired by the work on Kripke semantics in [1], for any $A \subseteq W$, let $\sim A = \{a \in W : \sim a \in A\}$.

Lemma 4.2. *Let $\mathcal{W} = (W, \vee, \wedge, 1, 0, \sim)$ be a bounded lattice with negation and let $A, B \subseteq W$.*
(a) (i)If $A \subseteq B$, then $\sim A \subseteq \sim B$; (ii) $\sim\sim A = A$.
(b) If A is a grill then $\sim A$ is a cogrill.
(c) If A is a cogrill, then $\sim A$ is a grill.
(d) If $x = (x_1, x_2)$ is a minimal pair, then $(\sim x_2, \sim x_1)$ is a minimal pair.

Proof. (a) (i) follows from the definition of $\sim A$ and (ii) follows from the definition of involution.

For (b): $1 \notin \sim A$ iff $\sim 1 \notin A$ iff $0 \notin A$. A is a grill so $1 \notin \sim A$. Let $b \in \sim A$ and $a \leq b$. Then $\sim b \in A$ and $\sim b \leq \sim a$. A is a grill so $\sim a \in A$; therefore $a \in \sim A$. Finally $a \wedge b \in \sim A$ iff $\sim (a \wedge b) \in A$ iff $\sim a \vee \sim b \in A$ iff $\sim a \in A$ or $\sim b \in A$ iff $a \in \sim A$ or $b \in \sim A$.

The proof of (c) is similar.

To prove (d), it remains to show (i) $\sim x_2 \cup \sim x_1 = W$ and (ii) $(\sim x_2, \sim x_1)$ is minimal. For (i) suppose there exists $a \in W$ such that $a \notin \sim x_2$. Then $\sim a \notin x_2$ so $\sim a \in x_1$; therefore $a \in \sim x_1$. For (ii) let y be a minimal pair such that (1)

$y_1 \subseteq\sim x_2$ and (2) $y_2 \subseteq\sim x_1$. We shall show that $y = (\sim x_2, \sim x_1)$. From (1), (2) by (a) we obtain: $\sim y_1 \subseteq x_2$ and $\sim y_2 \subseteq x_1$. Since x is a minimal pair this yields $\sim y_1 = x_2$ and $\sim y_2 = x_1$. Now by (a) we get: $y_1 =\sim\sim y_1 =\sim x_2$ and $y_2 =\sim\sim y_2 =\sim x_1$, hence $(y_1, y_2) = (\sim x_2, \sim x_1)$, which shows that $(\sim x_2, \sim x_1)$ is a minimal pair. ∎

Let $\mathcal{W}$ be bounded lattice with negation; define a unary operation $g : X_m(W) \rightarrow X_m(W)$ by: $g(x_1, x_2) = (\sim x_2, \sim x_1)$. The following is a consequence of Lemma 4.2.

Lemma 4.3. *Let $\mathcal{W}$ be a bounded lattice with negation and let $x, y \in X_m(W)$; then:*

(a) $g(g(x)) = x$.
(b) $x \leq_1 y \Leftrightarrow g(x) \leq_2 g(y)$.
(c) $x \leq_2 y \Leftrightarrow g(x) \leq_1 g(y)$.

By a *doubly ordered set with involution*, we mean a tuple $X = (X, \leq_1, \leq_2, g)$ where $(X, \leq_1, \leq_2)$ is a doubly ordered set and g is a unary operation on X satisfying (i) $g(g(x)) = x$ (ii) $x \leq_1 y \Rightarrow g(x) \leq_2 g(y)$ and (iii) $x \leq_2 y \Rightarrow g(x) \leq_1 g(y)$. From Lemma 4.3 we see that if $\mathcal{W}$ is a bounded lattice with negation, and g is the map defined above, then $(X_m(W), \leq_1, \leq_2, g)$ is a doubly ordered set with involution, called the *canonical $_m$frame of $\mathcal{W}$*.

Let $(X, \leq_1, \leq_2, g)$ be a doubly ordered set with involution; for $A \subseteq X$ define $\sim A = \{x \in X : g(x) \in \ulcorner_2 A\}$.

Lemma 4.4. *Let $X = (X, \leq_1, \leq_2, g)$ be a doubly ordered set with involution, and let $A \subseteq X$.*

(a) $x \in \ulcorner_2 A$ iff $g(x) \in\sim A$.
(b) If A is $\ulcorner_1$-stable, then $y \in \ulcorner_2 \sim A$ iff $g(y) \in A$.
(c) If A is $\ulcorner_1$-stable, then $\sim A$ is $\leq_1$-increasing.

Proof. (a) $x \in \ulcorner_2 A$ iff $g(g(x)) \in \ulcorner_2 A$ iff $g(x) \in\sim A$.

(b) Let A be $\ulcorner_2$-stable. Then:

($\subseteq$) $y \in \ulcorner_2 \sim A$ iff $\exists z(z \leq_2 y$ and $z \notin\sim A)$ iff $\exists z(g(z) \leq_1 g(y)$ and $g(z) \notin \ulcorner_2 A)$ iff $\exists z'(z' \leq_1 g(y)$ and $z' \notin \ulcorner_2 A)$. Then by the definition of $\ulcorner_1$ we get $g(y) \in \ulcorner_1 \ulcorner_2 A = A$, hence $g(y) \in A$.

($\supseteq$) $g(y) \in A$ iff $g(y) \in \ulcorner_1 \ulcorner_2 A$ iff $\exists z(z \leq_1 g(y)$ and $z \notin \ulcorner_2 A)$ iff (by (a)) $\exists z(g(z) \leq_2 y$ and $g(z) \notin\sim A)$. Hence $\exists z'(z' \leq_2 y$ and $z' \notin\sim A)$ and by the definition of $\ulcorner_2$ we get $y \in \ulcorner_2 \sim A$.

(c) Suppose A is $\ulcorner_1$-stable, $x \in\sim A$ and $x \leq_1 y$. Then $g(x) \in \ulcorner_2 A$ and $g(x) \leq_2 g(y)$. By Lemma 3.4(a), $\ulcorner_2 A$ is $\leq_2$-increasing, so $g(y) \in \ulcorner_2 A$ and by (a) $y = g(g(y)) \in \sim A$. ∎

Lemma 4.5. *Let $(X, \leq_1, \leq_2, g)$ be a doubly ordered set with involution and let $A, B \subseteq X$ be $\ulcorner_1$-stable sets; then:*

(a) $\sim A$ is $\ulcorner_1$-stable; (b) $\sim\sim A = A$; (c) If $A \subseteq B$ then $\sim B \subseteq \sim A$.

Proof. (a) Suppose A is $\ulcorner_1$-stable. $x \in \ulcorner_1 \ulcorner_2 \sim A$ iff $\exists y(y \leq_1 x$ and $y \notin \ulcorner_2 \sim A)$ iff (by Lemma 4.4(b)) $\exists y(y \leq_1 x$ and $g(y) \notin A)$ iff $\exists y(g(y) \leq_2 g(x)$ and

$g(y) \notin A)$ iff $\exists y'(y' \leq_2 g(x)$ and $y' \notin A)$ iff $g(x) \in \ulcorner_2 A$ iff (by Lemma 4.4(a)) $x = g(g(x)) \in \sim A$.

(b) $x \in \sim\sim A$ iff $g(x) \in \ulcorner_2 \sim A$ iff (by Lemma 4.4(b)) $g(g(x)) \in A$ iff $x \in A$.

(c) Suppose $A \subseteq B$ and $x \in \sim B$. Then $g(x) \in \ulcorner_2 B$; by Lemma 3.4(d), $g(x) \in \ulcorner_2 A$ so $x \in \sim A$. ■

Lemma 4.5 and Proposition 3.1 allow us to conclude:

Proposition 4.1. *Let $(X, \leq_1, \leq_2, g)$ be a doubly ordered set with involution, and let $L_m(X) = \{A \subseteq X : \ulcorner_1 \ulcorner_2 A = A\}$, the $\ulcorner_1$-stable subsets of X. For $A, B \in L_m(X)$ let $A \vee B = A \cup B$, $A \wedge B = \ulcorner_1(\ulcorner_2 A \cup \ulcorner_2 B)$, $0 = \emptyset$, $1 = X$ and $\sim A = \{x \in X : g(x) \in \ulcorner_2 A\}$. Then $(L_m(X), \vee, \wedge, 0, 1, \sim)$ is a bounded lattice with negation, called the complex $_m$algebra of the doubly ordered set with involution, $(X, \leq_1, \leq_2, g)$.*

Theorem 4.1. *(a) Let $\mathcal{W}$ be a bounded lattice with negation and let $(X_m(W), \leq_1, \leq_2, g)$ be its canonical $_m$frame. Then for any $x \in X_m(W)$, $\sim a \in x_1$ iff $a \in (g(x))_2$.*

(b) If h is the lattice embedding of Definition 3.6, then $h(\sim a) = \sim h(a)$.

Proof. (a) $\sim a \in x_1$ iff $a \in \sim x_1 = (g(x))_2$. (b) $x \in h(\sim a)$ iff $\sim a \in x_1$ iff $a \in (g(x))_2$ iff (by Theorem 3.1(b)) $g(x) \in \ulcorner_2 h(a)$ iff (by Lemma 4.4(a)) $x \in \sim h(a)$. ■

Theorems 4.1 and 3.2 allow us to conclude:

Theorem 4.2. *(Representation theorem) Every bounded lattice with negation is isomorphic to a subalgebra of the complex $_m$algebra of its canonical $_m$frame.*

Finally, paraconsistent Kripke models for LAT can be extended to models for NLAT by considering models on doubly ordered sets with involution with the following clauses for the negation:

- $x \models_{\mathbf{t}} \sim a$ iff $g(x) \models_{\mathbf{f}} a$.
- $x \models_{\mathbf{f}} \sim a$ iff $g(x) \models_{\mathbf{t}} a$.

Lemma 3.6 can be extended to formulas of NLAT, so NLAT is a 3-valued logic with negation that is nonexplosive - a paraconsistent logic in the strict sense.

The following proposition is a straightforward corollary of the representation theorem.

Proposition 4.2. *The following conditions are equivalent for all sequents $a \vdash b$ in NLAT:*

(i) $a \vdash b$ is true in all lattices with negation.
(ii) $a \vdash b$ is true in all lattices of the form $L_m(X)$ over doubly ordered sets X with involution.
(iii) $a \vdash b$ is true in all paraconsistent models $\mathcal{M}$ of NLAT.

5 Concluding Remarks

Following the results in Allwein and MacCaull [2] and MacCaull and Orłowska [9] a description of a sound and complete tableau-style deduction system corresponding to 3-valued semantics related to Urquhart and dual Urquhart representation of lattices and lattices with negation is forthcoming.

Currently under investigation are nonclassical versions of the modal operators $\Diamond$ and $\Box$. Without complementation, the $\Diamond$-modality and its associated $\Box$-modality resolve into 4 modalities. Orłowska and Vakarelov [11] considered non-distributive lattices without complements, developing Urquhart-style representation theorems for lattices with operators corresponding to each of the 4 modalities. Work is in progress to prove representation theorems using minimal pairs, giving paraconsistent Kripke-style semantics for NLAT with modal operators.

Acknowledgements. The authors wish to thank the anonymous referees for their helpful comments and for finding a number of typographical errors in the original manuscript.

References

[1] Allwein G. and M. Dunn. Kripke models for linear logic, Journal of Symbolic Logic, 58 (1993) 514-545.

[2] Allwein G. and W. MacCaull. A Kripke semantics for the logic of Gelfand quantales, Studia Logica, 61 (2001) 1-56.

[3] Arieli, O. and A. Avron. A model-theoretic approach for recovering consistent data from inconsistent knowledge-bases, Journal of Automated Reasoning, 22 (1999) 263-309.

[4] Belnap, N. A useful four-valued logic. In: G. Epstein and J.M.Dunn, (eds), Modern Uses of Multiple-Valued Logic, Reidel, pp 7-37, 1977.

[5] da-Costa, N. On the theory of inconsistent formal systems, Notre Dame Journal of Formal Logic, 15 (1974), 497-510.

[6] Dunn, M. Gaggle Theory: An abstraction of Galois connections and residuation, with applications to negation, implication, and various logical operators. In: J. van Eijk (ed), Logics in AI, European Workshop JELIA'90, Amsterdam, The Netherlands, September 1990, Proceedings. Lecture Notes in Artificial Intelligence No 478. 31-51, Springer-Verlag.

[7] Fitting, M. Bilattices and the semantics of logic programming, Journal of Logic Programming, 11 (1991) 91-116.

[8] Ginsberg, M. Multivalued logics: A uniform approach to reasoning in AI, Computer Intelligence, 4 (1988) 256-316.

[9] MacCaull, W. and E. Orłowska. Correspondence results for relational proof systems with application to the Lambek Calculus, Studia Logica, 71 (2002) 389-414.

[10] MacCaull, W. and D. Vakarelov, Lattice-based paraconsistent logic, In: I.Düntsch and M.Winter (eds) Proceedings of RelMiCS 8, the 8th International Seminar in Relational Methods in Computer Science, 155-162, 2005.

[11] Orłowska, E. and D. Vakarelov. Lattices with modal operators and lattice based modal logics, to appear.

[12] Rauszer, C. Semi-Boolean algebras and their applications to intuitionistic logic with dual operations, Fundamenta Informaticae, 83 (1974) 219-250.
[13] Urquhart, A. A topological representation for lattices, Algebra Universalis, 8 (1978) 45-58.
[14] Vakarelov, D. Semi-Boolean algebras and semantics for HB-predicate logic, Bull. Acad. Polon. Sci. Ser. Math. Phys. 22 (1974) 1087–1095.
[15] Vakarelov, D. Consistency, completeness and negation. In: G.Priest, R.Routley and J.Norman (eds), Paraconsistent Logic, Essays on the Inconsistent, Philosophia Verlag, 1989.
[16] Vakarelov, D. Intuitive semantics for some three-valued logics connected with information, contrariety and subcontrariety, Studia Logica 48 (1989) 565–575.

Verification of Pushdown Systems Using Omega Algebra with Domain*

Vincent Mathieu and Jules Desharnais

Département d'informatique et de génie logiciel,
Université Laval, Québec, QC, G1K 7P4 Canada
{Vincent.Mathieu, Jules.Desharnais}@ift.ulaval.ca

Abstract. We present a framework for the verification of pushdown systems. These systems can model the interprocedural control flow of computer programs. The framework is based on an extension of Kleene algebra called omega algebra with domain. This allows to formulate behavioural properties that refer to both actions and states.

1 Introduction

Computer technologies are increasingly used and software flaws can have dramatic consequences. This is why we are interested in analyzing programs statically (i.e., before execution) to uncover potential problems, in particular those related to security. Our research goal is to develop a verification framework based on Kleene algebra (KA) or extensions like omega algebra with domain (OAD) in which both program models and behavioural properties can be expressed and compared. Since large systems are not monolithic, we are aiming at verifying programs composed of multiple procedures, including recursive ones.

In this paper, we show how to use OAD to represent *pushdown systems* [9], which can model the control flow of programs with recursive procedures, and properties about them. We also show how to verify algebraically whether a pushdown system satisfies a property. Because models and properties are described with the same algebraic formalism, the principles of verification can also be expressed with this formalism. The focus here is on the description of the algebraic verification framework and implementation issues are not discussed. Indeed, much more research is required before an efficient model-checking tool can be developed.

Basic concepts on OAD and matrices are presented in Sect. 2. In Sect. 3, we introduce pushdown systems. Finally, we formalize the problem of representing and verifying pushdown systems with OAD in Sect. 4, where we also give examples. We conclude in Sect. 5 with future targets for research.

* This research was supported by NSERC (Natural Sciences and Engineering Research Council of Canada).

W. MacCaull et al. (Eds.): RelMiCS 2005, LNCS 3929, pp. 188–199, 2006.

2 Omega Algebra with Domain

Kleene algebra [3, 5] is an abstract representation of many structures used in computer science, such as regular languages. Omega algebra [2] is an extension of KA axiomatizing the concept of languages that can contain finite or infinite sequences. Another extension of KA is KA with tests [6], which is appropriate for the representation of regular programs. It can itself be extended to KA with domain [4]. In this section, we present OAD, which is a combination of KA with domain and omega algebra, and the notion of matrices over an OAD.

Definition 1. *An* omega algebra with domain *is an algebraic structure of type* $\langle O, B, +, \cdot, ^*, ^\ulcorner, ^\omega, 0, 1, ^- \rangle$ *such that* O *is a set of elements called* programs *and* $B \subseteq O$ *is a set of elements called* tests. *The result of the application of the domain operator* ($^\ulcorner$) *is a test. The operator* $^-$ *can only be applied to tests. The structure* $\langle B, +, \cdot, 0, 1, ^- \rangle$ *is a Boolean algebra and, finally, the algebra satisfies the following axioms, where* $b \in B$ *and* $x, y, z \in O$. *We write* xy *instead of* $x \cdot y$. *The increasing precedence of the operators is* $+, \cdot, (^*, ^\ulcorner, ^\omega)$.

$$
\begin{array}{lll}
x + (y + z) = (x + y) + z & \left.\begin{array}{l} 1 + xx^* \leqslant x^* \\ 1 + x^*x \leqslant x^* \end{array}\right\} & \textit{(*-unfolding)} \\
x + y = y + x & & \\
x + x = x & \left.\begin{array}{l} yx + z \leqslant x \Rightarrow y^*z \leqslant x \\ xy + z \leqslant x \Rightarrow zy^* \leqslant x \end{array}\right\} & \textit{(*-induction)} \\
x + 0 = x & & \\
x(yz) = (xy)z & x^\omega = xx^\omega & \textit{(ω-unfolding)} \\
x0 = 0 = 0x & x \leqslant yx + z \Rightarrow x \leqslant y^\omega + y^*z & \textit{(ω-induction)} \\
x1 = x = 1x & \left.\begin{array}{l} x \leqslant {}^\ulcorner x x \\ {}^\ulcorner(bx) \leqslant b \end{array}\right\} & \textit{(definition of domain)} \\
x(y + z) = xy + xz & & \\
(x + y)z = xz + yz & {}^\ulcorner(x {}^\ulcorner y) = {}^\ulcorner(xy) & \textit{(locality of composition)}
\end{array}
$$

The ordering $\leqslant$ *used in the axioms is defined by* $x \leqslant y \Leftrightarrow x + y = y$, *for all* $x, y \in O$.

The axiom of locality of composition is not included in the basic definition of OAD [4]. It is however needed for our applications. Fortunately, this is not a major constraint, since all the usual models (languages, traces, relations) satisfy it. The axiom means that the domain of a composition xy does not depend on the internal structure of the second argument (y), but only on its domain.

Here are some consequences of Definition 1.

Theorem 1. *1.* ${}^\ulcorner x = 0 \Leftrightarrow x = 0$.
2. If $ab = ba$, *then* (a) $ab^* = b^*a$, (b) $a^*b^* = b^*a^*$, (c) $(ab)^* \leqslant a^*b^*$.
3. $ab \leqslant 1 \Rightarrow a^*b^* \leqslant a^* + b^*$.

Proof. See [4] for 1 and [5] for 2(a,b). Property 2(c) is a consequence of *-induction, $1 \leqslant a^*b^*$ and 2(b). Property 3 follows by *-induction and *-unfolding applied to a^*.

Let $\langle O, B, +, \cdot, ^*, ^\ulcorner, ^\omega, 0, 1, ^- \rangle$ be an OAD. An *ω-expression* is an OAD term, i.e., an expression constructed with the OAD operators and producing an element of O. An *ω-regular expression* [2] is an ω-expression such that no $^\omega$ appears in the argument of $^\omega$ or in the left argument of $\cdot$. A *standard ω-regular expression* is an ω-regular expression which is a sum of expressions of the form xy^ω such that for no test b except 0, $b \leqslant y$. Under a standard interpretation in the model of guarded sequences [7], extended to include infinite sequences, an ω-regular expression as just defined maps to a regular language of infinite sequences. For a test b, the expression b^ω maps to a language that contains finite sequences, and this is the reason for excluding $b \leqslant y$ [8].

We will use standard ω-regular expressions to specify properties. In the sequel, we use $\mathsf{P} \subseteq O$ and $\mathsf{B} \subseteq B$ to denote sets of *atomic programs* and *atomic tests*, respectively. The following abbreviations refer to the indicated sets of expressions.

$^{\ulcorner\omega}\text{-SExp}_{\mathsf{P},\mathsf{B}}$:	standard ω-regular expressions on P and B that can contain domain,
$^{\ulcorner\omega}\text{-SExp}_{\mathsf{P}}$:	standard ω-regular expressions on P that can contain domain,
$^{\omega}\text{-SExp}_{\mathsf{P},\mathsf{B}}$:	standard ω-regular expressions on P and B without domain,
$^{\omega}\text{-SExp}_{\mathsf{P}}$:	standard ω-regular expressions on P without domain.

OAD has many interesting models, including languages of finite and infinite guarded strings, binary relations, and the model of finite and infinite traces on a Kripke frame. We present here the model of binary relations. This model is important for us because we define, in Sect. 4, an algebra for stack manipulations (pushing and popping) and the intended model of this algebra is a relational one. The language model is also useful because it is a model for control actions done by pushdown systems.

Let X be a set. The set of all binary relations on X, $\mathcal{P}(X \times X)$, is the set of elements of the algebra and the set of tests is the set of all subidentities, that is, $\mathcal{P}\{(x,x) \mid x \in X\}$. Let R, R_1 and R_2 be relations. The constants and operators are defined by

$$
\begin{aligned}
&0 \overset{\text{def}}{=} \{\}, \qquad 1 \overset{\text{def}}{=} \{(x,x) \mid x \in X\}, \qquad R_1 + R_2 \overset{\text{def}}{=} R_1 \cup R_2,\\
&R_1 \cdot R_2 \overset{\text{def}}{=} \{(x,x') \mid \exists x''.((x,x'') \in R_1 \wedge (x'',x') \in R_2)\},\\
&R^* \overset{\text{def}}{=} \textstyle\bigcup_{i\geqslant 0} R^i, \quad \text{where } R^0 = 1 \text{ and } R^{n+1} = R \cdot R^n,\\
&R^\omega \text{ is the greatest fixed point of } \lambda Z.R \cdot Z,\\
&{}^{\ulcorner}R \overset{\text{def}}{=} \{(x,x) \mid \exists x'.(x,x') \in R\},\\
&\overline{R} \overset{\text{def}}{=} \{(x,x) \mid (x,x) \notin R\} \ .
\end{aligned}
$$

The expression ${}^{\ulcorner}R$ is a subidentity that represents the domain (in the usual sense) of relation R. The expression $\overline{R}$ is defined only if R is a subidentity. Any omega algebra has a top element which is 1^ω. For the relational model, $1^\omega = X \times X$. It is easy to show that the relation R^ω is a vector, i.e., $R^\omega = R^\omega 1^\omega$. An element $x \in X$ is in the domain of R^ω iff there exists an infinite path starting from x in the graph of relation R.

The set of square $n \times n$ matrices on an OAD also forms an OAD in which the operations are defined as follows. The notation $\mathbf{A}[i,j]$ refers to the entry in row i and column j of matrix $\mathbf{A}$.

1. $\mathbf{0}$: matrix whose entries are all 0, i.e., $\mathbf{0}[i,j] = 0$.
2. $\mathbf{1}$: identity matrix (square), i.e., $\mathbf{1}[i,j] = \begin{cases} 1 \text{ if } i = j \\ 0 \text{ if } i \neq j. \end{cases}$
3. $(\mathbf{A}+\mathbf{B})[i,j] = \mathbf{A}[i,j] + \mathbf{B}[i,j]$.
4. $(\mathbf{AB})[i,j] = \sum_k \mathbf{A}[i,k]\mathbf{B}[k,j]$.
5. $\mathbf{A}^*$ is defined recursively [3, 5]. If $\mathbf{A} = (a)$, then $\mathbf{A}^* = (a^*)$. If $\mathbf{A}$ is a matrix with more than one entry, $\mathbf{A}$ can be divided into four submatrices such that $\mathbf{A} = \begin{pmatrix} \mathbf{B} & \mathbf{C} \\ \mathbf{D} & \mathbf{E} \end{pmatrix}$, where $\mathbf{B}$ and $\mathbf{E}$ are square. Then
$$\mathbf{A}^* = \begin{pmatrix} \mathbf{F}^* & \mathbf{F}^*\mathbf{C}\mathbf{E}^* \\ \mathbf{E}^*\mathbf{D}\mathbf{F}^* & \mathbf{E}^* + \mathbf{E}^*\mathbf{D}\mathbf{F}^*\mathbf{C}\mathbf{E}^* \end{pmatrix},$$
where $\mathbf{F} = \mathbf{B} + \mathbf{C}\mathbf{E}^*\mathbf{D}$.
6. Like $\mathbf{A}^*$, $\mathbf{A}^\omega$ is defined recursively:
$$(a)^\omega = (a^\omega) \quad \text{and} \quad \begin{pmatrix} \mathbf{B} & \mathbf{C} \\ \mathbf{D} & \mathbf{E} \end{pmatrix}^\omega = \begin{pmatrix} \mathbf{F}^\omega + \mathbf{B}^*\mathbf{C}\mathbf{G}^\omega & \mathbf{F}^\omega + \mathbf{B}^*\mathbf{C}\mathbf{G}^\omega \\ \mathbf{E}^*\mathbf{D}\mathbf{F}^\omega + \mathbf{G}^\omega & \mathbf{E}^*\mathbf{D}\mathbf{F}^\omega + \mathbf{G}^\omega \end{pmatrix},$$
where $\mathbf{B}$ and $\mathbf{E}$ are square, $\mathbf{F} = \mathbf{B} + \mathbf{C}\mathbf{E}^*\mathbf{D}$ and $\mathbf{G} = \mathbf{E} + \mathbf{D}\mathbf{B}^*\mathbf{C}$. It can be shown that this definition of $\mathbf{A}^\omega$ is a consequence of the ω-axioms [8]. This is done as for the definition of $\mathbf{A}^*$ [3, 5]. Since by this definition all columns of $\mathbf{A}^\omega$ are identical, it is often possible to use a single column vector instead of the full matrix for $\mathbf{A}^\omega$; the only constraint is to respect typing according to standard rules for matrices. The following equation can then be used at the recursive step:
$$\begin{pmatrix} \mathbf{B} & \mathbf{C} \\ \mathbf{D} & \mathbf{E} \end{pmatrix}^\omega = \begin{pmatrix} \mathbf{F}^\omega + \mathbf{B}^*\mathbf{C}\mathbf{G}^\omega \\ \mathbf{E}^*\mathbf{D}\mathbf{F}^\omega + \mathbf{G}^\omega \end{pmatrix} .$$
7. $(\ulcorner\mathbf{A})[i,i] = \sum_k \ulcorner(\mathbf{A}[i,k])$ and $(\ulcorner\mathbf{A})[i,j] = 0$ if $i \neq j$. For a matrix $\mathbf{A}$, only the diagonal of $\ulcorner\mathbf{A}$ contains non-zero entries. This matrix represents the set of states from which there exists a transition to another state (or the same state) in the transition diagram corresponding to matrix $\mathbf{A}$. If $\mathbf{A}$ is a binary relation, $\ulcorner\mathbf{A}$ is the domain of this relation (in the usual sense).
8. For tests only: $\overline{\mathbf{A}}[i,i] = \overline{\mathbf{A}[i,i]}$ and $\overline{\mathbf{A}}[i,j] = 0$ if $i \neq j$.

3 Pushdown Systems

Finite state systems can model the control flow of a regular program, but not the control flow of programs with recursive procedures. A model for the control flow of programs with recursive calls is a *pushdown system* (PDS). Informally, it is a finite system along with a stack data structure of unbounded length, and so the associated transition system may have an infinite state space.

Definition 2. *A* pushdown system (PDS) *is a quintuple* $\mathcal{P} = (L, A, S, \hookrightarrow, i)$, *where* L *is a finite set of* locations, A *is a set of* actions *and* S *is a* stack alphabet. *A* stack configuration *is a finite word on S (an element of S^*). A* state *of a PDS is a pair (l, w) such that $l \in L$ and $w \in S^*$. The relation $\hookrightarrow$ is a finite subset of* $(L \times S^*) \times A \times (L \times S^*)$. *Finally, $i \in L \times S^*$ is the initial state.*

We also write $(l_1, w_1) \overset{a}{\hookrightarrow} (l_2, w_2)$ if $((l_1, w_1), a, (l_2, w_2)) \in \hookrightarrow$. This means that it is possible to go from location l_1 to location l_2 by popping the sequence of stack symbols w_1, pushing the sequence of stack symbols w_2 and doing action a.

A transition in a PDS can pop and push zero, one or many elements on the stack. This is more general than the definition in [9], where it is only possible to pop exactly one symbol. This is done for the sake of conciseness and does not add expressiveness to the model. In addition, there are no actions in the definition of a PDS in [9] (but there are in the definition of Burkart [1]).

As described in [1, 9], the model checking problem on PDSs consists in verifying that a given PDS satisfies a given formula of a temporal logic. Model-checking the modal mu-calculus is decidable on PDSs [1]. So, this is also the case for many temporal logics like CTL, CTL*, LTL and the linear-time mu-calculus. The model-checking problem on PDSs for branching-time logics does not have an efficient solution. For example, there exists a formula of CTL for which the model-checking problem is PSPACE-complete in the size of the model. On the other hand, linear-time logics are more interesting. For example, the model-checking problem on PDSs for LTL and the linear-time mu-calculus has polynomial running time in the size of the model. So it is realistic to use these linear-time logics in practice.

4 Verifying Pushdown Systems with OAD

In this section, we describe how to model PDSs using matrices on an OAD and how to check whether they satisfy properties about states or actions specified with standard ω-regular expressions.

4.1 OAD Matrix Representation of Pushdown Systems

Suppose a PDS $\mathcal{P} \overset{\text{def}}{=} (L, A, S, \hookrightarrow, i)$. To represent it in OAD, we need a way to describe both system actions and stack manipulations. For this reason, we will use two sets of primitive (atomic) actions: a set of *control actions* $\mathsf{A} \overset{\text{def}}{=} A$ and a set of *stack actions*

$$\mathsf{S} \overset{\text{def}}{=} S \cup \{s^{\smile} \mid s \in S\} \cup \{\varepsilon\}$$

(note that $s^{\smile}$ is a stack action symbol, not the application of an operator $^{\smile}$ to stack action s). The OAD to be used for the representation of the PDS is the algebra $\mathcal{O} \overset{\text{def}}{=} \langle O, B, +, \cdot, ^*, ^{\ulcorner}, ^{\omega}, 0, 1, ^{\neg} \rangle$ generated by A and S under the axioms of OAD and the following additional laws satisfied by atomic control and stack actions, where $a \in \mathsf{A}$ and $s, t \in S$:

$$ss^{\smile} = 1, \qquad st^{\smile} = 0 \text{ if } s \neq t, \qquad \varepsilon + \sum_{s \in S} s^{\smile} s = 1, \qquad \varepsilon s^{\smile} = 0, \tag{1}$$

$$as = sa, \qquad as^{\smile} = s^{\smile} a, \qquad a\varepsilon = \varepsilon a. \tag{2}$$

A stack action $s \in \mathsf{S}$ corresponds to pushing a symbol $s \in S$ on the stack —we are here overloading the symbol s— and $s^{\smile} \in \mathsf{S}$ corresponds to popping $s \in S$ (returning the resulting stack). The action ε tests if the stack is empty. A relational model for these laws is obtained by using the interpretation R defined for $s \in S$ by $R(s) \stackrel{\text{def}}{=} \{(x, xs) \mid x \in S^*\}$, $R(s^{\smile})$ is the converse of $R(s)$ and $R(\varepsilon) \stackrel{\text{def}}{=} \{(\epsilon, \epsilon)\}$. The first axiom in (1) says that pushing followed by popping is like doing nothing. The second one says that it is impossible to push a symbol and then pop a different one. Consequences of the third axiom are that ε and $s^{\smile}s$ for $s \in S$ are tests. They correspond respectively to the test that the stack is empty and the test that the symbol $s \in S$ is on top of the stack (the test $s^{\smile}s$ pops s if present and then pushes it back). The third axiom itself says that the stack is either empty or has a symbol of S on its top. This axiom is useful for complementing a test. For example, if $S = \{s, t\}$, it is possible to show that $\overline{t^{\smile}t} = \varepsilon + s^{\smile}s$ by showing $\varepsilon + s^{\smile}s + t^{\smile}t = 1$, which is the present axiom, and $(\varepsilon + s^{\smile}s)t^{\smile}t = 0$. Finally, the last axiom says that it is impossible to pop a symbol if the stack is empty. By the axioms in (2), control actions and stack actions commute. Some consequences of this commutativity are given in Theorem 1(2).

The representation of a PDS $(L, A, S, \hookrightarrow, i)$ by a matrix over the OAD $\mathcal{O}$ is similar to the representation of a finite state automaton by a matrix over a KA [3, 5]. The PDS is represented by a pair of matrices $(\mathbf{I}, \mathbf{M})$ corresponding respectively to the initial state and the transition relation. Now, let $w = s_1 \ldots s_k$ be a sequence of stack actions; we use $w^{\smile}$ as an abbreviation for $s_k^{\smile} \ldots s_1^{\smile}$. If the PDS has n locations $l_1, \ldots, l_n$, then $\mathbf{M}$ is an $n \times n$ matrix such that

$$\mathbf{M}[i, j] \stackrel{\text{def}}{=} \sum \{a w_1^{\smile} w_2 \mid a \in \mathsf{A} \wedge w_1, w_2 \in S^* \wedge (l_i, w_1) \stackrel{a}{\hookrightarrow} (l_j, w_2)\}. \quad (3)$$

Matrix $\mathbf{I}$ is a row vector with n columns. If the initial state is $(l_i, s_1 \ldots s_n)$, the entry in column i of $\mathbf{I}$ is $s_n^{\smile} \ldots s_1^{\smile} \varepsilon s_1 \ldots s_n$ and the other entries are 0. Thus, all entries of $\mathbf{I}$ are tests. In particular, $s_n^{\smile} \ldots s_1^{\smile} \varepsilon s_1 \ldots s_n$ tests if the stack contains $s_1 \ldots s_n$ (having ε in the expression ensures there is nothing under s_1).

By defining $T_{i,j,a} \stackrel{\text{def}}{=} \sum \{a w_1^{\smile} w_2 \mid w_1, w_2 \in S^* \wedge (l_i, w_1) \stackrel{a}{\hookrightarrow} (l_j, w_2)\}$ and using distributivity on (3), we get

$$\mathbf{M}[i, j] = \sum_{a \in \mathsf{A}} a T_{i,j,a}. \quad (4)$$

Note that $T_{i,j,a} = 0$ if there is no transition by a from l_i to l_j.

4.2 Specifying Properties

A property is specified by a non-interpreted expression which gives, when interpreted with respect to a PDS, the set of sequences of actions corresponding to executions of the PDS that satisfy the property. In the sequel, we use expressions E and E^- to describe executions that respectively satisfy or violate the desired property. So, E and E^- are implicitly related.

4.3 Verification

Assume we are given a PDS $\mathbf{P} \stackrel{\text{def}}{=} (\mathbf{I}, \mathbf{M})$ and an expression E in $\ulcorner^{\omega}\text{-SExp}_{\mathsf{A},At}$. We want to verify if the PDS "satisfies" E, i.e., whether all executions of $\mathbf{P}$, when projected on the control actions (ignoring the stack) are executions allowed by E. Before explaining how this is done, we define the concept of *interpretation of an expression with respect to* $\mathbf{M}$.

Definition 3. *Let $\mathcal{M}_{\mathcal{O}}$ denote the set of matrices with entries from an OAD $\mathcal{O}$ defined as in Sect. 4.1. Let $\mathbf{M}$ be the matrix of a PDS $(\mathbf{I}, \mathbf{M})$ on $\mathcal{O}$. The interpretation of an expression in $\ulcorner^{\omega}\text{-SExp}_{\mathsf{A},At}$ with respect to $\mathbf{M}$ is a function $(\cdot)_{\mathbf{M}} : \ulcorner^{\omega}\text{-SExp}_{\mathsf{A},At} \to \mathcal{M}_{\mathcal{O}}$ that is defined inductively on the structure of expressions as follows: $p_{\mathbf{M}}$ for an atomic proposition $p \in At$ is a matrix that is a test and that represents the set of states that satisfy p with respect to the desired valuation (an example follows shortly; see also Sect. 4.5) and*

$$
\begin{array}{ll}
\textit{for } a \in \mathsf{A},\ a_{\mathbf{M}}[i,j] = aT_{i,j,a} & \textit{(see (4))}, \\
(a+b)_{\mathbf{M}} = a_{\mathbf{M}} + b_{\mathbf{M}}, & (ab)_{\mathbf{M}} = a_{\mathbf{M}} b_{\mathbf{M}}, \\
(a^*)_{\mathbf{M}} = (a_{\mathbf{M}})^*, & (a^{\omega})_{\mathbf{M}} = (a_{\mathbf{M}})^{\omega}, \\
(\ulcorner a)_{\mathbf{M}} = \ulcorner(a_{\mathbf{M}}), & \overline{p}_{\mathbf{M}} = \overline{p_{\mathbf{M}}}, \\
0_{\mathbf{M}} = \mathbf{0}, & 1_{\mathbf{M}} = \mathbf{1}.
\end{array}
$$

This interpretation of an expression over the model defined by matrix $\mathbf{M}$ shares much with the usual interpretation of modal expressions over a Kripke frame. In particular, the matrix $a_{\mathbf{M}}$ in which the elements of A are replaced by 1 represents the relation associated to an action a in a Kripke frame.

We now give an example of interpretation of an atomic proposition. Suppose given a PDS $(\{l_1, l_2\}, A, \{s, t\}, \hookrightarrow, i)$. The fact that the stack is empty or has an s on top is represented by $\varepsilon + s^{\smile}s$. The fact that the stack contains the symbol s is represented by the test $\ulcorner(t^{\smile *}s^{\smile})$. This set of stack configurations cannot be expressed without the domain operator. Because the possible number of t symbols to pop before reaching the first s is unbounded, the only way to express the set of stack configurations without using the domain operator would be to use an infinite expression $s^{\smile}s + t^{\smile}s^{\smile}st + t^{\smile}t^{\smile}s^{\smile}stt + \ldots$. Now suppose that the atomic proposition p is satisfied in location l_1 when the stack is empty or s is on top of the stack, and satisfied in l_2 when the stack contains s; then

$$
p_{\mathbf{M}} = \begin{pmatrix} \varepsilon + s^{\smile}s & 0 \\ 0 & \ulcorner(t^{\smile *}s^{\smile}) \end{pmatrix}.
$$

This illustrates that the domain operator is useful for expressing additional properties about states.

We consider two cases for the definition of satisfaction of an expression by a PDS.

Definition 4 (satisfaction).

1. *Let E be an expression of* $^{\omega}$-$\mathsf{SExp}_{\mathsf{A}}$ *specifying sequences of actions that the system may execute. Then the PDS* $(\mathbf{I}, \mathbf{M})$ *satisfies E iff* $\mathbf{IM}^{\omega} \leqslant \mathbf{I}E_{\mathbf{M}}$.
2. *Let E^{-} be an expression in* $^{\ulcorner\omega}$-$\mathsf{SExp}_{\mathsf{A},At}$ *specifying sequences of actions that the system must not execute. Then the PDS* $(\mathbf{I}, \mathbf{M})$ *satisfies E iff* $\mathbf{I}E^{-}_{\mathbf{M}} = \mathbf{0}$.

Note that $\mathbf{I}E_{\mathbf{M}}$ represents all sequences of actions specified by E that the system can execute from its initial state (the same can be said of E^{-}). The inequation of the first definition means that the set of sequences of actions that the PDS can execute is included in the set of sequences of actions corresponding to the executions that satisfy the desired property. The equation of the second definition means that the set of sequences of actions corresponding to the executions that violate the desired property is empty.

The expression E in the first definition is more restricted than E^{-} in the second one. The restriction to $^{\omega}$-$\mathsf{SExp}_{\mathsf{A}}$ (rather than $^{\ulcorner\omega}$-$\mathsf{SExp}_{\mathsf{A},At}$) ensures that the expression does not describe a property of the states of the PDS. This is because states are represented by tests and tests can only be introduced using atomic propositions (elements of At) and the domain operator. Multiplication of matrices shares with the composition of relations the property that intermediate states (locations) are lost. The same thing happens with the composition of stack actions, due to laws like $ss^{\smile} = 1$. This means that $\mathbf{IM}^{\omega}$ may contain a sequence of actions that is constructed in two different ways by going through different states, and there is no way to tell if this is happening or not. The second definition does not need the restriction: if a sequence can be constructed by going through undesired states, it will be in $\mathbf{I}E^{-}_{\mathbf{M}}$ (i.e., $\mathbf{I}E^{-}_{\mathbf{M}} \neq \mathbf{0}$).

4.4 Example 1

This example shows how to verify an LTL-like property on a system whose action traces have the form ca^{ω} or $ca^{n}cb^{n}cd^{\omega}$, for $n \in \mathbb{N}$. The PDS that models this system is $\mathcal{P} = \{\{l_1, l_2, l_3, l_4\}, \{a, b, c, d\}, \{s, t\}, \hookrightarrow, (l_1, \epsilon)\}$, where ϵ is the empty word on S and $\hookrightarrow$ contains the following transitions:

$$\begin{array}{lll} (l_1, \epsilon) \overset{c}{\hookrightarrow} (l_2, s), & (l_2, \epsilon) \overset{a}{\hookrightarrow} (l_2, t), & (l_2, \epsilon) \overset{c}{\hookrightarrow} (l_3, \epsilon), \\ (l_3, t) \overset{b}{\hookrightarrow} (l_3, \epsilon), & (l_3, s) \overset{c}{\hookrightarrow} (l_4, \epsilon), & (l_4, \epsilon) \overset{d}{\hookrightarrow} (l_4, \epsilon). \end{array}$$

Using matrices, the PDS is represented by $(\mathbf{I}, \mathbf{M})$, where

$$\mathbf{I} = (\,\varepsilon \quad 0 \quad 0 \quad 0\,) \text{ and } \mathbf{M} = \begin{pmatrix} 0 & cs & 0 & 0 \\ 0 & at & c & 0 \\ 0 & 0 & bt^{\smile} & cs^{\smile} \\ 0 & 0 & 0 & d \end{pmatrix}.$$

The property to verify on the PDS is that whenever an action a is executed, then either a is executed forever or eventually b is executed. This property can be stated in LTL as $\mathsf{G}((a)\mathsf{true} \Rightarrow \mathsf{G}(a)\mathsf{true} \vee \mathsf{F}(b)\mathsf{true})$. The negation of this property is represented by the standard ω-regular expression

$$E^- \stackrel{\text{def}}{=} (a+b+c+d)^* a (a+c+d)^* (c+d)(a+c+d)^\omega,$$

which says that to violate the property, an action a must eventually be executed ($(a+b+c+d)^*a$) and be followed by steps where no b is executed ($(a+c+d)^*$), followed by a step where no a and no b is executed ($c+d$); at this point, the only way to satisfy the property is to eventually execute b, but E^- says it cannot happen ($(a+c+d)^\omega$).

Let us check that E is satisfied, i.e., $\mathbf{I}E^-_{\mathrm{M}} = \mathbf{0}$. In the following derivation, expressions of the form F^*_{M} and F^ω_{M} mean $(F_{\mathrm{M}})^*$ and $(F_{\mathrm{M}})^\omega$, respectively.

$$\begin{aligned}
& \mathbf{I}E^-_{\mathrm{M}} = \mathbf{0} \\
\Leftrightarrow\ & \mathbf{I}((a+b+c+d)^* a (a+c+d)^* (c+d)(a+c+d)^\omega)_{\mathrm{M}} = \mathbf{0} \\
\Leftrightarrow\ & \qquad \langle \text{ Definition 3 } \rangle \\
& \mathbf{I}(a+b+c+d)^*_{\mathrm{M}}\ a_{\mathrm{M}}\ (a+c+d)^*_{\mathrm{M}}\ (c+d)_{\mathrm{M}}\ (a+c+d)^\omega_{\mathrm{M}} = \mathbf{0} \\
\Leftrightarrow\ & \qquad \langle\ \omega\text{-unfolding for } (a+c+d)^\omega_{\mathrm{M}}\ \rangle \\
& \mathbf{I}(a+b+c+d)^*_{\mathrm{M}}\ a_{\mathrm{M}}\ (a+c+d)^*_{\mathrm{M}}\ (c+d)_{\mathrm{M}}\ (a+c+d)_{\mathrm{M}}(a+c+d)^\omega_{\mathrm{M}} = \mathbf{0} \\
\Leftarrow\ & a_{\mathrm{M}}\ (a+c+d)^*_{\mathrm{M}}\ (c+d)_{\mathrm{M}}\ (a+c+d)_{\mathrm{M}} = \mathbf{0}
\end{aligned}$$

We now prove the last line of the previous derivation.

$$\begin{aligned}
& a_{\mathrm{M}}\ (a+c+d)^*_{\mathrm{M}}\ (c+d)_{\mathrm{M}}\ (a+c+d)_{\mathrm{M}} \\
=\ & \qquad \langle \text{ Definition 3 } \rangle \\
& a_{\mathrm{M}}\ (a_{\mathrm{M}} + c_{\mathrm{M}} + d_{\mathrm{M}})^*\ (c+d)_{\mathrm{M}}\ (a+c+d)_{\mathrm{M}} \\
=\ & \qquad \langle \text{ Definition 3, (3) and (4) } \rangle \\
& a_{\mathrm{M}}\ (\begin{pmatrix} 0 & 0 & 0 & 0 \\ 0 & at & 0 & 0 \\ 0 & 0 & 0 & 0 \\ 0 & 0 & 0 & 0 \end{pmatrix} + \begin{pmatrix} 0 & cs & 0 & 0 \\ 0 & 0 & c & 0 \\ 0 & 0 & 0 & cs^\smile \\ 0 & 0 & 0 & 0 \end{pmatrix} + \begin{pmatrix} 0 & 0 & 0 & 0 \\ 0 & 0 & 0 & 0 \\ 0 & 0 & 0 & 0 \\ 0 & 0 & 0 & d \end{pmatrix})^*\ (c+d)_{\mathrm{M}}\ (a+c+d)_{\mathrm{M}} \\
=\ & a_{\mathrm{M}} \begin{pmatrix} 0 & cs & 0 & 0 \\ 0 & at & c & 0 \\ 0 & 0 & 0 & cs^\smile \\ 0 & 0 & 0 & d \end{pmatrix}^* (c+d)_{\mathrm{M}}\ (a+c+d)_{\mathrm{M}} \\
=\ & \begin{pmatrix} 0 & 0 & 0 & 0 \\ 0 & at & 0 & 0 \\ 0 & 0 & 0 & 0 \\ 0 & 0 & 0 & 0 \end{pmatrix} \begin{pmatrix} 1 & cs(at)^* & cs(at)^*c & cs(at)^*ccs^\smile d^* \\ 0 & (at)^* & (at)^*c & (at)^*ccs^\smile d^* \\ 0 & 0 & 1 & cs^\smile d^* \\ 0 & 0 & 0 & d^* \end{pmatrix} (c+d)_{\mathrm{M}}\ (a+c+d)_{\mathrm{M}} \\
=\ & \begin{pmatrix} 0 & 0 & 0 & 0 \\ 0 & at(at)^* & at(at)^*c & at(at)^*ccs^\smile d^* \\ 0 & 0 & 0 & 0 \\ 0 & 0 & 0 & 0 \end{pmatrix} \begin{pmatrix} 0 & cs & 0 & 0 \\ 0 & 0 & c & 0 \\ 0 & 0 & 0 & cs^\smile \\ 0 & 0 & 0 & d \end{pmatrix} (a+c+d)_{\mathrm{M}}
\end{aligned}$$

$$= \begin{pmatrix} 0 & 0 & 0 & 0 \\ 0 & 0 & at(at)^*c & at(at)^*ccs^{\smile}d^* \\ 0 & 0 & 0 & 0 \\ 0 & 0 & 0 & 0 \end{pmatrix} \begin{pmatrix} 0 & cs & 0 & 0 \\ 0 & at & c & 0 \\ 0 & 0 & 0 & cs^{\smile} \\ 0 & 0 & 0 & d \end{pmatrix}$$

$$= \begin{pmatrix} 0 & 0 & 0 & 0 \\ 0 & 0 & 0 & at(at)^*ccs^{\smile}d^* \\ 0 & 0 & 0 & 0 \\ 0 & 0 & 0 & 0 \end{pmatrix}$$

$$= \quad \langle \text{ using (2) and (1), } at(at)^*ccs^{\smile} = (at)^*atccs^{\smile} = (at)^*accts^{\smile} = 0 \ \rangle$$

$$\mathbf{0}$$

This shows that none of the infinite executions specified by E^- can be executed by the system modelled by the given PDS.

Because the property verified in this example is not about states and does not contain the domain operator, it is possible to use the first definition of satisfaction (Definition 4). A suitable expression E (complementary to E^-) is

$$E \stackrel{\text{def}}{=} ((b+c+d)^*a(a+c+d)^*b)^\omega + \\ ((b+c+d)^*a(a+c+d)^*b)^* \, ((b+c+d)^\omega + (b+c+d)^*a^\omega) \ .$$

It satisfies $\mathbf{IM}^\omega \leqslant \mathbf{I}E_{\mathbf{M}}$, but the verification is much more tedious than the verification involving E^-.

4.5 Example 2

We present another example which uses the domain operator and the same PDS as the previous example. The property to verify is specified by an LTL formula based on states in which an atomic proposition represents a regular set of states; expressing this requires a domain operator. The LTL formula is $\neg \mathsf{F}p$. The proposition p characterizes states whose location is l_4 and stack contains the symbol s. The formula states that it is impossible to reach location l_4 and have symbol s in the stack (when the control is in this location). Because an atomic proposition $p \in At$ is involved, we must use the second definition of satisfaction (Definition 4). We have

$$E^- \stackrel{\text{def}}{=} (a+b+c+d)^*p(a+b+c+d)^\omega \ ,$$

because the only way to violate the property is to satisfy p at a given moment during the execution. We also have

$$p_{\mathrm{M}} \stackrel{\text{def}}{=} \begin{pmatrix} 0 & 0 & 0 & 0 \\ 0 & 0 & 0 & 0 \\ 0 & 0 & 0 & 0 \\ 0 & 0 & 0 & \ulcorner(t^{\smile *}s^{\smile}) \end{pmatrix} .$$

We have to show that $\mathbf{I}E^-_{\mathbf{M}} = \mathbf{0}$.

$$
\begin{aligned}
& \mathbf{I}E^{-}_{\mathrm{M}} = \mathbf{0} \\
\Leftrightarrow\ & \mathbf{I}((a+b+c+d)^{*}p(a+b+c+d)^{\omega})_{\mathrm{M}} = \mathbf{0} \\
\Leftrightarrow\ & \mathbf{I}(a+b+c+d)^{*}_{\mathrm{M}}p_{\mathrm{M}}(a+b+c+d)^{\omega}_{\mathrm{M}} = \mathbf{0} \\
\Leftarrow\ & \mathbf{I}(a+b+c+d)^{*}_{\mathrm{M}}p_{\mathrm{M}} = \mathbf{0} \\
\Leftrightarrow\ & \begin{pmatrix} \varepsilon & 0 & 0 & 0 \end{pmatrix} \begin{pmatrix} 0 & cs & 0 & 0 \\ 0 & at & c & 0 \\ 0 & 0 & bt^{\smile} & cs^{\smile} \\ 0 & 0 & 0 & d \end{pmatrix}^{*} \begin{pmatrix} 0 & 0 & 0 & 0 \\ 0 & 0 & 0 & 0 \\ 0 & 0 & 0 & 0 \\ 0 & 0 & 0 & \ulcorner(t^{\smile *}s^{\smile}) \end{pmatrix} = \mathbf{0} \\
\Leftrightarrow\ & \begin{pmatrix} \varepsilon & 0 & 0 & 0 \end{pmatrix} \begin{pmatrix} 1 & cs(at)^{*} & cs(at)^{*}c(bt^{\smile})^{*} & cs(at)^{*}c(bt^{\smile})^{*}cs^{\smile}d^{*} \\ 0 & (at)^{*} & (at)^{*}c(bt^{\smile})^{*} & (at)^{*}c(bt^{\smile})^{*}cs^{\smile}d^{*} \\ 0 & 0 & (bt^{\smile})^{*} & (bt^{\smile})^{*}cs^{\smile}d^{*} \\ 0 & 0 & 0 & d^{*} \end{pmatrix} \begin{pmatrix} 0 & 0 & 0 & 0 \\ 0 & 0 & 0 & 0 \\ 0 & 0 & 0 & 0 \\ 0 & 0 & 0 & \ulcorner(t^{\smile *}s^{\smile}) \end{pmatrix} = \mathbf{0} \\
\Leftrightarrow\ & \varepsilon cs(at)^{*}c(bt^{\smile})^{*}cs^{\smile}d^{*}\ulcorner(t^{\smile *}s^{\smile}) = 0 \\
\Leftrightarrow\ & \qquad \langle \text{ by Theorem 1(1) and locality of composition (Definition 1),} \\
& \qquad\quad a\ulcorner b = 0 \Leftrightarrow \ulcorner(a\ulcorner b) = 0 \Leftrightarrow \ulcorner(ab) = 0 \Leftrightarrow ab = 0 \ \rangle \\
& \varepsilon cs(at)^{*}c(bt^{\smile})^{*}cs^{\smile}d^{*}t^{\smile *}s^{\smile} = 0 \\
\Leftarrow\ & \qquad \langle \text{ Theorem 1(2c) } \rangle \\
& \varepsilon csa^{*}t^{*}cb^{*}t^{\smile *}cs^{\smile}d^{*}t^{\smile *}s^{\smile} = 0 \\
\Leftrightarrow\ & \qquad \langle \text{ commutativity axioms (2) and Theorem 1(2a,2b) } \rangle \\
& ca^{*}cb^{*}cd^{*}\varepsilon st^{*}t^{\smile *}s^{\smile}t^{\smile *}s^{\smile} = 0 \\
\Leftarrow\ & \qquad \langle \text{ by (1) and Theorem 1(3), } t^{*}t^{\smile *} \leqslant t^{*} + t^{\smile *} \ \rangle \\
& \varepsilon s(t^{*} + t^{\smile *})s^{\smile}t^{\smile *}s^{\smile} = 0 \\
\Leftrightarrow\ & \qquad \langle \text{ *-unfolding for } t^{*} \text{ and } t^{\smile *} \ \rangle \\
& \varepsilon s(1 + t^{*}t + t^{\smile}t^{\smile *})s^{\smile}t^{\smile *}s^{\smile} = 0 \\
\Leftrightarrow\ & \varepsilon(ss^{\smile} + st^{*}ts^{\smile} + st^{\smile}t^{\smile *}s^{\smile})t^{\smile *}s^{\smile} = 0 \\
\Leftrightarrow\ & \qquad \langle \text{ (1) and *-unfolding for } t^{\smile *} \ \rangle \\
& \varepsilon(t^{\smile}t^{\smile *} + 1)s^{\smile} = 0 \\
\Leftrightarrow\ & \varepsilon t^{\smile}t^{\smile *}s^{\smile} + \varepsilon s^{\smile} = 0 \qquad \text{—this holds by (1)}
\end{aligned}
$$

This shows that none of the infinite executions specified by E^{-} can be executed by the system modelled by the given PDS.

5 Conclusion

We have shown how to use omega algebra with domain to describe pushdown systems and specify properties they should satisfy. The choice of this extension of Kleene algebra allows us to express properties about both states and actions of the model. This is only a starting point and it opens the way for more research.

One research topic consists in determining the expressivity of the various classes of expressions $^{\ulcorner\omega}\text{-}\mathsf{SExp}_{\mathsf{P,B}}$, $^{\ulcorner\omega}\text{-}\mathsf{SExp}_{\mathsf{P}}$, $^{\omega}\text{-}\mathsf{SExp}_{\mathsf{P,B}}$ and $^{\omega}\text{-}\mathsf{SExp}_{\mathsf{P}}$, and their relationship with modal logics (we have hinted in the paper to connections with LTL).

One important goal we have is to mechanize the verification principles described in this paper. This will require investigating decidability issues in omega algebra with domain.

We also plan to study an alternative representation of pushdown systems. Matrices can be eliminated by adding a new stack symbol for each location. These symbols can be pushed on top of the stack to define the current location [1]. The resulting formalism is more homogeneous and might be more appropriate for proving some properties.

Acknowledgements

We thank the anonymous referees for detailed comments, as well as Claude Bolduc and Peter Höfner for interesting discussions.

References

1. Burkart, O., Caucal, D., Moller, F., Steffen, B.: Verification on infinite structures. In: Handbook of Process Algebra. Elsevier (2001) 545–623
2. Cohen, E.: Separation and reduction. In Backhouse, R., Oliveira, J.N., eds.: Proceedings of the 5th International Conference on Mathematics of Program Construction. Volume 1837 of Lecture Notes in Computer Science, Springer-Verlag (2000) 45–59
3. Conway, J.H.: Regular Algebra and Finite Machines. Chapman and Hall, London (1971)
4. Desharnais, J., Möller, B., Struth, G.: Kleene algebra with domain. ACM Transactions on Computational Logic, to appear. Preliminary version: Report No. 2003-07, Universität Augsburg, Institut für Informatik (2003)
5. Kozen, D.: A completeness theorem for Kleene algebras and the algebra of regular events. Information and Computation **110** (1994) 366–390
6. Kozen, D.: Kleene algebra with tests. ACM Transactions on Programming Languages and Systems (TOPLAS) **19** (1997) 427–443
7. Kozen, D., Smith, F.: Kleene algebra with tests: Completeness and decidability. In van Dalen, D., Bezem, M., eds.: 10th Int. Workshop on Computer Science Logic (CSL'96). Volume 1258 of Lecture Notes in Computer Science., Utrecht, The Netherlands, Springer-Verlag (1996) 244–259
8. Mathieu, V.: Vérification des systèmes à pile au moyen des algèbres de Kleene. Forthcoming Master's thesis, Université Laval, Québec, Canada (2006)
9. Schwoon, S.: Model-Checking Pushdown Systems. Ph.D. thesis, Technische Universität München (2002)

wp Is wlp

Bernhard Möller[1] and Georg Struth[2]

[1] Institut für Informatik, Universität Augsburg, D-86135 Augsburg, Germany
moeller@informatik.uni-augsburg.de
[2] Fakultät für Informatik, Universität der Bundeswehr München, D-85577 Neubiberg, Germany
struth@informatik.unibw-muenchen.de

Abstract. Using only a simple transition relation one cannot model commands that may or may not terminate in a given state. In a more general approach commands are relations enriched with termination vectors. We reconstruct this model in modal Kleene algebra. This links the recursive definition of the do od loop with a combination of the Kleene star and a convergence operator. Moreover, the standard wp operator coincides with the wlp operator in the modal Kleene algebra of commands. Therefore our earlier general soundness and relative completeness proof for Hoare logic in modal Kleene algebra can be re-used for wp. Although the definition of the loop semantics is motivated via the standard Egli-Milner ordering, the actual construction does not depend on Egli-Milner-isotony of the constructs involved.

1 Introduction

Total correctness has been extensively studied, a.o. using relational methods. One line of research (see e.g. [3, 8, 9, 12, 24]) provides strongly demonic semantics for regular programs. There, however, one cannot model commands that may or may not terminate in a given state. A second line of research (e.g. [4, 5, 13, 23, 25]) provides a weakly demonic semantics that allows such more general termination behaviour. We reconstruct the latter approach in modal Kleene algebra. This provides a new connection between the recursive definition of the do od loop and a combination of the Kleene star with convergence algebra. Moreover, it turns out that the standard wp operator coincides with the wlp operator of a suitable modal algebra of commands. Therefore the general soundness and relative completeness proof for Hoare logic in modal Kleene algebra given in [21] can be re-used for wp (where now, of course, expressiveness has to cover termination). Although the definition of the loop semantics is motivated via the standard Egli-Milner ordering, its actual construction does not depend on Egli-Milner-isotony of the constructs involved. A number of simple proofs are omitted due to lack of space; they can be found in the technical report [22].

2 Weak and Modal Semirings

A *weak semiring* is a quintuple $(S, +, 0, \cdot, 1)$ such that $(S, +, 0)$ is a commutative monoid and $(S, \cdot, 1)$ is a monoid such that $\cdot$ distributes over $+$ and is *left-strict*,

W. MacCaull et al. (Eds.): RelMiCS 2005, LNCS 3929, pp. 200–211, 2006.

i.e., $0 \cdot a = 0$. S is *idempotent* if $+$ is. In this case the relation $a \le b \overset{\text{def}}{\Leftrightarrow} a+b = b$ is an order, called the *natural order* on S, with least element 0. Moreover, $\cdot$ is isotone w.r.t. $\le$. A *semiring* is a weak semiring where $\cdot$ is also right-strict, i.e, $a \cdot 0 = 0$.

An important idempotent semiring is REL, the algebra of binary relations under union and composition over a set. Other interesting examples of weak idempotent semirings can be found within the set of endofunctions on an upper semilattice $(L, \sqcup, \bot)$ with least element $\bot$, where addition is defined as $(f+g)(x) = f(x) \sqcup g(x)$ and multiplication by function composition. The set of disjunctive functions (satisfying $f(x \sqcup y) = f(x) \sqcup f(y)$) forms a weak idempotent semiring. The induced natural order is the pointwise order $f \le g \Leftrightarrow \forall\, x \,.\, f(x) \le g(x)$. The subclass of strict disjunctive functions (satisfying additionally $f(\bot) = \bot$) even forms an idempotent semiring. These types of semirings include predicate transformer algebras and are at the centre of von Wright's algebraic approach [27].

A *(weak) test semiring* is a pair $(S, \mathsf{test}(S))$, where S is a(weak) idempotent semiring and $\mathsf{test}(S) \subseteq [0,1]$ is a Boolean subalgebra of the interval $[0,1]$ of S such that $0, 1 \in \mathsf{test}(S)$ and join and meet in $\mathsf{test}(S)$ coincide with $+$ and $\cdot$. This definition corresponds to the one in [18]. In REL the tests are partial identity relations (also called *monotypes* or *coreflexives*), encoding sets of states. We use $a, b, \ldots$ for general semiring elements and $p, q, \ldots$ for tests. By $\neg p$ we denote the complement of p in $\mathsf{test}(S)$ and set $p \to q = \neg p + q$. Moreover, we sometimes write $p \wedge q$ for $p \cdot q$ and $p \vee q$ for $p + q$. We freely use the Boolean laws for tests. An important property is

$$p \cdot a \cdot q \le 0 \Leftrightarrow a \cdot q \le \neg p \cdot a \,. \tag{1}$$

For ($\Rightarrow$) we note $a \cdot q = (p + \neg p) \cdot a \cdot q = p \cdot a \cdot q + \neg p \cdot a \cdot q = \neg p \cdot a \cdot q \le \neg p \cdot a$ by $q \le 1$. For ($\Leftarrow$) we have $a \cdot q \le \neg p \cdot a \Rightarrow p \cdot a \cdot q \le p \cdot \neg p \cdot a = 0 \cdot a = 0$.

A *(weak) modal semiring* is a pair $(S, [\,])$, where S is a (weak) test semiring and the *box* $[\,] : S \to (\mathsf{test}(S) \to \mathsf{test}(S))$ satisfies

$$p \le [a]q \Leftrightarrow p \cdot a \cdot \neg q \le 0 \,, \qquad [(a \cdot b)]p = [a]([b]p) \,.$$

The diamond is the de Morgan dual of the box, i.e., $\langle a \rangle p = \neg [a] \neg p$.

The box generalises the notion of the *weakest liberal precondition* wlp to arbitrary weak modal semirings. When a models a transition relation, $[a]p$ models those states from which execution of a is impossible or guaranteed to terminate in a state in set q. In REL one has $(x, x) \in [R]q \Leftrightarrow \forall\, y : xRy \Rightarrow (y, y) \in q$. In arbitrary weak semirings the box need not exist; for more details see [10].

The box axioms are equivalent to the equational domain axioms of [10]. In fact the *domain* of element a is $\ulcorner a \overset{\text{def}}{=} \neg [a]0$. Hence $\ulcorner a$ provides an abstract characterisation of the starting states of a. Conversely, $[a]q = \neg \ulcorner (a \cdot \neg q)$. Most of the consequences of the box axioms shown originally for strict modal semirings in [10] still hold for weak modal semirings (see [20]), in particular,

$$[a](p \cdot q) = [a]p \cdot [a]q \,, \qquad \langle a \rangle (p + q) = [a]p + [a]q \,, \tag{2}$$

$$[a + b]p = [a]p \cdot [b]p \,, \qquad \langle a + b \rangle p = \langle a \rangle p + \langle b \rangle p \,, \tag{3}$$

$$[p]q = p \to q \,, \qquad \langle p \rangle q = p \cdot q \,. \tag{4}$$

The latter implies $[1]q = q = \langle 1\rangle q$ as well as $[0]p = 1$ and $\langle 0\rangle p = 0$. By (3) $[a]$ and $\langle a\rangle$ are isotone. Moreover, by (3) box is antitone and diamond is isotone, i.e., $a \leq b \Rightarrow [b] \leq [a] \wedge \langle a\rangle \leq \langle b\rangle$. In a semiring (i.e., assuming right-strictness of $\cdot$) we additionally get

$$[a]1 = 1 , \qquad \langle a\rangle 0 = 0 . \tag{5}$$

A (weak) modal semiring S is *extensional* if for all $a, b \in S$ we have $[a] \leq [b] \Rightarrow b \geq a$. For example, REL is extensional. However, we can completely avoid extensionality, which makes our results much more widely applicable.

3 Commands and Correctness

While the previous section showed how to model the wlp-semantics of partial correctness in modal semirings, we now turn to total correctness. This requires modelling the states from which termination of a command can be guaranteed. The basic idea in [4, 5, 13, 23, 25] is to model a command as a pair (a, p) consisting of a transition a between states and a set p of states from which termination is guaranteed. Parnas [25] requires p to be contained in the domain of a. This allows distinguishing the "must-termination" given by p from the "may-termination" given by the domain and excludes "miraculous" commands that terminate without producing a result state. However, this entails that there is no neutral element w.r.t. demonic choice, since the obvious candidate fail with empty transition but full termination set does not satisfy Parnas's restriction. So there is not even an additive monoid structure. Nelson [23] dropped this restriction; we will base our treatment on his more liberal approach.

Assume now a modal semiring S (i.e., right-strictness of $\cdot$). we define the set of *commands* over S as $\mathrm{COM}(S) \stackrel{\text{def}}{=} S \times \mathsf{test}(S)$. In a command (a, p) the element $a \in S$ describes the state transition behaviour and $p \in \mathsf{test}(S)$ characterises the states with guaranteed termination; all states in $\neg p$ have the "result" of looping besides any proper states that may be reached from them under a. In this view the weakest (liberal) precondition can be defined as

$$\mathsf{wlp}.(a,p).q \stackrel{\text{def}}{=} [a]q , \qquad \mathsf{wp}.(a,p).q \stackrel{\text{def}}{=} p \cdot \mathsf{wlp}.(a,p).q .$$

Then by (5) we get $p = \mathsf{wp}.(a,p).1$, and hence, for command k, Nelson's *pairing condition* $\mathsf{wp}.k.q = \mathsf{wp}.k.1 \cdot \mathsf{wlp}.k.q$. The *guard* of a command,

$$\mathsf{grd}.(a,p) \stackrel{\text{def}}{=} \neg\mathsf{wp}.(a,p).0 = p \rightarrow \ulcorner a .$$

characterises the set of states that, if non-diverging, allow a transition under a. A command is called *total* if its guard equals one. The above formula links Parnas's condition on termination constraints with totality:

$$\mathsf{grd}.(a,p) = 1 \Leftrightarrow p \leq \ulcorner a .$$

Nelson remarks that totality of command k is also equivalent to Dijkstra's law of the excluded miracle $\mathsf{wp}.k.0 = 0$.

We now define the basic non-iterative commands.

$$\mathsf{fail} \stackrel{\text{def}}{=} (0,1)\ , \quad \mathsf{skip} \stackrel{\text{def}}{=} (1,1)\ , \quad \mathsf{loop} \stackrel{\text{def}}{=} (0,0)\ ,$$
$$(a,p) \,[\!]\, (b,q) \stackrel{\text{def}}{=} (a+b, p\cdot q)\ ,$$
$$(a,p)\ ;\ (b,q) \stackrel{\text{def}}{=} (a\cdot b, p\cdot [a]q)\ .$$

The straightforward proof of the following theorem can be found in [22].

Theorem 3.1 *The structure* $\mathrm{COM}(S) \stackrel{\text{def}}{=} (\mathrm{COM}(S), [\!], \mathsf{fail}, ;, \mathsf{skip})$ *over a semiring* S *is an idempotent weak semiring, the* command semiring *over* S. *However, it is not a semiring. The associated natural order on* $\mathrm{COM}(S)$ *is*

$$(a,p) \le (b,q) \Leftrightarrow a \le b \wedge p \ge q\ . \tag{6}$$

By antitony of box we obtain for commands k, l

$$k \le l \Rightarrow \mathsf{wlp}.k \ge \mathsf{wlp}.l \wedge \mathsf{wp}.k \ge \mathsf{wp}.l\ ,$$

where $\ge$ is the pointwise order between test transformers. The second conjunct is the converse of the usual refinement relation. If the underlying semiring is extensional then the converse implication holds as well.

By standard order theory, if S is a complete lattice then $\mathrm{COM}(S)$ is a complete lattice again with

$$\bigsqcup\{(a_i,p_i) \,|\, i \in I\} = (\bigsqcup\{a_i \,|\, i\in I\}, \bigsqcap\{a_i \,|\, i \in I\}).$$

Likewise, if S has a greatest element $\top$ then $\mathsf{chaos} \stackrel{\text{def}}{=} (\top, 0)$ is the greatest element of $\mathrm{COM}(S)$, whereas $\mathsf{havoc} \stackrel{\text{def}}{=} (\top, 1)$ represents the most nondeterministic everywhere terminating command.

4 Modalities for Commands

We now want to make $\mathrm{COM}(S)$ into a weak *modal* semiring as well. From (6) and $p \le 1$ it is immediate that $(a,p) \le \mathsf{skip} \Leftrightarrow a \le 1 \wedge p = 1$. It is easy to check that the elements of this shape are closed under ; and $[\!]$. Therefore it seems straightforward to use the *test commands* $\underline{p} \stackrel{\text{def}}{=} (p,1)$ and to choose

$$\mathsf{test}(\mathrm{COM}(S)) \stackrel{\text{def}}{=} \{\underline{p} \,|\, p \in \mathsf{test}(S)\}\ .$$

Clearly, this yields a Boolean algebra with $\neg\underline{p} = \underline{\neg p}$, $\underline{0} = \mathsf{fail}$ and $\underline{1} = \mathsf{skip}$.

Using this, we can also introduce a guarded statement as

$$p \longrightarrow k = \underline{p}\ ;\ k\ . \tag{7}$$

To check the first box axiom we calculate, using the definitions and $[a]1 = 1$ (we assume a semiring, i.e., right-strictness of $\cdot$),

$$(p,1)\ ;\ (c,r)\ ;\ \neg(q,1) = (p\cdot c, p \to r)\ ;\ (\neg q, 1) = (p\cdot c\cdot \neg q, p \to r)\ ,$$

so that, by (6) and shunting,

$$(p,1)\,;(c,r)\,;\neg(q,1) \le (0,1) \Leftrightarrow p\cdot c\cdot\neg q \le 0 \wedge p \rightarrow r \ge 1$$
$$\Leftrightarrow p \le [c]q \wedge p \le r \Leftrightarrow p \le \mathsf{wp}.(c,r).q\ .$$

For the second box axiom we calculate, using the definitions, the second box axiom, conjunctivity of $[a]$ and the definitions again,

$$\begin{aligned}\mathsf{wp}.((a,p)\,;(b,q)).r &= p\cdot[a]q\cdot[a\cdot b]r = p\cdot[a]q\cdot[a]([b]r)\\ &= p\cdot[a](q\cdot[b]r) = \mathsf{wp}.(a,p).(\mathsf{wp}(b,q).r)\ .\end{aligned}$$

Altogether, we have shown

Theorem 4.1 *Setting* $[k]\underline{q} \stackrel{\text{def}}{=} \underline{\mathsf{wp}.k.q}$ *makes* COM(S) *a weak modal semiring.*

Hence the general definitions for modal semirings tie in nicely with the wp semantics. This equation explains the title of our paper: wp is nothing but wlp in the weak modal semiring of commands.

Now the usual properties of wlp and wp come for free, since both are box operators in modal semirings:

$$\begin{aligned}\mathsf{w(l)p.fail}.r = 1\ ,\quad &\mathsf{w(l)p.skip}.r = r\ ,\\ \mathsf{w(l)p}.(k \mathbin{[]} l).r &= \mathsf{w(l)p}.k.r \wedge \mathsf{w(l)p}.l.r\ ,\\ \mathsf{w(l)p}.(p \longrightarrow l).r &= p \rightarrow \mathsf{w(l)p}.l.r\ .\end{aligned}$$

The only command that does not have an abstract counterpart in all modal semirings is loop. For it the box operators behave asymmetrically:

$$\mathsf{wlp.loop}.r = 1\ ,\qquad \mathsf{wp.loop}.r = 0\ . \tag{8}$$

Theorem 4.1 implies, moreover, that for $k \in$ COM(S) we have $\ulcorner k = \underline{\mathsf{grd}.k}$, another pleasing connection with the general theory of weak modal semirings. From this observation we obtain the usual guard laws for free:

$$\begin{aligned}\mathsf{grd.fail} = 0\ ,\quad \mathsf{grd.skip} = 1\ ,\quad &\mathsf{grd}.(p \longrightarrow k) = p\cdot\mathsf{grd}.k\ ,\\ \mathsf{grd}.(k \mathbin{[]} l) = \mathsf{grd}.k + \mathsf{grd}.l\ ,\quad &\mathsf{grd}.(k\,;l) = \neg\mathsf{wp}.k.(\neg\mathsf{grd}.l)\ .\end{aligned}$$

Additionally, $\mathsf{grd.loop} = 1$.

Finally, we define Nelson's biased choice operator $\boxminus$ that will be used in the definition of the if fi command in the next section:

$$k \boxminus l \stackrel{\text{def}}{=} k \mathbin{[]} (\neg\mathsf{grd}.k \longrightarrow l)\ .$$

Then $\boxminus$ is the *overwrite* operation in COM(S) that in general weak modal semirings is defined as $a|b \stackrel{\text{def}}{=} a + \neg\ulcorner a\cdot b$. A corresponding operator is used in B [1] and Z [26], but also in calculating with pointer and object structures [14, 19]. This operation satisfies a number of useful laws from which we get three properties of biased choice for free:

$$\begin{aligned}a|0 = a = 0|a\ ,\qquad &k \boxminus \mathsf{fail} = k = \mathsf{fail} \boxminus k\ ,\\ a|(b|c) = (a|b)|c\ ,\qquad &k \boxminus (l \boxminus m) = (k \boxminus l) \boxminus m\ ,\\ \ulcorner(a+b) = \ulcorner a + \ulcorner b\ ,\qquad &\mathsf{grd}.(k \boxminus l) = \mathsf{grd}.k + \mathsf{grd}.l\ .\end{aligned}$$

To ease reading we will simply write p instead of $\underline{p}$ in the remainder; the context will make clear where the lifting would have to be filled in.

5 Loops, Kleene Algebra and the Egli-Milner Order

So far we have not dealt with iteration. We show now that the semantics of Dijkstra's do od loop can be defined in closed terms if we assume that the underlying modal semiring S is a *convergence algebra*, that is, has additional operations * of finite iteration and $\triangle$ that yields termination information.

Let us give the necessary definitions. A *weak left Kleene algebra* is a structure $(S, ^*)$ such that S is an idempotent weak semiring and the *star* * satisfies, for $a, b, c \in S$, the *left unfold* and *left induction axioms*

$$1 + a \cdot a^* \leq a^* , \qquad b + a \cdot c \leq c \Rightarrow a^* \cdot b \leq c .$$

Hence $a^* \cdot b$ is the least pre-fixpoint and the least fixpoint of the function $\lambda x . a \cdot x + b$. As a consequence, star is $\leq$-isotone. Symmetrically, a *weak right Kleene algebra* $(S, ^*)$ satisfies the *right unfold* and *right induction axioms*

$$1 + a^* \cdot a \leq a^* , \qquad b + c \cdot a \leq c \Rightarrow b \cdot a^* \leq c .$$

A *weak left (right) modal Kleene algebra* is a weak left (right) Kleene algebra in which S is modal. Finally, a left (right) modal Kleene algebra is a weak left (right) modal Kleene algebra with a full underlying semiring. The law

$$a \cdot c \leq c \cdot b \Rightarrow a^* \cdot c \leq c \cdot b^* \tag{9}$$

holds in every left Kleene algebra: For the right-hand side it suffices by left induction to show that $c + a \cdot c \cdot b^* \leq c \cdot b^*$. But $c + a \cdot c \cdot b^* \leq c + c \cdot b \cdot b^* = c \cdot (1 + b \cdot b^*) \leq c \cdot b^*$. Even in weak left modal Kleene algebras we have $p^* = 1$ for all $p \in \mathsf{test}(S)$ and the following induction law [10].

Lemma 5.1 *In a left modal Kleene algebra,* $q \leq p \cdot [a]q \Rightarrow q \leq p \cdot [a^*]p$.
Proof. Assume $q \leq p \cdot [a]q$, i.e., $q \leq p \wedge q \leq [a]q$. The claim is equivalent to $q \leq p \wedge q \leq [a^*]q$. The first conjunct is an assuption. For the second one we calculate $q \leq [a]q \Leftrightarrow a \cdot \neg q \leq \neg q \cdot a \Rightarrow a^* \cdot \neg q \leq \neg q \cdot a^* \Leftrightarrow q \leq [a^*]q$. The first and third steps follow from (1), the second one from (9). □

Now we are ready to show

Theorem 5.2 *The command semiring over a left Kleene algebra can be made into a left modal Kleene algebra by setting* $(a, p)^* \stackrel{\text{def}}{=} (a^*, [a^*]p)$.
Proof. For the left unfold axiom we calculate, using the definitions, the second box axiom, (3) and the left unfold axiom for S,

$$\begin{aligned}(1,1) \,[\!]\, (a,p) \,;\, (a^*, [a^*]p) &= (1 + a \cdot a^*, p \cdot [a]([a^*]p) \\ &= (1 + a \cdot a^*, [1 + a \cdot a^*]p) = (a^*, [a^*]p) .\end{aligned}$$

For the left induction axiom assume $(b, q) \,[\!]\, (a, p) \,;\, (c, r) \leq (c, r)$, i.e., $b + a \cdot c \leq c \wedge q \cdot p \cdot [a]r \geq r$, which by left star induction for S and Lemma 5.1 implies

$$a^* \cdot b \leq c \wedge [a^*](q \cdot p) \geq r . \tag{$*$}$$

Now we calculate, using the definitions, conjunctivity of $[a^*]$ and $(*)$,

$$(a^*, [a^*]p) \,;\, (b, q) = (a^* \cdot b, [a^*]p \cdot [a^*]q) = (a^* \cdot b, [a^*](p \cdot q)) \leq (c, r) . \qquad □$$

Analogously one shows that under the same definition of star the command semiring over a right Kleene algebra is a right Kleene algebra again.

A *weak Kleene algebra* is a structure $(S,^*)$ that is both a left and a right Kleene algebra over a weak semiring S; it is a *Kleene algebra* if S is a strict semiring. The notion of a *(weak) modal Kleene algebra* is defined analogously. Summarizing the above remarks we have

Theorem 5.3 *The command semiring over a (modal) Kleene algebra can again be made into a (modal) Kleene algebra by the above definition.*

Let us now look at the semantics x of the loop $\mathsf{do}\, k\, \mathsf{od}$. It is supposed to satisfy the recursion equation (cf. [23])

$$x = (k\,;x) \,[]\, \neg\mathsf{grd}.k \longrightarrow \mathsf{skip}\ . \tag{10}$$

Given the Kleene algebra structures of commands it is tempting to define the semantics of the loop $\mathsf{do}\, k\, \mathsf{od}$ as the $\leq$-least solution, viz. by the standard expression $k^*\,;\neg grd.k$. However, for $k = \mathsf{skip}$ we obtain $k^*\,;\neg grd.k = \mathsf{skip}\,;\mathsf{fail} = \mathsf{fail}$, whereas the semantics of $\mathsf{do\,skip\,od}$ should be loop.

So $\leq$ is not the adequate approximation order for recursions such as the one for loops; it is in a sense "too angelic". Instead, one uses the *Egli-Milner approximation relation* $\sqsubseteq$ over COM(S), given by (see [23])

$$k \sqsubseteq_{\mathrm{EM}} l \Leftrightarrow \mathsf{wp}.k \leq \mathsf{wp}.l \wedge \mathsf{wlp}.l \leq \mathsf{wlp}.k\ .$$

It is an order iff S is extensional. Equivalently, $k \sqsubseteq_{\mathrm{EM}} l \Leftrightarrow \mathsf{wp}.k.1 \leq \mathsf{wp}.l.1 \wedge \mathsf{wp}.k \leq \mathsf{wlp}.l \wedge \mathsf{wlp}.l \leq \mathsf{wlp}.k$. Thus, to allow S to be non-extensional, we define

$$(a,p) \sqsubseteq (b,q) \stackrel{\text{def}}{=} p \leq q \wedge \mathsf{wp}.(a,p) \leq \mathsf{wlp}.(b,q) \wedge a \leq b\ .$$

Lemma 5.4 *The relation* $\sqsubseteq$ *is an order with least element* loop.

Proof. Antisymmetry follows from that of $\leq$, while reflexivity is immediate from that of $\leq$ and $\mathsf{wp}.k \leq \mathsf{wlp}.k$. For transitivity, assume $(a,p) \sqsubseteq (b,q)$ and $(b,q) \sqsubseteq (c,r)$. From transitivity of $\leq$ we get $a \leq c$ and $p \leq r$. Consider now an arbitrary $s \in \mathsf{test}(S)$. First, $\mathsf{wp}.(a,p).s = p \cdot [a]s = q \cdot p \cdot [a]s = q \cdot \mathsf{wp}.(a,p).s$, since $p \leq q$. Now, $\mathsf{wp}.(a,p).s = q \cdot \mathsf{wp}.(a,p).s \leq q \cdot \mathsf{wlp}.(b,q).s = \mathsf{wp}.(b,q).s \leq \mathsf{wlp}.(c,r).s$. Finally, $\sqsubseteq$-leastness of loop follows from $\leq$-leastness of 0 and (8). □

The meaning of a recursive command then is the $\sqsubseteq$-least fixpoint of the associated function (provided it exists; $\sqsubseteq$ need not induce a cpo in general). A treatment of full recursion will be the subject of a later paper. To actually find a convenient representation of the $\sqsubseteq$-least solution of (10) we need an additional concept that captures termination information.

A *convergence algebra* [11] is a pair $(S, \triangle)$ where S is a left modal Kleene algebra and the *convergence* operation $\triangle : S \to \mathsf{test}(S)$ satisfies, for all $a \in S$ and $p, q \in \mathsf{test}(S)$, the unfold and coinduction laws

$$[a](\triangle a) \leq \triangle a, \qquad [a]p \cdot q \leq p \Rightarrow \triangle a \cdot [a^*]q \leq p\ .$$

This axiomatises $\triangle a \cdot [a^*]q$ as the least pre-fixpoint and least fixpoint of the function $\lambda p \,.\, [a]p \cdot q$; in particular, $\triangle a$ is the least pre-fixpoint and the least fixpoint of $[a]$. For the pre-fixpoints of $[a]$ we have $[a]p \le p \Leftrightarrow \neg p \le \langle a \rangle \neg p$. Since $q \le \langle a \rangle q$ means that every state in q has a successor in q, the complements of the pre-fixpoints consist of states with the possibility of nontermination under iterated execution of a. Hence the *least* pre-fixpoint $\triangle a$ characterises the states from which a does not admit infinite transition sequences. It corresponds to the *halting predicate* of the modal μ [16]). Hence we call an element a *Noetherian* if $\triangle a = 1$. For $p \in \mathsf{test}(S)$ we have $\triangle p = \neg p$.

For our treatment of loops we now assume a convergence algebra as the underlying semiring. First we extend the convergence operation to commands and define a particular command that captures termination information by setting, for $a \in S, p \in \mathsf{test}(S)$ and $k \in \mathrm{COM}(S)$,

$$\triangle(a,p) \stackrel{\text{def}}{=} \triangle a \;, \qquad \mathsf{trm}.k \stackrel{\text{def}}{=} (0, \triangle k) \;.$$

We define command k to be *Noetherian*, in signs $\mathrm{NOE}(k)$, if $\triangle k = 1$.

Lemma 5.5 *1.* $\mathsf{trm}.(a,p)$ *is the* $\sqsubseteq$*-least solution of the equation* $x = (a,1)\,;x$.
2. $\mathsf{trm}.(a,p)$ *(like all commands of the form* $(0,q)$*) is a left zero w.r.t.* $;$.

To tackle the semantics of the loop $\mathsf{do}\,k\,\mathsf{od}$, we slightly generalise and define the command $\mathsf{do}\,k\,\mathsf{exit}\,l\,\mathsf{od}$ as the $\sqsubseteq$-least solution of the recursion equation

$$x \;=\; (k\,;x) \;[]\; \neg\mathsf{grd}.k \longrightarrow l \;. \tag{11}$$

Let us calculate conditions for such a solution (y,t). Assume

$$(y,t) \;=\; ((a,p)\,;(y,t)) \;[]\; \neg g \longrightarrow (b,q)$$

where $g \stackrel{\text{def}}{=} \mathsf{grd}.(a,p)$. Plugging in the definitions, we have to satisfy

$$y \;=\; a \cdot y + \neg g \cdot b \;, \qquad t \;=\; [a]t \cdot p \cdot (\neg g \to q) \;.$$

To get a $\sqsubseteq$-least solution (y,t), we have to use the $\le$-least solutions of these equations, which by left star induction and convergence induction are

$$y \;=\; a^* \cdot \neg g \cdot b \;, \qquad t \;=\; \triangle a \cdot [a^*](p \cdot \neg g \to q) \;.$$

We show that (y,t) is indeed the $\sqsubseteq$-least solution of (11). Consider an arbitrary solution (z,u). In remains to verify that $\mathsf{wp}.(y,t) \le \mathsf{wlp}.(z,u)$. First, for arbitrary s, using the fixpoint property of (z,u), (3) and the second box axiom

$$\mathsf{wlp}.(z,u).s \;=\; [a \cdot z + \neg g \cdot b]s \;=\; [a]([z]s) \cdot [\neg g \cdot b]s \;=\; [a](\mathsf{wlp}.(z,u).s) \cdot [\neg g \cdot b]s \;.$$

Hence by the convergence induction axiom we have

$$\triangle a \cdot [a^*]([\neg g \cdot b]s) \le \mathsf{wlp}.(z,u).s \;.$$

Now, by definition and the second box axiom,

$$\begin{aligned}\mathsf{wp}.(y,t).s &= t \cdot [y]s \;=\; \triangle a \cdot [a^*](p \cdot \neg g \to q) \cdot [a^* \cdot \neg g \cdot b]s \\ &\le \triangle a \cdot [a^* \cdot \neg g \cdot b]s \;=\; \triangle a \cdot [a^*]([\neg g \cdot b]s) \;.\end{aligned}$$

Next, we bring our least solution (y,t) into somewhat nicer form:

$$
\begin{aligned}
& (a^* \cdot \neg g \cdot b, \triangle a \cdot [a^*](p \cdot \neg g \to q) \\
= & \quad \{\!\![\text{ definition of } [\,] \;]\!\!\} \\
& (a^* \cdot \neg g \cdot b, [a^*](p \cdot \neg g \to q)) \, [\,] \, (0, \triangle a) \\
= & \quad \{\!\![\text{ conjunctivity of } [a^*] \;]\!\!\} \\
& (a^* \cdot \neg g \cdot b, [a^*]p \cdot [a^*](\neg g \to q)) \, [\,] \, (0, \triangle a) \\
= & \quad \{\!\![\text{ definition of } ; \;]\!\!\} \\
& ((a^*, [a^*]p) \,;\, (\neg g \cdot b, \neg g \to q)) \, [\,] \, (0, \triangle a) \\
= & \quad \{\!\![\text{ definition of star and } \longrightarrow \;]\!\!\} \\
& ((a, p)^* \,;\, (\neg g \longrightarrow (b, q)) \, [\,] \, (0, \triangle a) \,.
\end{aligned}
$$

Altogether we have shown

Theorem 5.6 $\mathsf{do}\, k \,\mathsf{exit}\, l \,\mathsf{od} \;=\; (k^* \,;\, \neg\mathsf{grd}.k \longrightarrow l) \, [\,] \, \mathsf{trm}.k$.

Note that this theorem does not depend on completeness of the underlying semiring nor on Egli-Milner-isotony of the command-building operations involved. Moreover, the form of the expressions in the semantics has arisen directly from the star and convergence axioms.

For $l = \mathsf{skip}$ we obtain the semantics $\mathsf{do}\, k \,\mathsf{od} \;=\; (k^* \,;\, \neg\mathsf{grd}.k) \, [\,] \, \mathsf{trm}.k$. And now, indeed, $\mathsf{do}\,\mathsf{skip}\,\mathsf{od} \;=\; \mathsf{loop}$. We have the following connection.

Lemma 5.7 $\mathsf{do}\, k \,\mathsf{exit}\, l \,\mathsf{od} \;=\; \mathsf{do}\, k \,\mathsf{od} \,;\, l$.

Moreover, we obtain the semantics of the $\mathsf{if}\;\mathsf{fi}$ command which, according to [23], should be the $\sqsubseteq$-least solution of the equation $x \;=\; k \,[\!]\, x$. Plugging in the definition of $[\!]$ we can rewrite that into

$$x \;=\; (\neg\mathsf{grd}.k \,;\, x) \, [\,] \, \mathsf{grd}.k \longrightarrow k$$

and the above theorem and lemma yield

$$\mathsf{if}\, k \,\mathsf{fi} \;=\; \mathsf{do}\, \neg\mathsf{grd}.k \,\mathsf{exit}\, k \,\mathsf{od} \;=\; \mathsf{do}\, \neg\mathsf{grd}.k \longrightarrow \mathsf{skip}\,\mathsf{od} \,;\, k \;.$$

In particular, $\mathsf{if}\,\mathsf{fail}\,\mathsf{fi} \;=\; \mathsf{loop}$.

6 Hoare Calculus for WP

Since we have seen that wp is wlp in an appropriate weak modal semiring, we can use the general soundness and relative completeness proof for propositional Hoare logic from [21], since that proof nowhere uses strictness of the underlying semiring. This yields fairly quickly a sound and relatively complete proof system for wp. In an arbitrary weak modal semiring, soundness of a Hoare triple $\{p\}\ a\ \{q\}$ with tests p, q is defined as $p \le [a]q$. The proof in [21], an abstract representation of the standard proof (see e.g. [2]) shows that relative completeness is achieved if the triple $\{[a]q\}\ a\ \{q\}$ is derivable for every command a and every test q (where one has to assume sufficient expressiveness, i.e., that the assertion logic is rich enough to express all tests $[a]q$). For the atomic commands this yields the axioms

$$\{1\}\ \mathsf{fail}\ \{q\} \qquad \{0\}\ \mathsf{loop}\ \{q\} \qquad \{q\}\ \mathsf{skip}\ \{q\} \qquad \{\triangle k\}\ \mathsf{trm}.k\ \{q\}$$

An appropriate rule for demonic choice is

$$\frac{\{p\}\ k\ \{r\} \qquad \{q\}\ l\ \{r\}}{\{p \cdot q\}\ k \,[\!]\, l\ \{r\}}$$

For the loop we observe that, except for the termination part, $\mathsf{do}\, k\, \mathsf{od}$ behaves like $\mathsf{while}\ \mathsf{grd}.k\ \mathsf{do}\ k$. For that, the usual while rule

$$\frac{\{q \wedge p\}\ k\ \{q\}}{\{q\}\ \mathsf{while}\ p\, \mathsf{do}\ k\ \{\neg p \wedge q\}}$$

is sound and relatively complete. Combining this with the rule for choice we obtain, after some simplification, the sound and relatively complete rule

$$\frac{\{p\}\ k\ \{p\}}{\{\triangle k \cdot p\}\ \mathsf{do}\, k\, \mathsf{od}\ \{p \cdot \neg \mathsf{grd}.k\}}$$

From that one can derive the rule

$$\frac{\{p\}\ k\ \{p\} \qquad \mathrm{NOE}(k)}{\{p\}\ \mathsf{do}\, k\, \mathsf{od}\ \{p \cdot \neg \mathsf{grd}.k\}}$$

7 Extensions: Angelic Choice and Infinite Iteration

In this section we give two extensions of the basic language of commands.

First, in $\mathrm{COM}(S)$ an angelic choice operator can be defined as

$$(a,p) \sqcup (b,q) \stackrel{\text{def}}{=} (a+b, p+q) \ .$$

It is clearly idempotent, associative and commutative.

Lemma 7.1 *The operators $\sqcup$ and $[\!]$ distribute over each other; in particular, $\sqcup$ is $\leq$-isotone. Moreover, $k \,[\!]\, l \leq k \sqcup l$ with $\mathsf{wlp}.(k \sqcup l) = \mathsf{wlp}.(k \,[\!]\, l)$ and*

$$\mathsf{wp}.(k \sqcup l).r = \mathsf{wp}.k.r \cdot \mathsf{wlp}.l.r + \mathsf{wp}.l.r \cdot \mathsf{wlp}.k.r \ .$$

The second extension concerns infinite iteration. A *weak omega algebra* [6, 20] is a structure $(S, {}^\omega)$ consisting of a left Kleene algebra S and a unary *omega* operation ${}^\omega$ that satisfies, for $a, b, c \in S$, the *unfold* and *coinduction laws*

$$a^\omega = a \cdot a^\omega \ , \tag{12}$$

$$c \leq a \cdot c + b \Rightarrow c \leq a^\omega + a^* \cdot b \ . \tag{13}$$

This axiomatises $a^\omega + a^* \cdot b$ as the greatest fixpoint of the function $\lambda x \,.\, a \cdot x + b$. In particular, a^ω is the greatest fixpoint of $\lambda x \,.\, a \cdot x$. Every weak omega algebra S has a greatest element $\top = 1^\omega$.

As in the case of Kleene algebras, we want to make the command semiring $\mathrm{COM}(S)$ over a weak omega algebra into a weak omega algebra, too. Let us find solutions to the recursion equation

$$(y, t) = ((a, p) ; (y, t)) \,[\!]\, (b, q) .$$

From the definitions we get the equations

$$y = a \cdot y + b , \qquad t = p \cdot [a]t \cdot q .$$

To get a $\leq$-greatest solution in $\mathrm{COM}(S)$ we have to take the $\leq$-greatest solution for y and the $\leq$-least solution for t, which are, by omega coinduction and convergence induction,

$$y = a^\omega + a^* \cdot b , \qquad t = \Delta a \cdot [a^*](p \cdot q) .$$

Setting $(b, q) = \mathsf{fail}$, we obtain

Lemma 7.2 *Over a weak omega algebra S that is also a convergence algebra, the semiring* $\mathrm{COM}(S)$ *can be made into a weak omega algebra by setting*

$$(a, p)^\omega \stackrel{\text{def}}{=} (a^\omega, \Delta a \cdot [a^*]p) .$$

8 Conclusion and Outlook

The modal view of the weakly demonic semantical model has led to a number of new insights. In particular, the possibility of combining the "angelic" semantics provided by the star operation with termination information through a demonic choice to get the appropriate demonic semantics seems to be novel.

The techniques of the present paper have in [15] been adapted to give an algebraic semantics for the normal designs as used in Hoare and He's Unifying Theories of Programming [17].

Future work will concern an analogous treatment of full recursion as well as applications to deriving new refinement laws.

Acknowledgements. We are grateful to J. Desharnais, W. Guttmann, P. Höfner and the anonymous referees for helpful discussions and remarks.

References

1. J.-R. Abrial: The B-Book. Cambridge University Press 1996
2. K.-R. Apt, E.-R. Olderog: Verification of Sequential and Concurrent Programs, 2nd edition. Springer 1997
3. R. C. Backhouse, J. van der Woude: Demonic operators and monotype factors. Mathematical Structures in Computer Science, 3(4):417–433 (1993)
4. R. Berghammer, H. Zierer: Relational algebraic semantics of deterministic and non-deterministic programs. Theoretical Computer Science, 43:123–147 (1986)
5. M. Broy, R. Gnatz, M. Wirsing: Semantics of nondeterministic and non-continuous constructs. In F.L. Bauer, M. Broy (eds.): Program construction. Lecture Notes in Computer Science **69**. Berlin: Springer 1979, 553–592
6. E. Cohen: Separation and reduction. In R. Backhouse, J. N. Oliveira(eds.): Mathematics of Program Construction. Lecture Notes in Computer Science **1837**. Berlin: Springer 2000, 45–59

7. J.H. Conway: Regular algebra and finite machines. London: Chapman and Hall 1971
8. J. Desharnais, N. Belkhiter, S.B.M. Sghaier, F. Tchier, A. Jaoua, A. Mili, and N. Zaguia: Embedding a demonic semilattice in a relation algebra. Theoretical Computer Science 149:333–360 (1995)
9. J. Desharnais, A. Mili, T.T. Nguyen: Refinement and demonic semantics. In C. Brink, W. Kahl, G. Schmidt (eds): Relational methods in computer science, Chapter 11. Springer 1997, 166–183
10. J. Desharnais, B. Möller, G. Struth: Kleene algebra with domain. ACM Transactions on Computational Logic (to appear)
11. J. Desharnais, B. Möller, G. Struth: Termination in modal Kleene algebra. In J.-J. Lévy, E. Mayr, and J. Mitchell, editors, Exploring new frontiers of theoretical informatics. IFIP International Federation for Information Processing Series **155**. Kluwer 2004, 653–666
12. J. Desharnais, B. Möller F. Tchier: Kleene under a modal demonic star. Journal on Logic and Algebraic Programming, Special Issue on Relation Algebra and Kleene Algebra, 2005 (to appear)
13. H. Doornbos: A relational model of programs without the restriction to Egli-Milner-monotone constructs. In E.-R. Olderog (ed.): Programming concepts, methods and calculi. North-Holland 1994, 363–382
14. T. Ehm: The Kleene algebra of nested pointer structures: theory and applications. Universität Augsburg, PhD Thesis, Dec. 2003
15. W. Guttmann, B. Möller: Modal design algebra. In: S. Dunne (ed.): Proc. First International Symposium on Unifying Theories of Programming, Walworth Castle, 5–7 Feb. 2006. To appear in LNCS. Preliminary version: Institut für Informatik, Universität Augsburg, Report 2005-15
16. D. Harel, D. Kozen, J. Tiuryn: Dynamic Logic. MIT Press 2000
17. C.A.R. Hoare, J. He: Unifying theories of programming. Prentice Hall 1998
18. D. Kozen: Kleene algebras with tests. ACM TOPLAS 19:427–443 (1997)
19. B. Möller: Towards pointer algebra. Science of Computer Programming 21:57–90 (1993)
20. B. Möller: Lazy Kleene algebra. In D. Kozen (ed.): Mathematics of Program Construction. Lecture Notes in Computer Science **3125**. Berlin: Springer 2004, 252–273
21. B. Möller, G. Struth: Modal Kleene algebra and partial correctness. In C. Rattray, S. Maharaj, C. Shankland (eds.): Algebraic methods and software technology. Lecture Notes in Computer Science **3116**. Berlin: Springer 2004, 379–393
22. B. Möller, G. Struth: WP is WLP. Institut für Informatik, Universität Augsburg, Report 2004-14
23. G. Nelson: A generalization of Dijkstra's calculus. ACM Transactions on Programming Languages and Systems 11:517–561 (1989)
24. T.T. Nguyen: A relational model of nondeterministic programs.International J. Foundations Comp. Sci. 2:101–131 (1991)
25. D. Parnas: A generalized control structure and its formal definition. Commun. ACM 26:572–581 (1983)
26. J. M. Spivey: Understanding Z. Cambrigde University Press 1988
27. J. von Wright: Towards a refinement algebra. Science of Computer Programming 51:23–45 (2004)

Relational Representability for Algebras of Substructural Logics*

Ewa Orłowska[1] and Anna Maria Radzikowska[2]

[1] National Institute of Telecommunications,
Szachowa 1, 04–894 Warsaw, Poland
orlowska@itl.waw.pl

[2] Faculty of Mathematics and Information Science,
Warsaw University of Technology,
Plac Politechniki 1, 00–661 Warsaw, Poland
annrad@mini.pw.edu.pl

Abstract. Representation theorems for the algebras of substructural logics FL, FLe, FLc, and FLw are presented. The construction of the representation algebras is an extension of the constructions from Urquhart ([25]) and Allwein and Dunn ([1]). Namely, the representation algebras are built from the frames, which are appropriately associated to substructural logics. As a by–product we obtain a Kripke–style frame semantics for these logics.

1 Introduction

In this paper we present representation theorems for the algebras of basic substructural logics, FL, FLe, FLc, and FLw, within a framework of Urquhart [20] and Allwein–Dunn [1] representability theory (UAD–framework), generalized to a duality between the algebras and abstract frames (relational systems).

Let *Alg* be a class of algebras and let *Alg*–frames be a class of the corresponding frames. With every algebra from *Alg* we associate a canonical frame (a dual space), and with every *Alg*–frame we associate a complex algebra. We show that canonical frames belong to the class of *Alg*–frames and that complex algebras belong to the class *Alg*. Then a representation theorem states that every algebra of *Alg* is embeddable into the complex algebra of its canonical frame. The recent examples of applications of a generalized UAD–framework can be found in [6], [7], [21], and in a survey [8].

In the majority of representability results that can be found in the literature the class of *Alg*–frames consists of canonical frames. Often they are additionally equipped with a topology.

For distributive lattices this kind of representability is straightforward and well known. Let $(W, \wedge, \vee)$ be a distributive lattice. A *dl*–frame is a system

* The work was carried on in the framework of COST Action 274/TARSKI on *Theory and Applications of Relational Structures as Knowledge Instruments* (www.tarski.org).

W. MacCaull et al. (Eds.): RelMiCS 2005, LNCS 3929, pp. 212–224, 2006.

$(X, \leqslant)$, where X is a non–empty set and $\leqslant$ is a partial order on X. The canonical frame of the lattice W is the structure $(X(W), \subseteq)$, where $X(W)$ is the family of prime filters of W and $\subseteq$ is set inclusion. The complex algebra of a *dl*–frame X is an algebra $(C(X), \cap, \cup)$, where $C(X)$ is the family of $\leqslant$–increasing subsets of X with the operations of set intersection and union. The representation theorem says that W is isomorphic to a subalgebra of $C(X(W))$.

For non–distributive lattices the corresponding frames are of the form $(X, \leqslant_1, \leqslant_2)$, where $\leqslant_1$ and $\leqslant_2$ are quasi orders such that $\leqslant_1 \cap \leqslant_2 =$ identity on X. The universe of the canonical frame of a non–distributive lattice $(W, \wedge, \vee)$ is a binary relation on 2^W, where 2^W stands for the powerset of W. The universe of the complex algebra of the canonical frame consists of subrelations of that relation. The representation theorem for non–distributive lattices is presented in [25] in a topological framework.

In this paper we will not consider topological aspects of representation, so in our approach the algebras and the frames are treated at the same level of abstraction and the frames are not equipped with a topology. The reason being that our aim is to apply the representation theorems to a development of relational semantics which will be equivalent to algebraic semantics. Such a relationship between a class of algebras and a class of frames is referred to as a duality via truth in [22]. In that paper we also consider some other representation theorems ([2]) for general lattices which may lead to duality via truth. A representation theorem presented in [13] uses the frames which are two sorted, they consist of a pair of sets X and Y and of a binary relation which is a subset of $X \times Y$. Therefore, they are not so well suited to providing a relational semantics which is most often defined in terms of the frames with a single universe. In the Urquhart representation the "provable" and "refutable" are modeled with two quasi order relations on a set, while in Hartonas–Dunn representation they are modeled with two sets, which in a canonical frame are the set of filters and the set of ideals of a lattice, respectively. Moreover, the Urquhart representation can be viewed as a relational representation. A universe of the canonical frame of a lattice is a binary relation on the powerset of the universe of the lattice, and the embedding guaranteed by the representation theorem assigns subrelations of that relation to the elements of the lattice. In [23] we present the classical Stone–like representation of modal algebras in the framework of duality via truth and we extend these results to some classes of information algebras presented in [4].

Surveys of algebraic approaches to substructural logics can be found in [17] and [18].

2 Preliminaries

Let X be a non–empty set and let $\leqslant_1$ and $\leqslant_2$ be two quasi orders on X. A structure $(X, \leqslant_1, \leqslant_2)$ is called a *doubly ordered set* iff for all $x, y \in X$, if $x \leqslant_1 y$ and $x \leqslant_2 y$ then $x = y$.

For a doubly ordered set $(X, \leqslant_1, \leqslant_2)$, a subset $A \subseteq X$ is called $\leqslant_1$*–increasing* (resp. $\leqslant_2$*–increasing*) iff for all $x, y \in X$, $x \in A$ and $x \leqslant_1 y$ (resp. $x \leqslant_2 y$) imply $y \in A$.

Given a doubly ordered set $(X, \leqslant_1, \leqslant_2)$, we define two mappings $l, r : 2^X \to 2^X$ by: for every $A \subseteq X$,

$$l(A) = \{x \in X : (\forall y \in X)\ x \leqslant_1 y \Rightarrow y \notin A\} \tag{1}$$
$$r(A) = \{x \in X : (\forall y \in X)\ x \leqslant_2 y \Rightarrow y \notin A\}. \tag{2}$$

Observe that $l(A) = [\leqslant_1] - A$ and $r(A) = [\leqslant_2] - A$, where for a binary relation R on a set X and for a subset $A \subseteq X$, $[R]A$ is an ordinary necessity operator defined as $[R]A = \{x \in X : \forall y \in X\ xRy \Rightarrow y \in A\}$.

For a doubly ordered set $(X, \leqslant_1, \leqslant_2)$, $A \subseteq X$ is called *l–stable* (resp. *r–stable*) iff $l(r(A)) = A$ (resp. $r(l(A)) = A$). The family of all l-stable (resp. r–stable) subsets of X will be denoted by $L(X)$ (resp. $R(X)$).

Lemma 1. ([6],[21]) *Let $(X, \leqslant_1, \leqslant_2)$ be a doubly ordered set. Then for every subset $A \subseteq X$,*
(i) *$l(A)$ is $\leqslant_1$–increasing and $r(A)$ is $\leqslant_2$–increasing*
(ii) *if A is $\leqslant_1$–increasing, then $A \subseteq l(r(A))$*
(iii) *if A is $\leqslant_2$–increasing, then $A \subseteq r(l(A))$.* ∎

Let $(X, \leqslant_1, \leqslant_2)$ be a doubly ordered set. Define the two binary operations $\sqcap$ and $\sqcup$ in 2^X and two constants $\mathbf{0}$ and $\mathbf{1}$ as follows: for all $A, B \subseteq X$,

$$A \sqcap B = A \cap B \tag{3}$$
$$A \sqcup B = l(r(A) \cap r(B)) \tag{4}$$
$$\mathbf{0} = \emptyset \tag{5}$$
$$\mathbf{1} = X. \tag{6}$$

Observe that $\sqcup$ is defined from $\sqcap$ resembling a De Morgan law with two different negations.

In [25] it was shown that the system $(L(X), \sqcap, \sqcup, \mathbf{0}, \mathbf{1})$ is a lattice. It is called the ***complex algebra of*** X.

Let $(W, \wedge, \vee, 0, 1)$ be a bounded lattice. By a *filter-ideal pair* of W we mean a pair $x = (x_1, x_2)$ such that x_1 is a filter of W, x_2 is an ideal of W and $x_1 \cap x_2 = \emptyset$. The family of all filter–ideal pairs of a lattice W will be denoted by $FIP(W)$. Let us define the following three quasi ordering relations on $FIP(W)$: for any $(x_1, x_2), (y_1, y_2) \in FIP(W)$,

$$(x_1, x_2) \preccurlyeq_1 (y_1, y_2) \quad \text{iff} \quad x_1 \subseteq y_1 \tag{7}$$
$$(x_1, x_2) \preccurlyeq_2 (y_1, y_2) \quad \text{iff} \quad x_2 \subseteq y_2 \tag{8}$$
$$(x_1, x_2) \preccurlyeq (y_1, y_2) \quad \text{iff} \quad (x_1, x_2) \preccurlyeq_1 (y_1, y_2)\ \&\ (x_1, x_2) \preccurlyeq_2 (y_1, y_2). \tag{9}$$

We say that $(x_1, x_2) \in FIP(W)$ is *maximal* iff it is maximal wrt $\preccurlyeq$. By $X(W)$ we will denote the family of all maximal filter–ideal pairs of the lattice W. Note that $X(W)$ is a binary relation on 2^W. It was shown ([25]) that for any $(x_1, x_2) \in FIP(W)$ there exists $(y_1, y_2) \in X(W)$ such that $(x_1, y_1) \preccurlyeq (y_1, y_2)$. If $(y_1, y_2) \in X(W)$ then it is referred to as the *extension of* (x_1, x_2).

Definition 1. *Let* $(W, \wedge, \vee, 0, 1)$ *be a bounded lattice. The* ***canonical frame of*** W *is the structure* $(X(W), \preccurlyeq_1, \preccurlyeq_2)$. □

Consider the complex algebra $(L(X(W)), \sqcap, \sqcup, \mathbf{0}, \mathbf{1})$ of the canonical frame of a lattice $(W, \wedge, \vee, 0, 1)$. Note that $L(X(W))$ is an algebra of subrelations of $X(W)$. Let us define a mapping $h : W \to 2^{X(W)}$ as follows: for every $a \in W$,

$$h(a) = \{x \in X(W) : a \in x_1\}. \tag{10}$$

Proposition 1. [25] *For every bounded lattice* $(W, \wedge, \vee, 0, 1)$, h *is a lattice embedding.*

The following theorem is a weak version of the Urquhart result.

Theorem 1 (Representation theorem for lattices). *Every bounded lattice is isomorphic to a subalgebra of the complex algebra of its canonical frame.* ■

3 FL Algebras and Frames

Definition 2. *An* ***FL algebra*** *is a structure* $(W, \wedge, \vee, \odot, \to, \leftarrow, \lambda, \delta)$ *such that*

(FL.1) $(W, \wedge, \vee)$ *is a lattice*

(FL.2) $(W, \odot, \lambda)$ *is a monoid*

(FL.3) δ *is an arbitrary, fixed element of* W

(FL.4) $\to$ *and* $\leftarrow$ *are binary operations in* W *satisfying the following residuation conditions: for all* $a, b, c \in W$,

$$a \odot b \leqslant c \quad \textit{iff} \quad b \leqslant a \to c$$
$$a \odot b \leqslant c \quad \textit{iff} \quad a \leqslant c \leftarrow b.$$

The operation $\odot$ *of an FL algebra is called product and the operations* $\leftarrow$ *and* $\to$ *are called the left and right residuum of* $\odot$. □

In the rest of the paper we consider FL algebras based on bounded lattices, in order to conform to the requirements of the Urquhart representation of lattices which is a basis for our results. Observe that in every FL algebra $a \odot 0 = 0$ and $0 \odot a = 0$. This can be proved from the residuation axioms. Therefore, it is sufficient to assume that the least element 0 exists, the existence of the greatest element 1 can then be proved. Moreover, we assume that λ is not a zero element of the lattice $(W, \wedge, \vee)$. This assumption is equivalent to non–triviality of the algebra. It is essential for proving the representation theorems presented in this paper. Since for all the classes of algebras considered in the paper, we always assume that in the complex algebras of frames 0 is defined as the empty set and 1 is defined as the universe of the underlying frame, in the remaining sections we will not write explicitly the lattice bounds 0 and 1.

Note that an FL algebra is an extension of a residuated lattice by an element $\delta \in W$. For recent developments on residuated lattices, see [3], [9], [10], [11], [12], [14], [15], [16], [20] and [24].

The following lemma provides some basic properties of FL algebras.

Lemma 2. *Let $(W, \wedge, \vee, \odot, \rightarrow, \leftarrow, \lambda, \delta)$ be an FL algebra. For all $a, b, c \in W$ and for every indexed family $(b_i)_{i \in I}$ of elements of W,*

(i) $b \leqslant a \rightarrow a \odot b$

(i') $b \leqslant b \odot a \leftarrow a$

(ii) $a \odot (a \rightarrow b) \leqslant b$

(ii') $(b \leftarrow a) \odot a \leqslant b$

(iii) *if* $a \leqslant b$, *then*
$c \rightarrow a \leqslant c \rightarrow b$
$b \rightarrow c \leqslant a \rightarrow c$

(iii') *if* $a \leqslant b$, *then*
$a \leftarrow c \leqslant b \leftarrow c$
$c \leftarrow b \leqslant c \leftarrow a$

(iv) *if* $a \leqslant b$, *then*
$c \odot a \leqslant c \odot b$
$a \odot c \leqslant b \odot c$

(v) *if* $\sup_i b_i$ *exists, then*
$a \odot \sup_i b_i = \sup_i (a \odot b_i)$
$\sup_i b_i \odot a = \sup_i (b_i \odot a)$

(vi) *if* $\inf_i b_i$ *exists, then*
$a \rightarrow \inf_i b_i = \inf_i (a \rightarrow b_i)$

(vi') *if* $\inf_i b_i$ *exists, then*
$\inf_i b_i \leftarrow a = \inf_i (b_i \leftarrow a)$

(vii) *if* $\sup_i b_i$ *exists, then*
$\sup_i b_i \rightarrow a = \inf_i (b_i \rightarrow a)$

(vii') *if* $\sup_i b_i$ *exists, then*
$a \leftarrow \sup_i b_i = \inf_i (a \leftarrow b_i)$. □

FL frames are LP frames of [6] expanded by a unary relation D which is a counterpart to a distinguished element δ of FL algebras.

Definition 3. *An **FL frame** is a relational system $(X, \leqslant_1, \leqslant_2, R, S, Q, I, D)$ such that $(X, \leqslant_1, \leqslant_2)$ is a doubly ordered set, R, S and Q are ternary relations on X and $I \subseteq X$, $D \subseteq X$ are unary relations on X such that the following conditions are satisfied: for all $x, x', y, y', z, z' \in X$,*

A. Monotonicity conditions

(M.1) $R(x,y,z)$ & $x' \leqslant_1 x$ & $y' \leqslant_1 y$ & $z \leqslant_1 z' \Rightarrow R(x',y',z')$
(M.2) $S(x,y,z)$ & $x \leqslant_2 x'$ & $y' \leqslant_1 y$ & $z' \leqslant_2 z \Rightarrow S(x',y',z')$
(M.3) $Q(x,y,z)$ & $x' \leqslant_1 x$ & $y \leqslant_2 y'$ & $z' \leqslant_2 z \Rightarrow Q(x',y',z')$
(M.4) $I(x)$ & $x \leqslant_1 x' \Rightarrow I(x')$
(M.5) $D(x)$ & $x \leqslant_1 x' \Rightarrow D(x')$

B. Stability conditions

(S.1) $R(x,y,z) \Rightarrow \exists x'' \in X\ (x \leqslant_1 x''$ & $S(x'',y,z))$
(S.2) $R(x,y,z) \Rightarrow \exists y'' \in X\ (y \leqslant_1 y''$ & $Q(x,y'',z))$
(S.3) $S(x,y,z) \Rightarrow \exists z'' \in X\ (z \leqslant_2 z''$ & $R(x,y,z''))$
(S.4) $Q(x,y,z) \Rightarrow \exists z'' \in X\ (z \leqslant_2 z''$ & $R(x,y,z''))$
(S.5) $\exists u \in X(R(x,y,u)$ & $Q(x',u,z)) \Rightarrow \exists w \in X(R(x',x,w)$ & $S(w,y,z))$
(S.6) $\exists u \in X(R(x,y,u)$ & $S(u,z,z')) \Rightarrow \exists w \in X(R(y,z,w)$ & $Q(x,w,z'))$
(S.7) $I(x)$ & $(R(x,y,z)$ *or* $R(y,x,z)) \Rightarrow y \leqslant_1 z$
(S.8) $\exists u \in X(I(u)$ & $S(u,x,x))$
(S.9) $\exists u \in X(I(u)$ & $Q(x,u,x))$. □

Let $(X, \leqslant_1, \leqslant_2, R, S, Q, I, D)$ be an FL frame. Define four mappings $\otimes_S$, $\otimes$, $\multimap$, $\multimapinv : 2^X \times 2^X \to 2^X$ as follows: for all $A, B \subseteq X$,

$$\begin{aligned} A \otimes_S B &= \{z \in X : \forall x, y \in X\ (S(x,y,z)\ \&\ y \in B \Rightarrow x \in r(A))\} && (11)\\ A \otimes B &= l(A \otimes_S B) && (12)\\ A \multimap B &= \{x \in X : (\forall y, z \in X)(R(y,x,z)\ \&\ y \in A \Rightarrow z \in B)\} && (13)\\ B \multimapinv A &= \{x \in X : (\forall y, z \in X)(R(x,y,z)\ \&\ y \in A \Rightarrow z \in B)\}. && (14) \end{aligned}$$

Moreover, let Λ and Δ be defined as:

$$\Lambda = l(r(I)) \quad (15)$$
$$\Delta = l(r(D)). \quad (16)$$

Lemma 3. ([7]) *For any $A, B \subseteq X$,*
 (i) *$A \otimes_S B$ is r–stable*
 (ii) *Λ and Δ are l–stable*
 (iii) *if A and B are l–stable, then so are $A \otimes B$, $A \multimap B$ and $A \multimapinv B$.* ■

Definition 4. *The* ***complex algebra*** *of an FL frame $(X, \leqslant_1, \leqslant_2, R, S, Q, I, D)$ is the structure $(L(X), \sqcap, \sqcup, \otimes, \multimap, \multimapinv, \Lambda, \Delta)$ with the operations defined by (3), (4), (12), (13), (14) and the constants (15) and (16).* □

Proceeding as in [7] one can show the following fact.

Proposition 2. *The complex algebra of an FL frame is an FL algebra such that Λ is not a zero element.* □

Let $(W, \wedge, \vee, \odot, \to, \leftarrow, \lambda, \delta)$ be an FL algebra. By a *filter* (resp. *ideal*) of W we mean a filter (resp. ideal) of the underlying lattice $(W, \wedge, \vee)$. We will write $X(W)$ to denote the family of all maximal filter–ideal pairs of the lattice reduct of W. Define the following ternary relations on $X(W)$ ([1],[6]): for all $x, y, z \in X(W)$,

$$\begin{aligned} R^\star(x,y,z) &\quad \text{iff} \quad (\forall a, b \in W)\ a \in x_1\ \&\ b \in y_1 \Rightarrow a \odot b \in z_1\\ S^\star(x,y,z) &\quad \text{iff} \quad (\forall a, b \in W)\ a \odot b \in z_2\ \&\ b \in y_1 \Rightarrow a \in x_2\\ Q^\star(x,y,z) &\quad \text{iff} \quad (\forall a, b \in W)\ a \odot b \in z_2\ \&\ a \in x_1 \Rightarrow b \in y_2. \end{aligned}$$

Moreover, define

$$I^\star = \{x \in X(W) : \lambda \in x_1\} \quad (17)$$
$$D^\star = \{x \in X(W) : \delta \in x_1\}. \quad (18)$$

Definition 5. *Let an FL algebra $(W, \wedge, \vee, \odot, \to, \leftarrow, \lambda, \delta)$ be given. The structure $(X(W), \preccurlyeq_1, \preccurlyeq_2, R^\star, S^\star, Q^\star, I^\star, D^\star)$ is the* ***canonical frame of*** *W.* □

The following proposition can be proved as in [1].

Proposition 3. *The canonical frame of an FL algebra is an FL frame.* ■

Let $(W, \wedge, \vee, \odot, \rightarrow, \leftarrow, \lambda, \delta)$ be an FL algebra, $(X(W), \preccurlyeq_1, \preccurlyeq_2, R^\star, S^\star, Q^\star, I^\star, D^\star)$ be the canonical frame of W, and let $(L(X(W)), \sqcap, \sqcup, \otimes, \twoheadrightarrow, \twoheadleftarrow, \Lambda, \Delta)$ be the complex algebra of $X(W)$. Let the mapping $h : W \to 2^{X(W)}$ be defined as in (10), i.e. for every $a \in W$, $h(a) = \{x \in X(W) : a \in x_1\}$. We show that W is isomorphic to a subalgebra of $L(X(W))$.

Proceeding in the analogous way as in [6] we can prove the following:

Lemma 4. *Let $(W, \wedge, \vee, \odot, \rightarrow, \leftarrow, \lambda, \delta)$ be an FL algebra, let $(L(X(W)), \sqcap, \sqcup, \otimes, \twoheadrightarrow, \twoheadleftarrow, \Lambda, \Delta)$ be the complex algebra of its canonical frame, and let h be defined by (10). Then*

(i) $h(\lambda) = \Lambda$

(ii) $h(\delta) = \Delta$

(iii) $h(a \odot b) = h(a) \otimes h(b)$. ■

For any $A, B \subseteq W$, denote

$$A \leftarrow B = \{a \leftarrow b : a \in A \ \&\ b \in B\}$$
$$A \rightarrow B = \{a \rightarrow b : a \in A \ \&\ b \in B\}.$$

Lemma 5. *Let $(W, \wedge, \vee, \odot, \rightarrow, \leftarrow, 1, \lambda, \delta)$ be an FL algebra. Let F and F' be filters of W and let I be an ideal of W. Define the following subsets of W:*

$$U = \{a \in W : F \cap (I \leftarrow \{a\}) \neq \emptyset\}, \qquad V = \{a \in W : F \cap (\{a\} \leftarrow F') \neq \emptyset\},$$
$$U' = \{a \in W : F \cap (\{a\} \rightarrow I) \neq \emptyset\}, \qquad V' = \{a \in W : F \cap (F' \rightarrow \{a\}) \neq \emptyset\}.$$

Then U and U' are ideals of W and V and V' are filters of W.

Proof. By way of example we show that U is an ideal of W. Let $a, b \in W$ be such that **(i)** $a \in U$ and **(ii)** $b \leqslant a$. By the definition of U, **(i)** implies that there exists $c \in I$ such that **(iii)** $c \leftarrow a \in F$. By Lemma 2**(iii')** we get from **(ii)** that $c \leftarrow a \leqslant c \leftarrow b$. Hence, by **(iii)**, we get **(iv)** $c \leftarrow b \in F$, since F is a filter. Therefore, for some $c \in I$ **(iv)** holds, which implies $b \in U$.

Assume that **(v)** $a, b \in U$. We show that $a \vee b \in U$. From **(v)**, there are $c, d \in I$ such that **(vi)** $c \leftarrow a \in F$ and **(vii)** $d \leftarrow b \in F$. Since $c \leqslant c \vee d$ and $d \leqslant c \vee d$, by Lemma 2**(iii)** we get $c \leftarrow a \leqslant (c \vee d) \leftarrow a$ and $d \leftarrow b \leqslant (c \vee d) \leftarrow b$. Hence, by **(vi)** and **(vii)** it follows that $(c \vee d) \leftarrow a \in F$ and $(c \vee d) \leftarrow b \in F$, so $((c \vee d) \leftarrow a) \wedge ((c \vee d) \leftarrow b) \in F$. By Lemma 2**(vii')**, $((c \vee d) \leftarrow a) \wedge ((c \vee d) \leftarrow b) = (c \vee d) \leftarrow (a \vee b)$. Then $(c \vee d) \leftarrow (a \vee b) \in F$. Since $c, d \in I$, $c \vee d \in I$. So we get that for some $e = c \vee d \in I$, $e \leftarrow (a \vee b) \in F$, which means that $a \vee b \in U$. ■

Let $(X(W), \preccurlyeq_1, \preccurlyeq_2, R^\star, S^\star, Q^\star, I^\star, D^\star)$ be the canonical frame of an FL algebra. Define the following auxiliary ternary relations on $X(W)$: for all $x, y, z \in X(W)$,

$$R^\star_{\leftarrow}(x, y, z) \quad \text{iff} \quad (\forall a, b \in W)\ b \leftarrow a \in x_1 \ \&\ a \in y_1 \Rightarrow b \in z_1 \tag{19}$$
$$R^\star_{\rightarrow}(x, y, z) \quad \text{iff} \quad (\forall a, b \in W)\ a \in x_1 \ \&\ a \rightarrow b \in y_1 \Rightarrow b \in z_1. \tag{20}$$

It is easy to show that $R^\star = R^\star_{\leftarrow} = R^\star_{\rightarrow}$

Theorem 2 (Representation theorem for FL algebras). *Every FL algebra is isomorphic to a subalgebra of the complex algebra of its canonical frame.*

Proof. By Proposition 1 and Lemma 4 it suffices to show that **(i)** $h(a \leftarrow b) = h(a) \Leftarrow h(b)$ and **(ii)** $h(a \rightarrow b) = h(a) \Rightarrow h(b)$.

(i) ($\subseteq$) Let $x \in h(a \leftarrow b)$. By (10) this means that **(i.1)** $a \leftarrow b \in x_1$. Assume that $x \notin h(a) \Leftarrow h(b)$. Then there exist $y, z \in X(W)$ such that **(i.2)** $R^\star(x, y, z)$, **(i.3)** $y \in h(b)$, and **(i.4)** $z \notin h(a)$. From **(i.3)** we get **(i.5)** $b \in y_1$. Since $R^\star = R^\star_{\leftarrow}$, from **(i.1)**, **(i.2)**, **(i.5)**, and (19), $a \in z_1$, i.e. $z \in h(a)$, which contradicts **(i.4)**.

($\supseteq$) Assume that **(i.6)** $x \notin h(b \leftarrow a)$. We will show that $x \notin h(b) \Leftarrow h(a)$. From **(i.6)** we have **(i.7)** $b \leftarrow a \notin x_1$. Put $U = \{c \in W : x_1 \cap ((b] \leftarrow \{c\}) \neq \emptyset\}$, where $(b]$ stands for the ideal generated by b. By Lemma 5, U is an ideal. Suppose that $a \in U$. Then there exists $b' \in W$ such that **(i.8)** $b' \leqslant b$ and **(i.9)** $b' \leftarrow a \in x_1$. By Lemma 2(iii') and **(i.8)** we get **(i.10)** $b' \leftarrow a \leqslant b \leftarrow a$. Since x_1 is a filter, **(i.9)** and **(i.10)** imply $b \leftarrow a \in x_1$, which contradicts **(i.7)**. Hence $a \notin U$. Let $[a)$ be the filter generated by a. Then $[a) \cap U = \emptyset$, so $([a), U)$ is a filter–ideal pair. Let (y_1, y_2) be its extension to a maximal filter–ideal pair. Then $[a) \subseteq y_1$ and $U \subseteq y_2$. Since $a \in y_1$, we have **(i.11)** $y \in h(a)$.

Now, consider the set $V = \{c \in W : x_1 \cap (\{c\} \leftarrow y_1) \neq \emptyset\}$. By Lemma 5, V is a filter of W. Suppose that $b \in V$. Then there is $c' \in W$ such that **(i.12)** $c' \in y_1$ and **(i.13)** $b \leftarrow c' \in x_1$. By the definition of U, **(i.13)** implies $c' \in U \subseteq y_2$ – a contradiction with **(i.12)**. Hence $b \notin V$. Then $(V, (b])$ is a filter–ideal pair. Let (z_1, z_2) be its extension to a maximal filter–ideal pair. Then **(i.14)** $V \subseteq z_1$ and $(b] \subseteq z_2$. Since $b \in z_2$, we get $b \notin z_1$, so **(i.15)** $z \notin h(b)$.

Finally, consider $c, d \in W$ such that $c \leftarrow d \in x_1$ and $d \in y_1$. Then $c \in V$, so $c \in z_1$ by **(i.14)**. By the definition (19), $R^\star_{\leftarrow}(x, y, z)$ holds, and so **(i.16)** $R^\star(x, y, z)$, since $R^\star_{\leftarrow} = R^\star$. Therefore, we have shown that for some $y, z \in X(W)$, **(i.11)**, **(i.15)**, and **(i.16)** hold, which means by (14) and Proposition 3 that $x \notin h(b) \Leftarrow h(a)$.

The proof of **(ii)** is similar. ■

4 FLe Algebras and Their Representability

Definition 6. *Let $(W, \wedge, \vee, \odot, \rightarrow, \leftarrow, \lambda, \delta)$ be an FL algebra. We say that W is an **FLe algebra** iff $a \odot b = b \odot a$ for all $a, b \in W$.* □

It is easily noted that commutativity of $\odot$ implies that $\rightarrow$ and $\leftarrow$ coincide. Then any FLe algebra will be just written $(W, \wedge, \vee, \odot, \rightarrow, \lambda, \delta)$.

Definition 7. *An **FLe frame** is an FL frame $(X, \leqslant_1, \leqslant_2, R, S, Q, I, D)$ satisfying the following additional stability condition for all $x, y, z \in X$:*

(Se) $R(x, y, z) \Rightarrow R(y, x, z)$. □

From the definitions (13) and (14) we get by the condition **(Se)** that $\Leftarrow\; = \;\Rightarrow$.

The *complex algebra of an FLe frame* is a structure of the form $(L(X), \sqcap, \sqcup, \otimes, \Rightarrow, \Leftarrow, \Lambda, \Delta)$ with the operations defined by (3), (4), (12), (13), (14) and the constants (15) and (16).

Lemma 6. *The complex algebra of an FLe frame is an FLe algebra.*

Proof. Let $(X, \leqslant_1, \leqslant_2, R, S, Q, I, D)$ be an *FLe* frame. We show that for all l–stable sets $A, B \subseteq X$, it holds $A \otimes B = B \otimes A$. Then, in view of Proposition 2 we get the result.

Note that it suffices to show that $A \otimes_S B = B \otimes_S A$. Then $l(A \otimes_S B) = l(B \otimes_S A)$, so $A \otimes B = B \otimes A$.

($\subseteq$) Let $z \in X$ be such that $z \notin B \otimes_S A$. Then there exist $x, y \in X$ such that **(i)** $S(x, y, z)$, **(ii)** $y \in A$, and **(iii)** $x \notin r(B)$. From **(iii)** it follows that there exists $x' \in X$ such that **(iv)** $x \leqslant_2 x'$ and **(v)** $x' \in B$. By **(M.2)**, from **(i)** and **(iv)** we get $S(x', y, z)$, which by **(S.3)** implies that there exists $z' \in X$ such that **(vi)** $z \leqslant_2 z'$ and **(vii)** $R(x', y, z')$. From **(vii)**, by **(Se)** we have $R(y, x', z')$. Whence by **(S.1)**, there exists $y' \in X$ such that **(viii)** $y \leqslant_1 y'$ and **(ix)** $S(y', x', z')$. By assumption, A is l–stable, so it is $\leqslant_1$–increasing by Lemma 1**(i)**. Then, from **(ii)** and **(viii)** we get $y' \in A$, hence **(x)** $y' \notin r(A)$. By the definition of $\otimes_S$, **(v)**, **(ix)**, and **(x)** imply **(xi)** $z' \notin A \otimes_S B$. In order to show that $z \notin A \otimes_S B$, assume on the contrary that $z \in A \otimes_S B$. From Lemma 3**(i)**, $A \otimes_S B$ is r–stable, so $\leqslant_2$–increasing. Hence, by **(vi)**, we get $z' \in A \otimes_S B$ – a contradiction with **(xi)**.
($\supseteq$) can be proved in the similar way. ■

Lemma 7. *The canonical frame of an FLe algebra is an FLe frame.*

Proof. It suffices to show that the canonical frame of FLe algebra satisfies **(Se)**. Let $x, y, z \in X(W)$ and assume that $R^\star(x, y, z)$ holds. Then for all $a, b \in W$, $a \in x_1$ and $b \in y_1$ imply $a \odot b \in z_1$. Since $a \odot b = b \odot a$, $R^\star(y, x, z)$ holds. ■

We complete this section with the following representation theorem.

Theorem 3 (Representation theorem for FLe algebras). *Every FLe algebra is isomorphic to a subalgebra of the complex algebra of its canonical frame.*
Proof. Follows from Theorem 2 and Lemma 7. ■

5 FLc Algebras and Their Representability

Definition 8. *Let $(W, \wedge, \vee, \odot, \to, \leftarrow, \lambda, \delta)$ be an FL algebra. We say that W is an* ***FLc algebra*** *iff it satisfies the following contraction condition for every $a \in W$, $a \leqslant a \odot a$.* □

Definition 9. *An* ***FLc frame*** *is an FL frame $(X, \leqslant_1, \leqslant_2, R, S, Q, I, D)$ satisfying the following stability condition for any $x \in X$:*
(Sc) $R(x, x, x)$. □

The *complex algebra of an FLc frame* is defined in the same way as the complex algebra of an FL frame.

Lemma 8. *The complex algebra of an FLc frame is an FLc algebra.*

Proof. It suffices to show that for any l–stable set $A \subseteq W$, $A \subseteq A \otimes A$.

Let **(i)** $x \in A$ and assume that $x' \in W$ is such that **(ii)** $x \leqslant_1 x'$. Since A is $\leqslant_1$–increasing, from **(i)** and **(ii)** we get **(iii)** $x' \in A$. Furthermore, by **(Sc)**, $R(x', x', x')$ holds, which by **(S.1)** implies that there exists $x" \in W$ such that **(iv)** $x' \leqslant_1 x"$ and **(v)** $S(x", x', x')$. By the transitivity of $\leqslant_1$, from **(ii)** and **(iv)** we get **(vi)** $x \leqslant_1 x"$. Since A is l–stable, **(i)** and **(vi)** give **(vii)** $x" \notin r(A)$. Then for $x' \in W$, **(iii)**, **(v)**, and **(vii)** hold, which means that **(viii)** $x' \notin A \otimes_S A$. Since **(ii)** implies **(viii)** for an arbitrary $x' \in W$, we get $x \in l(A \otimes_S A) = A \otimes A$. ∎

Lemma 9. *The canonical frame of an FLc algebra is an FLc frame.*

Proof. It suffices to show that **(i)** for every $x \in X(W)$, $R^*(x, x, x)$ holds. Since $a \wedge b \leqslant a$ and $a \wedge b \leqslant b$, from Lemma 2**(iv)** we have **(ii)** $(a \wedge b) \odot (a \wedge b) \leqslant a \odot b$. Moreover, from **(FLc)**, $a \wedge b \leqslant (a \wedge b) \odot (a \wedge b)$, so by **(ii)** we get **(iii)** $a \wedge b \leqslant a \odot b$. Let $x \in X(W)$ and let **(iv)** $a, b \in x_1$. Since x_1 is a filter, $a \wedge b \in x_1$, and also $a \odot b \in x_1$ by **(iii)**. Hence, by **(iv)** and the definition of $R^\star$, we get **(i)**. ∎

Theorem 4 (Representation theorem for FLc algebras). *Every FLc algebra is isomorphic to a subalgebra of the complex algebra of its canonical frame.*

Proof. Follows from Theorem 2 and Lemmas 8 and 9. ∎

6 FLw Algebras and Their Representability

Definition 10. *Let* $(W, \wedge, \vee, \odot, \rightarrow, \leftarrow, \lambda, \delta)$ *be an FL algebra. We say that* W *is an* ***FLw algebra*** *iff it satisfies for all* $a, b \in W$:

(FLw.1) $a \odot b \leqslant a$
(FLw.2) $a \odot b \leqslant b$
(FLw.3) δ *is the bottom element of the underlying lattice* $(W, \wedge, \vee)$. □

An FLw algebra will be written $(W, \wedge, \vee, \odot, \rightarrow, \leftarrow, \lambda, 0)$.

Definition 11. *An* ***FLw frame*** *is an FL frame* $(X, \leqslant_1, \leqslant_2, R, S, Q, I, D)$ *satisfying the following stability conditions: for all* $x, y, z \in X$:

(Sw.1) $S(x, y, z) \Rightarrow z \leqslant_2 x$
(Sw.2) $Q(x, y, z) \Rightarrow z \leqslant_2 y$
(Sw.3) $D = \emptyset$. □

The ***complex algebra of an FLw frame*** is defined in the same way as the complex algebra of an FL frame.

Lemma 10. *The complex algebra of an FLw frame is an FLw algebra.*

Proof. We have to show that for all l–stable subsets $A, B \subseteq W$, **(i)** $A \otimes B \subseteq A$ and **(ii)** $A \otimes B \subseteq B$. Clearly, **(Flw.3)** holds in the complex algebra due to **(Sw.3)**. **(i)** Let **(i.1)** $x \in A \otimes B$ and let $x' \in W$ be an arbitrary element satisfying **(i.2)** $x \leqslant_1 x'$. Hence, by **(i.1)**, we get that $x' \notin A \otimes_S B$, which by the definition (11) of $\otimes_S$ means that there exist $y, z \in W$ such that **(i.3)** $S(y, z, x')$, **(i.4)** $z \in B$

and **(i.5)** $y \notin r(A)$. By **(Sw.1)**, from **(i.3)** we get **(i.6)** $x' \leqslant_2 y$. Furthermore, **(i.5)** implies that there exists $y' \in W$ such that **(i.7)** $y \leqslant_2 y'$ and **(i.8)** $y' \in A$. From **(i.6)** and **(i.7)** we get $x' \leqslant_2 y'$, which together with **(i.8)** gives **(i.9)** $x' \notin r(A)$. So we have shown that for any $x' \in W$ satisfying **(i.2)**, **(i.9)** holds, which means that $x \in l(r(A))$. Since A is l–stable, we finally get $x \in A$.

Using **(Sw.2)** and proceeding in the analogous way **(ii)** can be proved. ∎

Lemma 11. *The canonical frame of an FLw algebra is an FLw frame.*

Proof. We have to show that **(i)** $D^* = \emptyset$, and for all $x, y, z \in X(W)$, **(ii)** $S^*(x, y, z)$ implies $z_2 \subseteq x_2$, and **(iii)** $Q^*(x, y, z)$ implies $z_2 \subseteq y_2$.
(i) Follows from the definition (18) of D^*.
(ii) Let $x, y, z \in X(W)$ and let **(ii.1)** $a \in z_2$. Assume that **(ii.2)** $S^*(x, y, z)$ holds. Since $a \odot 1 \leqslant a$ and z_2 is an ideal, by **(ii.1)** we get **(ii.3)** $a \odot 1 \in z_2$. Since y_1 is a filter, we have **(ii.4)** $1 \in y_2$. From **(ii.2)**, **(ii.3)** and **(ii.4)** it follows that $a \in x_2$. Hence $z_2 \subseteq x_2$.
(iii) can be proved in the similar way. ∎

Concluding the discussion we have:

Theorem 5 (Representation theorem for FLw algebras). *Every FLw algebra is isomorphic to a subalgebra of the complex algebra of its canonical frame.*

Proof. Follows from Theorem 2 and Lemmas 10 and 11. ∎

7 Conclusions

We have presented relational representation theorems for the classes of algebras of basic substructural logics FL, FLe, FLc, and FLw, within a generalized UAD–framework. Representability for the algebras FLec, FLew, and FLecw obtained from the basic classes by making the relevant joins of algebras can be obtained by adequately joining (and possibly simplifying) the axioms of the corresponding frames. Observe that the definitions of the relations of canonical frames are the same for all the classes FL, FLe, FLc, and FLw. The only difference is in the properties that these relations satisfy. Therefore the representability results can be obtained in a modular way. Joining the properties often leads to a simplification. For example, since the algebras of FLec are distributive, in the corresponding frames $\leqslant_2$ is the converse of $\leqslant_1$.

Further work on applications of the generalized UAD–framework to various logics based on not–necessarily distributive lattices and to various classes of lattice–based relation algebras is planned.

Acknowledgements

We would like to thank the anonymous referees for very helpful and constructive comments.

References

1. Allwein G. and Dunn J. M. (1993). Kripke models for linear logic. J. Symb. Logic 58, 514–545.
2. Bimbo K. (1999). Substructural logics, combinatory logic, and λ–calculus, PhD thesis, Indiana University, Bloomington, USA.
3. Blount K. and Tsinakis C. (2003). The structure of residuated lattices. Int. J. of Algebra Comput. 13(4), 437–461.
4. Demri S. and Orłowska E. (2002). *Incomplete Information: Structure, Inference, Complexity.* EATCS Monographs in Theoretical Computer Science, Springer.
5. Dilworth R. P. and Ward N. (1939) Residuated lattices. Transactions of the American Mathematical Society 45, 335–354.
6. Düntsch I., Orłowska E., and Radzikowska A. M. (2003). Lattice–based relation algebras and their representability. In: de Swart C. C. M. et al (eds), *Theory and Applications of Relational Structures as Knowledge Instruments*, Lecture Notes in Computer Science 2929, Springer–Verlag, 234–258.
7. Düntsch I., Orłowska E. and Radzikowska A. M. (2003). Lattice–based relation algebras II. Preprint.
8. Düntsch I., Orłowska E., Radzikowska A. M., and Vakarelov D. (2004). Relational representation theorems for some lattice–based structures. Journal of Relation Methods in Computer Science JoRMiCS, vol.1, Special Volume, ISSN 1439-2275, 132-160.
9. Esteva F. and Godo L. (2001). Monoidal t–norm based logic: towards a logic for left–continuous t–norms. Fuzzy Sets and Systems 124, 271–288.
10. Flondor P., Georgescu G., and Iorgulecu A. (2001). Psedo–t–norms and pseudo–BL algebras. Soft Computing 5, No 5, 355–371.
11. Hajek P. (1998). *Metamathematics of Fuzzy Logic*, Kluwer, Dordrecht.
12. Hart J. B., Rafter L., and Tsinakis C. (2002). The structure of commutative residuated lattices. Internat. J. Algebra Comput. 12, no. 4, 509–524.
13. Hartonas C. and Dunn J. M. (1997). Stone duality for lattices. Algebra Universalis 37, 391–401.
14. Höhle U. (1996). Commutative, residuated l-monoids. In: Höhle U., Klement U. P. (eds), *Non–Classical Logics and their Applications to Fuzzy Subsets*, Kluwer, Dordrecht, 53–106.
15. Jipsen P. (2001). A Gentzen system and decidability for residuated lattices. Preprint.
16. Jipsen P. and Tsinaksis C. (2003). A Survey of Residuated Lattices. In: Martinez J. (ed), *Ordered Algebraic Structures*, Kluwer Academic Publishers, Dordrecht, 19–56.
17. Ono H. (1993). Semantics for substructural logics. In: Dosen K. and Schroeder–Heister P. (eds), *Substructural Logics*, Oxford University Press, 259–291.
18. Ono H. (2003). Substructural logics and residuated lattices – an introduction. In: Hendricks V. and Malinowski J. (eds), *Trends in Logic: 50 Years of Studia Logica*, Trends in Logic vol. 20, Kluwer, 177–212.
19. Orłowska E. and Radzikowska A. M. (2001). Information relations and operators based on double residuated lattices. In: de Swart H. C. M. (ed), Proceedings of the 6th Seminar on Relational Methods in Computer Science *RelMiCS'2001*, 185–199.
20. Orłowska E. and Radzikowska A. M. (2002). Double residuated lattices and their applications. In: de Swart H. C. M. (ed), Relational Methods in Computer Science, Lecture Notes in Computer Science 2561, Springer–Verlag, Heidelberg, 171–189.

21. Orłowska E. and Vakarelov D. (2003). Lattice–based modal algebras and modal logics. In: Hajek P., Valdes L., and Westerstahl D. (eds), Proceedings of the 12th International Congress of Logic, Methodology and Philosophy of Science, Oviedo, August 2003, Elsevier, in print. Abstract in the Volume of Abstracts, 22–23.
22. Orłowska E. and Rewitzky I. (2005). Duality via Truth: Semantic frameworks for lattice–based logics. Logic Journal of the IGPL 13, 467–490.
23. Orłowska, E., Rewitzky I., and Düntsch I. (2005). Relational semantics through duality. Lecture Notes in Computer Science, Springer–Verlag, to appear.
24. Turunen E. (1999) *Mathematics Behind Fuzzy Logic*, Springer–Verlag.
25. Urquhart A. (1978). A topological representation theorem for lattices. Algebra Universalis 8, 45–58.

Knuth-Bendix Completion as a Data Structure

Georg Struth

Fakultät für Informatik, Universität der Bundeswehr München
struth@informatik.unibw-muenchen.de

Abstract. We propose a cooperating Knuth-Bendix completion procedure for transitive relations and equivalences and apply it as a data structure for novel dynamic strongly connected component algorithms. Benefits are separation of declarative and procedural concerns, simple generic specifications and flexible optimisation via execution strategies.

1 Introduction

A fully dynamic graph algorithm is a data structure which implements an on-line sequence of updates for edge insertion and deletion and which maintains and answers queries about graph properties. Such algorithms have recently received considerable attention (cf. [3, 2, 5]), but only few instances exist. They depend on complex data structures and low-level techniques. A declarative approach seems desirable for developing further instances. But even static efficient graph algorithms are difficult outside the imperative paradigm.

We propose a cooperating Knuth-Bendix completion procedure for transitive relations and equivalences for developing efficient dynamic algorithms for strongly connected components (SCCs). The two parts of the procedure maintain information about reachability and equivalences in an inherently dynamic way. Efficiency is obtained by rewriting techniques and execution strategies.

The main idea behind our novel SCC algorithms is simple: Cycle detection uses the fact that all cycles contain critical pairs. Information about cycles is propagated from the relational part of the Knuth-Bendix procedure to the equational part to construct a canonical rewrite system; a canonical representative for each cycle and thereby each SCC. Information is then propagated back to reduce edges to canonical representatives, thus building the SCC-graph. Completion is inherently incremental: edge insertions do not affect the precompiled data structure. Also the effects of deletions can be traced, to a certain extent, locally. Queries to the data structure are handled efficiently by the rewrite-based decision procedure. For a graph $(V, \to)$, this yields running time $O(|V|^2)$ by adding one single rule to the completion procedure. Simple execution strategies lead to $O(|V| + |{\to}|)$ running time, which is that of the best static algorithms. Update and query times of the dynamic cases depend on concrete implementations. We can recover, for instance, the best known dynamic cycle detection algorithm [7].

The approach offers further benefits. Equivalence classes can be maintained by a declarative union-find data structure. Our algorithms are highly non-deterministic and generic: most implementation details need not concern the developer. Completeness proofs are simple and modular relative to completion.

W. MacCaull et al. (Eds.): RelMiCS 2005, LNCS 3929, pp. 225–236, 2006.

Completion for transitive relations and preorderings has been proposed in [8]. The idea of applying *equational* completion in declarative programming appears in [6]. A declarative reconstruction of Shostak's congruence closure algorithm by equational completion appears in [4], implicitly using declarative union-find.

In this extended abstract, we can only sketch the approach. Background material can be found at the author's web-page.

2 Preliminaries

We assume basic knowledge about terms, relations and graphs. We restrict our attention to ground, that is, variable-free terms. Let T_Σ be a set of (ground) terms with signature Σ. As usual, we identify terms with Σ-labelled trees with nodes or positions in $\mathbb{N}^*$. We write $r[t]_p$ $(r[s/t]_p)$ to denote the term obtained by replacing in r the subterm $(s = r|_p)$ at position p by t. We lift this notation to other expressions.

Let $\rightarrow$ be a binary relation. We write $\leftarrow$ for its converse, $\leftrightarrow$ for its symmetric closure, $\rightarrow^+$ for its transitive closure and $\rightarrow^*$ for its reflexive transitive closure. $\rightarrow$ is *transitive*, if $\rightarrow\ =\ \rightarrow^+$, a *preordering*, if $\rightarrow\ =\ \rightarrow^*$ and *noetherian*, if all $\rightarrow$-chains are finite.

Let $\rightarrow$ be a binary relation on a term algebra A with associated set of ground terms T_Σ. $\rightarrow$ is *compatible* if $s \rightarrow t$ implies $r[s]_p \rightarrow r[t]_p$ for all $r, s, t \in T_\Sigma$ and positions p. A (ground) *rewrite rule* is a pair of (ground) terms. The (ground) *rewrite relation* $\rightarrow_R$ induced by a set R of (ground) rewrite rules is the smallest compatible relation containing R. To orient rewrite rules, we also consider syntactic orderings $\prec$ on (ground) terms. A *reduction ordering* is a noetherian compatible preordering. We always assume that $\prec$ is linear on ground terms and contains the proper subterm relation.

A finite binary relation $\rightarrow$ can be represented as a digraph $G = (V, \rightarrow)$. As usual, a *k-path* in G is a sequence of $k + 1$ vertices along edges. It is *simple* if all vertices are distinct. A *k-cycle* in G is a k-path with $v_0 = v_k$. Vertex v is *reachable* from vertex u if $u \rightarrow^+ v$. The *strongly connected components* (SCCs) of G are the equivalence classes of the mutual reachability relation $\approx\ =\ \leftarrow^* \cap \rightarrow^*$. This definition implies that every vertex is reachable from itself. The *component graph* (SCC-graph) $G_{SCC} = G/\!\approx$ of G is defined on $\approx$-equivalence classes in the standard way.

3 Rewriting for Transitive Relations

We presuppose the basic concepts and notation of rewriting for non-symmetric transitive relations and congruences ([8, 9]) and sketch only the main ideas. Let $\rightarrow_I$ be a binary relation and $\prec$ some reduction ordering on T_Σ. Then $\rightarrow_I$ can be partitioned into a decreasing part $\rightarrow_R$, an increasing part $\rightarrow_S$ and a non-orientable part $\rightarrow_\Delta$ with respect to $\prec$. By this ordering, relational chains now show peaks and valleys, just like equational proofs do in equational rewriting.

In further analogy, relational rewriting asks for conditions to obtain valleys. These are useful for deciding reachability by reducing terms along $\prec$, which is noetherian, as we will see. Arbitrary paths can be transformed into valleys when all $\to_R$-steps can be permuted to the left of $\to_S$-steps.

Lemma 1. *Let $\to_R$ and $\to_S$ be binary relations on some set A. Let $\to_R \cup \leftarrow_S$ be noetherian. Then $\to_S\to_R \subseteq \to_R^+\to_S^* \cup \to_S^+$ iff $(\to_R \cup \to_S)^+ \subseteq \to_R^+\to_S^* \cup \to_S^+$.*

Formally, a *valley* is a path in $\to_R^+\to_S^* \cup \to_S^+$; a *peak* a path in $\to_S\to_R$. $\to_R$ *semicommutes* over $\to_S$ when every peak can be replaced by a valley. Lemma 1 generalises Newman's lemma. Setting $\to_R$ and $\leftarrow_S$ to $\to_T$, its left-hand side is *local confluence* of $\to_T$, $\leftarrow_T\to_T \subseteq \to_T^*\leftarrow_T^*$; the right-hand side is the *Church-Rosser property*, $\leftrightarrow_T^* \subseteq \to_T^*\leftarrow_T^*$. When $\to_T$ is a noetherian rewrite relation that has the Church-Rosser property, every term has a normal form representing its $\leftrightarrow_T^*$-equivalence class. In relational rewriting, this need not be the case.

Semicommutation and local confluence are further refined to critical pairs, using the relative positions of rewrite steps in terms. Intuitively, these are pairs of terms that are connected via a peak and may be irreplaceable by a valley. Their number is finite when R and S are finite. See [8] for a formal definition.

Combining an equational rewrite relation $\to_T$ with $\to_R$ and $\to_S$ yields peaks and critical pairs in $\leftarrow_T\to_R$, $\to_S\to_T$ and $\leftarrow_T\to_T$ and valleys in $\to_T^+\to_R^*\to_S^*\leftarrow_T^*$ $\cup \to_R^+\to_S^*\leftarrow_T^* \cup \to_S^+\leftarrow_T^* \cup \leftarrow_T^+$. When $\to_T$ is confluent and noetherian, it computes unique normal forms for equivalent terms. Then, valleys operate solely on T-normal forms.

In absence of critical pairs and when $\to_R$, $\leftarrow_S$ and $\to_T$ are noetherian, we obtain a search procedure for reachability along valleys. For a non-symmetric transitive relation, $s \to_I^+ t$ holds if either the T-normal form of s and the T-normal form of t are in the reflexive part of $\to_I^+$ or if some $\to_R$-path from the T-normal form of s and some $\leftarrow_S$-path from the T-normal form of t are joinable. This is again a decision procedure if $\to_R$ and $\leftarrow_S$ are finitely branching. When A is just a set of constants (e.g. vertices of a graph), the decision procedure takes time linear in $|A \cup \to_R \cup \to_S|$, for instance by depth-first search. When A carries a ground term structure, memoisation yields polynomial time complexity.

We call a pair (R, S) of sets of rewrite rules that supports the decision procedure for reachability of relational rewriting a *normal system*. By definition, all critical peaks can be joined by a rewrite proof, $\to_R$ and $\leftarrow_S$ are noetherian and no rule from R or S can be deleted without changing the theory. A set T of equational rewrite rules is *canonical* if the induced term rewrite system is confluent and noetherian.

4 Cooperating Completion

Equational Knuth-Bendix completion procedures (KB-procedures) are presented, e.g., in [9]; relational KB-procedures in [8]. Their combination allows an interleaved construction of a canonical term rewrite system and a normal system, roughly by critical pair computations and deletion of redundant expressions. The cooperating KB-procedure consists of the inference rules of the two procedures for equations and relations plus rules that compute critical pairs between relational and equational rules. Also the equational simplification rules are extended such that equations can also simplify relational rules. Since this procedure is interesting in its own right and has further applications, we present it for term structures. However, only constants are needed for SCC algorithms.

A KB-procedure implements a transition system, using a syntactic reduction ordering $\prec$ on terms, (oriented) equational and relational rules and paths. States are tuples of sets of equational, relational and rewrite rules. The transition relation is specified by transition rules of two kinds. Deductive rules add certain consequences to a state that correspond to critical pair computations. Simplification rules combine deduction steps with deletions implementing an (approximate) notion of redundancy: An (oriented) equation or inequality is *redundant*, if it can be replaced by a smaller path.

We denote the combined KB-procedure for a non-symmetric transitive relation by C. Its states are of the form $q_i = (U_i, \Delta_i, O_i)$, $i \in \mathbb{N}$. The first component $U_i = I_i \cup E_i$, where I_i is a set of (unoriented) relational pairs and E_i is a set of (unoriented) equational pairs on T_Σ. The second component $\Delta_i \subseteq \{(t,t) : t \in T_\Sigma\}$. This set stores the reflexive part of U_i^+. This is necessary since U_i^+ is not reflexive in general. The third component $O = R_i \cup S_i \cup T_i$, where R_i, S_i and T_i are sets of (oriented) rewrite rules for inequalities and equations on T_Σ.

In the initial state q_0, the sets Δ_0 and O_0 are empty. U_0 contains the given set of unoriented pairs to be processed. A *run* of C is a (finite or infinite) sequence $q_0, q_1, q_2, \ldots$ of states such that q_0 is the initial state and consecutive states are related by transitions that apply some inference rule in C in a forward way. We also define *limit states* q_∞ of C, where $\Delta_\infty = \bigcup_{i=0}^{\infty} \Delta_i$ and $X_\infty = \bigcup_{i=0}^{\infty}(\bigcap_{j=i}^{\infty} X_j)$, for X ranging over U, O, R, S and T.

A run *succeeds*, if (i) U_∞ is empty, (ii) T_∞ yields a canonical system for E and the equational part of I and (iii) if (R_∞, S_∞) is a normal system on E-equivalence classes. A run is *fair*, if every enabled transition is eventually executed. Then every relevant critical pair will eventually be computed. A completion procedure is *correct* if every successful run produces a limit rewrite system that is equivalent to the initial specification.

We now define the deduction and simplification rules of C. They are standard, except for the combination of equational and relational reasoning.

$$(U, \Delta, O) \Rightarrow (U \cup \{sIt\}, \Delta, O) \qquad \text{(DEDUCE)}$$

if (s,t) is a critical pair involving R or S. Note that critical pairs from T are subsumed by simplification rules. Rewriting into T-rules occurs at strict subterms.

$$\begin{aligned}(U \cup \{sIt\}, \Delta, O) &\Rightarrow (U, \Delta, O \cup \{sRt\}), \qquad \text{(ORIENT)}\\ (U \cup \{tIs\}, \Delta, O) &\Rightarrow (U, \Delta, O \cup \{tSs\}),\\ (U \cup \{sEt\}, \Delta, O) &\Rightarrow (U, \Delta, O \cup \{sTt\}),\\ (U \cup \{tEs\}, \Delta, O) &\Rightarrow (U, \Delta, O \cup \{sTt\}),\end{aligned}$$

where $s \succ t$. Moreover, we store the reflexive part of $\rightarrow_{I\cup E}^{+}$ in Δ:

$$(U \cup \{s(I \cup E)s\}, \Delta, O) \Rightarrow (U, \Delta \cup \{s\Delta s\}, U), \qquad \text{(DIAGONAL)}$$

DIAGONAL can be restricted to unoriented expressions when eagerly applied.

The simplification rules use equational rewrite rules to simplify oriented and unoriented pairs:

$$(U[l], \Delta, O \cup \{lTr\}) \Rightarrow (U[l/r], \Delta, O \cup \{lTr\}), \qquad \text{(SIMPLIFY)}$$

$$(U, \Delta \cup \{s[l]\Delta s[l]\}, O \cup \{lTr\}) \Rightarrow (U, \Delta \cup \{s[l/r]\Delta s[l/r]\}, O \cup \{lTr\}). \qquad \text{(DSIMPLIFY)}$$

DSIMPLIFY could be restricted to one side. The result of simplification should then be stored in I and be shifted back into Δ after applying SIMPLIFY.

$$\begin{aligned} &(U, \Delta, O \cup \{s[l]Tt, lTr\}) \Rightarrow (U \cup \{s[l/r]Et\}, \Delta, O \cup \{lTr\}),\\ &(U, \Delta, O \cup \{s[l]Rt, lTr\}) \Rightarrow (U \cup \{s[l/r]It\}, \Delta, O \cup \{lTr\}),\\ &(U, \Delta, O \cup \{sSt[l], lTr\}) \Rightarrow (U \cup \{sIt[l/r]\}, \Delta, O \cup \{lTr\}). \end{aligned} \qquad \text{(COLLAPSE)}$$

In the first case, either rewriting occurs at a strict subterm, or else $s = l$ and $t \succ r$.

$$\begin{aligned} &(U, \Delta, O \cup \{sTt[l], lTr\}) \Rightarrow (U, \Delta, O \cup \{sTt[l/r], lTr\}),\\ &(U, \Delta, O \cup \{sSt[l], lTr\}) \Rightarrow (U, \Delta, O \cup \{sSt[l/r], lTr\}),\\ &(U, \Delta, O \cup \{s[l]St, lTr\}) \Rightarrow (U, \Delta, O \cup \{s[l/r]St, lTr\}). \end{aligned} \qquad \text{(COMPOSE)}$$

There are also search-based rules for simplifying relational pairs [8].

C specialises to an equational KB-procedure by forgetting all rules involving I, R and S and to a KB-procedure for non-symmetric transitive relations [8], forgetting all rules involving E and T. Conversely, as an extension of these procedures, the combined procedure adds just a few more simple cases to well-known rules. The proofs of soundness and correctness of C are quite similar to those of equational and relational KB-completion (c.f. [9, 8]).

Theorem 2. *Every fair implementation of* C *is correct.*

There are three main differences between relational and equational completion. First, even in the ground case, DEDUCE is not subsumed by the relational simplification rules [8]. Second, straightforward implementations of C admit infinite runs, (c.f. [8]). The procedure can be forced to terminate in polynomial time, when the input expressions are appropriately transformed. Third, the generalisation to the non-ground case is not straightforward (c.f. [8]). Extensions of C are also discussed in Section 8.

5 Strongly Connected Components

We now apply C to develop static SCC algorithms. Dynamic aspects are discussed in Section 8. Readers might look at the example in Section 7 in parallel.

Consider a digraph $G = (V, \rightarrow_I)$ together with a linear well-founded ordering $\prec$, also called *precedence* on V. Thus there is no term structure for vertices.

Lemma 3. *All runs of* C *on a digraph* G *terminate in* $O(|V|^2)$ *steps.*

Proof. The C-rules add, relabel and delete edges. Since no deleted edge must ever be added again (because it is and remains redundant) and every edge is relabelled at most once, the procedure terminates after at most $3|V|^2$ steps. □

Every edge in G can be oriented. A path is *oriented*, if it contains only edges labelled with R or S. We call an SCC *trivial*, if it contains one single vertex. We use $\leftrightarrow_T^*$ for representing the mutual reachability relation $\approx$. Since it is a global property of $\rightarrow_I$ and therefore of G, we introduce the local variant $\approx_I = \leftarrow_I \cap \rightarrow_I$ and model $\approx_1$ by the rule

$$(U, \Delta, O \cup \{l \rightarrow_R r, r \rightarrow_S l\}) \Rightarrow (U, \Delta \cup \{r \rightarrow_\Delta r\}, O \cup \{l \rightarrow_T r\}). \qquad \text{(SCC)}$$

Adding SCC to C yields the state transition system C_{SCC}. In the initial state, we assume that the input digraph is presented by I. E, Δ and O are empty. ORIENT and DIAGONAL steps will be left implicit.

We now motivate that adding just SCC to C suffices for detecting all SCCs and constructing the component graph G_{SCC}. First, all runs of C_{SCC} obviously still terminate after $O(|V|^2)$ steps. Soundness is evident, since all runs of C_{SCC} preserve reachability and mutual reachability. Also, the term and proof measure can easily be extended to integrate SCC into the proof transformation.

The main ideas behind correctness are also simple. Roughly, the relational rewrite system is used to detect cyclic structures, the equational one to collapse cycles and represent equivalence classes. The SCC rule couples the two KB-procedures and establishes their cooperation.

C_{SCC} has two main phases (cf. the example in Section 7). For the first phase, observe that every SCC is a cluster of simple cycles.

Lemma 4. *Every simple oriented cycle contains at least one* R*- and one* S*-step.*

This holds, since unlike at Escher's picture, one can neither walk up nor walk down forever on an oriented cycle. So, by Lemma 4, every simple oriented k-cycle contains a critical pair $\rightarrow_S\rightarrow_R$ and DEDUCE adds a new edge that generates a $k-1$-cycle. By iteration, C_{SCC} eventually generates a 2-cycle, for which SCC introduces a T-rule. This alone suffices for cycle detection.

Lemma 5. *Every fair implementation of* C_{SCC} *is correct for cycle detection.*

Now to the second phase. Once a T-rule $l \rightarrow_T r$ has been introduced, SIMPLIFY and COLLAPSE replace all occurrences of l in an edge by r. Hence, to recapitulate, every i-cycle, $3 \leq i \leq k$, obtained from DEDUCE-steps on a k-cycle is collapsed to an $i-1$-cycle. Also, a new 2-cycle appears and consequently, a new T-rule is introduced by SCC. By iteration, the initial k-cycle is collapsed into a 2-cycle and then to a single vertex by SCC. All in- and outgoing edges of the initial k-cycle are now in- and outgoing edges to this surviving vertex. The remaining vertices of the cycle do no longer contribute to rewrite rules. They are connected by T-rules to the surviving vertex instead. Algebraically, C_{SCC} collapses the cycle into an equivalence class of the component graph, which is now represented by T-rules pointing from each member of the class to its canonical representative. The relational rules apply solely at the canonical representative and thus rewrite modulo the $\approx$-equivalence class. Consequently, every SCC is eventually collapsed to a canonical representative by T, based on a canonical term rewrite system. This system is an abstract declarative data structure that maintains information about the equivalence classes of the SCC-graph. The normal system that results from the completion process is an abstract declarative data structure that maintains reachability information in the SCC-graph. This informal argument can be formally rephrased as follows.

Theorem 6. *Every fair implementation of* C_{SCC} *is correct.*

(i) *Let* v *be the* $\prec$*-minimal element of* $[v]_\approx$*. Then there exists a rule* $v \rightarrow_{\Delta_\infty} v$ *and for all* $v' \neq v$ *in* $[v]_\approx$ *there exists a rewrite rule* $v' \rightarrow_{T_\infty} v$*. Thus* $\rightarrow_{T_\infty}$ *computes canonical elements for and represents every non-trivial SCC of* G*.*
(ii) $\leftrightarrow^+_{T_\infty} \cup id_V = \approx$.
(iii) *For all* $v, v' \in V$*,* v' *is reachable from* v *in the graph* $G' = (V, \rightarrow_{R_\infty}, \rightarrow_{S_\infty}, \rightarrow_{T_\infty})$ *by a path in* $\rightarrow^{\leq 1}_{T\infty} (\rightarrow^+_{R_\infty} \rightarrow^*_{S_\infty} \cup \rightarrow^+_{S_\infty}) \leftarrow^{\leq 1}_{T_\infty}$ *if and only if* $[v]_\approx \rightarrow^+_\approx [v']_\approx$ *in* $G/\approx$*.* $\rightarrow^{\leq 1}$ *denotes that there is at most one such rewrite step.*

The proof uses correctness of C and is therefore modular.

The statement in Theorem 6(iii) is not entirely as expected. G' is not G_{SCC}, it only induces the same reachability relation. The reason is that DEDUCE steps may add persisting new edges to the graph and therefore G' can have different edges than G_{SCC}. Additional work is required to recover G_{SCC} from G', but in many applications, including transitive query evaluation, constraint solving problems and model checking, reachability equivalence suffices.

C_{SCC} is a simple and robust extension of C, but its running time cannot compete with the $O(|V| + |\rightarrow|)$ of standard algorithms [1].

Corollary 7. C_{SCC} *eliminates isolated simple* k*-cycles in* $O(k)$ *transitions. All runs terminate after* $O(|V|^2)$ *transitions.*

Proof. There are $k-2$ DEDUCE transitions, $k-1$ SCC and COLLAPSE transitions and $2k-3$ ORIENT transitions; thus $5k-7$ transitions in the algorithm suffice for eliminating a cycle. The overall running time follows from the proof of Theorem 6(i). □

So $O(|V|^2)$ is also a very loose upper bound for SCC-elimination, but in fact the number of steps in C_{SCC} depends strongly on the clustering of cycles.

C_{SCC} is declarative, rule-based non-deterministic, thus specifies a whole class of algorithms. The following section shows that efficient algorithms can be obtained by eliminating non-determinism via execution strategies. For running time analysis we assume—following arguments in [4] for similar procedures—that all implementations of transitions in C and C_{SCC} can be executed in constant time.

6 Strategic Refinements

We now show that refinements via execution strategies yield efficient static SCC algorithms. First, two questions must be answered. How to translate the output graph G' of C_{SCC} into G_{SCC}? How to improve the running time of C_{SCC}?

We define a strategy that restricts DEDUCE steps to cycles and answers both questions at once. Let C^s_{SCC} be the state transition system C_{SCC} together with the following strategy.

- Orient inequalities using depth-first search on I_0,
- Construct $\prec$ on vertices on-the-fly, using the discovery times of vertices.
- Give successor vertices a smaller weight with respect to $\prec$ as long as possible.

This strategy orients inequalities into R-rules as long as possible. We call *R_f-edge* an R-edge and *S_f-edge* an S-edge whose right-hand vertex is finished by depth-first search. We call *R_d-edge* an R-edge and *S_d-edge* an S-edge whose right-hand vertex is discovered but not finished by depth-first search.

Lemma 8. *C^s_{SCC} has the following properties.*

(i) Every S-edge is either in S_f or S_d.
(ii) Every SCC contains a simple cycle with exactly one S_d-edge.

Proof. We use simple properties of depth-first search (cf. [1]). (i) is obvious. For (ii), perform depth-first search on a SCC. Obviously, it contains a simple cycle with at least one S-edge. Assume further that no cycle has yet been detected. Let the S-edge on the cycle be an S_f-edge $v \to_{S_f} v'$. Then v is discovered, but not finished and v' is finished. Since we are on a cycle, some predecessor u from which v has been discovered must be reachable from v'. Since v' is finished, u must also be finished. But then v must also be finished, since it is reachable from u, a contradiction. Thus the S-edge must be an S_d-edge. Moreover, this S_d-edge must be the step closing the cycle, since v' must have been discovered. □

We can therefore refine the strategy in C^s_{SCC} further by restricting DEDUCE steps to R_d and S_d. These DEDUCE steps must be applied eagerly as soon as an S_d-edge has been discovered. Also the succeeding simplifications must be applied eagerly. Vertices that have been discovered or finished must keep their labels under the simplifications. This guarantees that precisely those critical pairs needed for collapsing cycles are computed. But all these critical pairs finally disappear in the T-rules that characterise the SCCs. Outside the SCCs, no new edge is added. Moreover, the outside edges are simplified by T-rules and finally become edges of the component graph. This implies the following result.

Theorem 9. *Every fair implementation of C^s_{SCC} is correct. For every input graph G, the algorithm computes the component graph G_{SCC}. The SCCs of G are represented by T_∞. The edges of G_{SCC} are represented by R_∞ and S_∞.*

Corollary 10. *Cycle detection with C^s_{SCC} means detection of S_d-edges. It is possible with* ORIENT *rules only, that is by pure depth-first search.*

We have thus reconstructed the standard algorithm for cycle detection in digraphs, using solely an execution strategy. The main work in the algorithm is shifted almost entirely to this strategy. Therefore, the running time of cycle detection is determined by depth-first search and becomes that of the best known algorithms. As a side effect, a spanning tree of the input graph is computed. C^s_{SCC} can be further refined.

Lemma 11. *For C^s_{SCC}, every T-tree for a simple k-cycle has $k-1$ branches, that is one T-rule per vertex to the minimal one (except from the latter itself).*

By Lemma 8(ii), every simple cycle in a SCC contains exactly one S-edge. The particular T-rules are computed backwards from the smallest node to the entry vertex of the SCC. One can therefore collect the vertices of the T-tree as follows: Search backwards (using depth-first search) all those vertices that are discovered, but not finished, starting with right-hand sides of S-rules and stopping at their left-hand sides. This saves some intermediate DEDUCE steps and yields a variant of the standard SCC algorithm [1].

In contrast to the standard algorithm, however, which performs two global searches on the input graph, C^s_{SCC} eliminates cycles in one full sweep. This may be an advantage in practice. Moreover, the construction of G_{SCC} is included in C^s_{SCC}. As a further benefit from the locality and non-determinism of C_{SCC} and C^s_{SCC}, edges can easily be inserted to a graph at running time; the algorithm is per se incrementally dynamic.

Prohibiting COMPOSE can even further improve the running time of C_{SCC} and C^s_{SCC}. A representation of SCCs by T-rules then corresponds precisely to a disjoint-set forest in a union-find data structure [1]. The effect of COMPOSE is precisely path-compression. Modelling the other heuristics of union-find, namely union by rank, is more involved. Choosing the representative of the bigger partial equivalence class as a new representative when merging two partial classes may violate the ordering constraints. The solution is to introduce a rank function and assign a weight $(r(v), o(v))$ to each vertex v, where $r(v)$, $r : V \to \mathbb{N}$, is the rank of v and $o(v)$ its size according to the usual precedence $\prec$. The components of this weight function are compared lexicographically. $r(v)$ is initialised with 0 and incremented, whenever v becomes the smaller vertex of a new T-rule. We therefore use the following variants of SCC and ORIENT for E-rules

$$(U, \Delta, O \cup \{l \to_R r, r \to_S l\}) \Rightarrow (U \cup \{l \to_E r\}, \Delta \cup \{r \to_\Delta r\}, O),$$
$$(U \cup \{s \to_E t\}, \Delta, O) \Rightarrow (U, \Delta, O \cup \{s \to_T t\}.$$

ORIENT is applied if $r(s) \leq r(t)$. Then $r(t)$ is incremented by one. Otherwise, its conclusion contains $t \to_T s$ and $r(s)$ is incremented by one. If moreover SIMPLIFY is applied eagerly, then only canonical representatives of equivalence classes are compared by SCC. It is well-known that $r(s)$ approximates the logarithm of the size of the T-subtree rooted at s and is also an upper bound for the height of each vertex in the T-tree [1]. Therefore for every vertex there are only logarithmically many smaller vertices. This heuristics may have considerable impact on the running time of C_{SCC} and associated search procedures. Note also that the union-find data-structure has been described in a purely declarative way.

7 An Example

For further illustration of our previous arguments, consider the following example. We eliminate the simple isolated oriented 4-cycle in the first diagram of Figure 1, using the precedence defined by $v_i \succ v_j$ iff $i < j$.

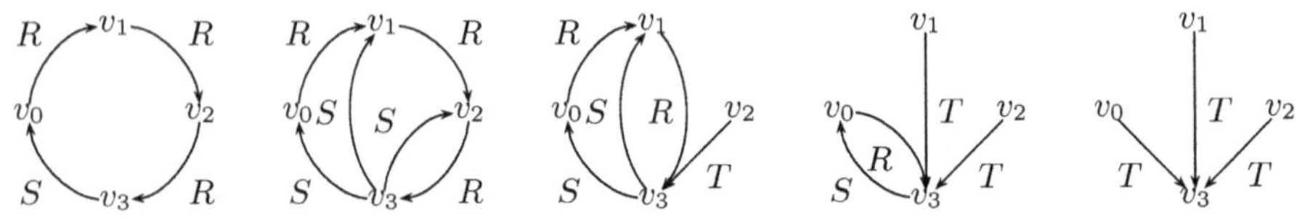

Fig. 1. Eliminating a simple cycle with C_{SCC}

Obviously, there is the critical pair $v_3 \rightarrow_S \rightarrow_R v_1$, for which Deduce and Orient compute the edge $v_3 \rightarrow_S v_1$. This yields the 3-cycle $v_1 \rightarrow_R v_2 \rightarrow_R v_3 \rightarrow_S v_1$ and the critical pair $v_3 \rightarrow_S \rightarrow_R v_2$. Deduce and Orient compute the edge $v_3 \rightarrow_S v_2$. This yields the 2-cycle $v_2 \rightarrow_R v_3 \rightarrow_S v_2$. This situation is shown in the second diagram.

Now SCC comes into play. It adds the edges $v_2 \rightarrow_T v_3$ and $v_3 \rightarrow_\Delta v_3$ and discards $v_2 \rightarrow_R v_3$ and $v_3 \rightarrow_S v_2$. Compose then replaces $v_1 \rightarrow_R v_2$ by $v_1 \rightarrow_R v_3$. All k-cycles are now collapsed into $k-1$-cycles and there is the 2-cycle $v_1 \rightarrow_R v_3 \rightarrow_S v_1$. This is shown in the third diagram. Iteration of this transformation leads to the fourth diagram with a final 2-cycle $v_0 \rightarrow_R v_3 \rightarrow_S v_0$. After its elimination by SCC, v_3 is the canonical representative of the cycle; the equivalence class is completely described by T.

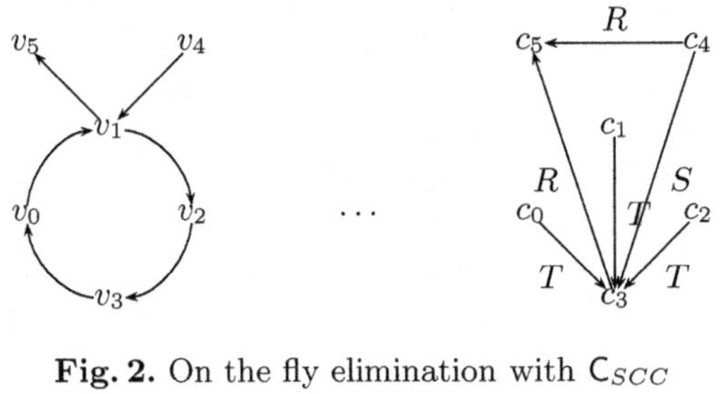

Fig. 2. On the fly elimination with C_{SCC}

Let us add the edge $v_4 \rightarrow_S v_1$ to the original cycle. Then Deduce and Orient add also $v_4 \rightarrow_S v_2$ and $v_4 \rightarrow_S v_3$. This critical pair computation is unnecessary for cycle elimination. These rules are replaced by $v_4 \rightarrow_S v_3$—an edge on the equivalence class—in the final graph, using Collapse. Let us now add also $v_1 \rightarrow_R v_5$ to the original graph. Then Deduce and Orient add also $v_4 \rightarrow_R v_5$. It does not belong to the component graph. Note that it does not matter at which stage of the process we add a new edge to the graph.

Now assume that we orient the nodes of the graph on the fly with C^s_{SCC}. Assume that we start with v_0 and v_1, but then continue with v_5. Then, when we continue with v_2, v_5 is already finished, hence need not be considered by

DEDUCE. We then disregard all critical pairs that are unnecessary for cycle elimination. All nodes on the cycle are discovered, but not finished and all rules on the cycle are R_d- and S_d-rules. We have detected the cycle (by mere depth-first search) as soon as we have oriented $v_3 \rightarrow_I v_1$ into S_d. For collecting all nodes on the cycle into one equivalence class, it suffices to depth-first search backwards from v_0 all the discovered, but non-finished nodes, until we return to v_1, the right-hand side of the S-rule. This yields the component graph via the standard SCC algorithm by strategic refinement from C_{SCC}.

8 Discussion

In this extended abstract, we can discuss only some aspects of our method. We now sketch two extensions. First, the combined KB-procedure, but also the cycle detection and SCC algorithms easily extend to preorderings. Then, the SCC algorithm constructs the associated partial ordering. Now, Δ is not needed, all inequalities of the form $s \rightarrow_I s$ can instead be deleted by the KB-procedure. The SCC-rule can accordingly be simplified. Second, there are various feasible extensions to the non-ground case. These include the case of linear variables and in particular a combined ground KB-completion modulo associativity and commutativity. These procedures can also easily be extended to non-ground variants of C_{SCC}. This yields (possibly non-terminating) semi-decision procedures that might be interesting for constraint systems.

In Section 5 and Section 6, we have restricted our attention to the static aspects of SCC algorithms. We now discuss dynamic aspects. We interpret C_{SCC} as an abstract declarative data structure that maintains information about reachability and cycles of a graph, hence of a component graph. In the basic version, inserting new edges is very simple. Since C_{SCC} is rule-based non-deterministic, it does not affect previous computations. New critical pairs must only be computed between the new edge and the previously compiled data structure. Also deletion of an edge can be tracked locally, for instance by computing all critical pairs with the deleted edge and deleting them successively. In a second step, the graph must then be closed under C_{SCC}, which is again local. This shows that C_{SCC} is inherently dynamic. A formal investigation is left for future work.

Using C^s_{SCC} instead of C_{SCC}, we can again increase efficiency. Dynamic cycle detection and edge insertion with C^s_{SCC}, for instance, is quite simple, since, by Lemma 8(iii) and Corollary 10, it suffices to check for cycles, when the new edge is in S. We determine whether an S-step is in S_d or S_f by recording the discovery and finishing times of all vertices. If it is in S_d (that is, when the finishing time of its right-hand vertex is greater than the discovery time of its left-hand vertex), it must be closing a cycle. If it is in S_f (that is, when the finishing time of its right-hand vertex is smaller than the discovery time of its left-hand vertex), it could also be the case that two paths are joined that do not form a cycle. One can then use depth-first search to test for cyclicity along a valley. Note that we cannot use the search procedure of non-symmetric rewriting, since we did not compute critical pairs involving S_f. This strategy reconstructs the best known dynamic cycle detection technique in [7]. But in contrast to previous approaches, ours is again local, which should pay in practice. Moreover, it leaves much space for heuristics and further refinement.

9 Conclusion

In this extended abstract, we have specified a cooperating Knuth-Bendix completion procedure for relations and equations. A main application is graph traversals modulo equivalence relations defined on vertices or edges. Here, we discussed incrementally dynamic cycle detection and SCC algorithms and briefly sketched techniques for efficient decrementation. All procedures are declarative and generic, implementation details can be hidden in the completion procedure. Efficiency can be obtained by refining with strategies. There is much space for approximations and other heuristics (cf. the union-find example). In contrast to the standard algorithms, ours are inherently dynamic. The approach is general enough to cover further applications, for instance declarative memoisation and termination analysis of ground rewriting systems. We also plan a more detailed treatment of the decrementally dynamic aspects of our algorithms, for instance on matrix representations of graphs. Dynamic updates might then be reducible to efficient updates in a system of linear equations.

Acknowledgements. I would like to thank Jesper Larsson Träff, Wolfram Kahl and the anonymous referees for helpful comments.

References

1. T. H. Cormen, C. E. Leiserson, and R. L. Rivest. *Introduction to Algorithms.* MIT Press, 1990.
2. C. Demetrescu and G. Italiano. Fully dynamic transitive closure: Breaking through the $O(n^2)$ barrier. In *Proc. 41st IEEE Symp. on Foundations of Computer Science*, pages 381–389. IEEE Computer Society, 2000.
3. M. Henzinger and V. King. Fully dynamic biconnectivity and transitive closure. In *Proc. 36th Symp. on Foundations of Computer Science*, pages 664–672. IEEE Computer Society, 1995.
4. D. Kapur. Shostak's congruence closure as completion. In H. Comon, editor, *Rewrite Techniques and Applications, 8th International Conference, RTA-97*, volume 1232 of *LNCS*, pages 23–37. Springer-Verlag, 1997.
5. V. King. Fully dynamic algorithms for maintaining all-time shortest paths and transitive closure in digraphs. In *Proc. 40st IEEE Symp. on Foundations of Computer Science*, pages 81–89. IEEE Computer Society, 1999.
6. G. Nelson. Techniques for program verification. Technical Report CSL-81-10, Xerox Palo Alto Research Center, 1981.
7. O. Shmueli. Dynamic cycle detection. *Information Processing Letters*, 17(4): 185–188, 1983.
8. G. Struth. Knuth-Bendix completion for non-symmetric transitive relations. In M. van den Brand and R. Verma, editors, *Second International Workshop on Rule-Based Programming (RULE2001)*, volume 59 of *Electronic Notes in Theoretical Computer Science*. Elsevier Science Publishers, 2001.
9. Terese, editor. *Term Rewriting Systems.* Cambridge University Press, 2003.

Quantifier Elimination in Elementary Set Theory*

Ewa Orłowska[1] and Andrzej Szałas[2,3]

[1] National Institute of Telecommunications, Warsaw, Poland
orlowska@itl.waw.pl
[2] The University of Economics and Computer Science, Olsztyn, Poland
[3] Department of Information and Computer Science
University of Linköping, Sweden
andsz@ida.liu.se

Abstract. In the current paper we provide two methods for quantifier elimination applicable to a large class of formulas of the elementary set theory. The first one adapts the Ackermann method [1] and the second one adapts the fixpoint method of [20]. We show applications of the proposed techniques in the theory of correspondence between modal logics and elementary set theory. The proposed techniques can also be applied in an automated generation of proof rules based on the semantic-based translation of axioms of a given logic into the elementary set theory.

1 Introduction

Second-order quantifier elimination, initiated probably in [1] in 1935, emerged in the last twenty years as a powerful technique in many applications, including the modal correspondence theory [37, 36, 12, 19, 20, 30, 31, 32, 33], relational, deductive and knowledge databases [6, 8, 13], complexity theory [34], knowledge representation, commonsense reasoning and approximate reasoning [7, 9, 16, 17]. All of the quoted results are based on elimination of second-order quantifiers and can be automatized, as done in [31, 7]. An online implementation of the algorithms based on [31, 7, 20] is available through http://www.ida.liu.se/labs/kplab/projects/dlsstar/.

On the other hand, there is a rich area of relational calculus, addressing many problems (see, e.g., [10, 11, 15, 18, 21, 23, 24, 25, 26, 27, 28, 29]), where second-order quantifier elimination techniques can substantially facilitate and/or automatize the reasoning.

The starting point of the current paper are the papers [4, 18, 21] and [20, 31, 32, 33]. More precisely, the paper [4] provides the translation of modal axioms into the formulas of elementary set theory. Then, in [18, 21] new methods for generating Rasiowa-Sikorski deductive rules for non-classical logics are introduced. These methods can be automatized except for applications of recursive rules and for dealing with conditions reflecting the second-order properties of translations from modal logics to the underlying set theory. On the other hand, a second-order quantifier elimination technique based on the Ackermann lemma of [1] has been proposed in the context of Kripke-based

* Partially supported by the EU COST Action 274 (TARSKI), INTAS project 04-77-7080 and the grant 3 T11C 023 29 of the Polish Ministry of Science and Information Society Technologies.

W. MacCaull et al. (Eds.): RelMiCS 2005, LNCS 3929, pp. 237–248, 2006.

translation, first in [31] and then, in a stronger context of the classical fixpoint calculus in [20], and modal fixpoint calculus in [33].

The techniques we develop here are quite general. However, in the current paper we mainly concentrate on applications of the proposed techniques in the field of modal correspondence theory. The methodology we develop here is then the following:

1. translate a given modal axiom into the elementary set theory using the method of [4];
2. use the natural deduction system for elementary set theory given in [21] in order to eliminate all occurrences of powerset operator $\mathcal{P}$ which is the set-theoretical counterpart of modal operator $\Box$;
3. consider fixpoint equations resulting from applications of recursive rules;
4. in all final sequents which do not close the proof tree try to eliminate second-order quantification using Lemma 4.1 (resulting in first-order conditions) or, if not successful, using Lemma 6.1 (resulting in fixpoint conditions, equivalent to a combination of infinite conjunctions and disjunctions);
5. derive Rasiowa-Sikorski style proof rules based on the obtained correspondences, using a generalization of the method proposed in [18].

In order to achieve the above goal we first recall elementary set theory Ω and the translation of modal formulas into the terms of Ω (based on [21]). Then we recall a particular constraint-rule correspondence of [18] and generalize it to an infinitary case. Next we formulate the Ackermann-like lemma (Lemma 4.1) and then we introduce fixpoint calculus into the elementary set theory and formulate a fixpoint quantifier elimination lemma (Lemma 6.1) in the framework of relational calculus for the theory. The fixpoint lemma reflects the fixpoint theorems given in [20, 33]. Here, however, it additionally allows us to eliminate quantifiers of any higher-order type.

We also demonstrate applications of the proposed technique in automated generation of Rasiowa-Sikorski style rules for modal logics, based on the elementary set theory.

2 Elementary Set Theory Ω

2.1 Syntax

The language is built over the following symbols:

- a denumerable set VAR of individual variables
- binary function symbols $\mathring{\cup}$, $\mathring{-}$ ("union" and "difference")
- unary function symbol $\mathcal{P}()$ ("powerset")
- binary predicates $\mathring{\in}$ and $\mathring{\subseteq}$, used in the infix notation ("membership" and "inclusion")
- propositional connectives $\neg, \vee, \wedge, \rightarrow, \equiv$
- quantifiers $\forall, \exists$.

The *set of terms*, TERMS, is defined as the least set satisfying the following conditions:

- VAR $\subseteq$ TERMS
- if $t, t' \in$ TERMS, then $t \mathring{\cup} t', t \mathring{-} t', \mathcal{P}(t) \in$ TERMS.

The *set of formulas*, FORMS, is defined as the least set satisfying the conditions:

- if $t, t' \in$ TERMS, then $t \mathring{\in} t', t \mathring{\subseteq} t' \in$ FORMS (*atomic formulas*)
- FORMS is closed under applications of propositional connectives and quantifiers.

2.2 Semantics

By a *model of* Ω we mean any model $\mathcal{S} = \langle U, \mathring{\cup}, \mathring{-}, \mathcal{P}, \mathring{\in}, \mathring{\subseteq} \rangle$ of the axioms

$$\begin{array}{l} x \mathring{\in} y \mathring{\cup} z \equiv x \mathring{\in} y \vee x \mathring{\in} z \\ x \mathring{\in} y \mathring{-} z \equiv x \mathring{\in} y \wedge x \mathring{\notin} z \\ x \mathring{\subseteq} y \equiv \forall z (z \mathring{\in} x \rightarrow z \mathring{\in} y) \\ x \mathring{\in} \mathcal{P}(y) \equiv x \mathring{\subseteq} y. \end{array} \tag{1}$$

The semantics of Ω wrt a model $\mathcal{S} = \langle U, \mathring{\cup}, \mathring{-}, \mathcal{P}, \mathring{\in}, \mathring{\subseteq} \rangle$ is defined in the standard way, by extending valuations VAR $\longrightarrow U$ in order to provide values for terms and truth values for formulas.

As usual, we use the same symbols for syntactic objects of the language and their corresponding interpretations in a model.

Given a model of Ω, $\mathcal{S} = \langle U, \mathring{\cup}, \mathring{-}, \mathcal{P}, \mathring{\in}, \mathring{\subseteq} \rangle$, a valuation v : VAR $\longrightarrow U$ and a formula $\Psi \in$ FORMS, we write $\mathcal{S}, v \models \Psi$ to mean that Ψ is satisfied in $\mathcal{S}$ under valuation v and $\mathcal{S} \models \Psi$ to mean that Ψ is valid in $\mathcal{S}$, i.e., for any valuation v, $\mathcal{S}, v \models \Psi$.

Table 1. Specific rules for Ω.

$(\mathring{\in}\mathring{\cup})\ \dfrac{K, x \mathring{\in} (y \mathring{\cup} z), H}{K, x \mathring{\in} y, x \mathring{\in} z, H}$	$(\mathring{\notin}\mathring{\cup})\ \dfrac{K, x \mathring{\notin} (y \mathring{\cup} z), H}{K, x \mathring{\notin} y, H \mid K, x \mathring{\notin} z, H}$
$(\mathring{\in}\mathring{-})\ \dfrac{K, x \mathring{\in} (y \mathring{-} z), H}{K, x \mathring{\in} y, H \mid K, x \mathring{\notin} z, H}$	$(\mathring{\notin}\mathring{-})\ \dfrac{K, x \mathring{\notin} (y \mathring{-} z), H}{K, x \mathring{\notin} y, x \mathring{\in} z, H}$
$(\mathring{\subseteq})\ \dfrac{K, x \mathring{\subseteq} y, H}{K, z \mathring{\notin} x, z \mathring{\in} y, H}$ (z is a new variable)	$(\mathring{\not\subseteq})\ \dfrac{K, x \mathring{\not\subseteq} y, H}{K, t \mathring{\in} x, x \mathring{\not\subseteq} y, H \mid K, t \mathring{\notin} y, x \mathring{\not\subseteq} y, H}$ (t is an arbitrary term)
$(\mathring{\in}\mathcal{P})\ \dfrac{K, x \mathring{\in} \mathcal{P}(y), H}{K, x \mathring{\subseteq} y, H}$	$(\mathring{\notin}\mathcal{P})\ \dfrac{K, x \mathring{\notin} \mathcal{P}(y), H}{K, x \mathring{\not\subseteq} y, H}$

2.3 A Rasiowa-Sikorski Deduction System for Ω

A proof system for Ω has been given in [21]. The proof system consists of the rules for propositional connectives and first-order quantifiers, including the cut rule, together with specific rules for Ω. The specific rules are provided in Table 1, where K, H are finite sets of formulas, $K, \phi, H \stackrel{\text{def}}{=} K \cup \{\phi\} \cup H$, branching in a proof tree is denoted by $\mid$ and axioms are sets of formulas containing both ϕ and $\neg\phi$, for some formula ϕ. Intuitively, a set of formulas represents the disjunction of its members and branching corresponds to the conjunction.

2.4 Some Definitions and Basic Facts

Let us now list some definitions and facts used throughout the paper.

Proposition 2.1. *Let $\mathcal{S} = \langle U, \mathring{\cup}, \mathring{-}, \mathcal{P}, \mathring{\in}, \mathring{\subseteq} \rangle$ be a model of Ω. Then the relation $\mathring{\subseteq}$ is reflexive and transitive. It is not guaranteed to be weakly antisymmetric.* ◁

As usual, for E being a term or a formula, we write $E(\bar{x}, z)$ to indicate that all the free variables in E are those from the sequence $\bar{x}$ and z.

Definition 2.2.

- *We say that model $\mathcal{S}\langle U, \mathring{\cup}, \mathring{-}, \mathcal{P}, \mathring{\in}, \mathring{\subseteq} \rangle$ is* partially ordered *if $\langle U, \mathring{\subseteq} \rangle$ is a partial order. If $\langle U, \mathring{\subseteq} \rangle$ is a complete partial order, we also say that $\mathcal{S}$ is a* complete partial order. *In such a case the least and the greatest element of U wrt $\mathring{\subseteq}$ exist and are denoted by $\bot$ and $\top$, respectively, and we often expand $\mathcal{S}$ by explicitly listing elements $\bot$ and $\top$.*
- *A term $F(\bar{x}, z) \in$ TERMS is* up-monotone *(respectively,* down-monotone*) wrt z in model $\mathcal{S}$ iff for any valuation v in $\mathcal{S}$,*
  $v(z') \mathring{\subseteq} v(z'')$ implies $\mathcal{S}, v \models F(\bar{x}, z') \mathring{\subseteq} F(\bar{x}, z'')$
  (respectively, implies $\mathcal{S}, v \models F(\bar{x}, z'') \mathring{\subseteq} F(\bar{x}, z')$).
- *A formula $\Psi(\bar{x}, z) \in$ FORMS is* up-monotone *(respectively,* down-monotone*) wrt z in model $\mathcal{S}$ iff for any any valuation v in $\mathcal{S}$,*
  $v(z') \mathring{\subseteq} v(z'')$ implies $\mathcal{S}, v \models \Psi(\bar{x}, z') \to \Psi(\bar{x}, z'')$
  (respectively, implies $\mathcal{S}, v \models \Psi(\bar{x}, z'') \to \Psi(\bar{x}, z')$). ◁

Example 2.3. By the third axiom of (1) we have that formula $x \mathring{\in} z$ is up-monotone wrt z and that formula $x \mathring{\notin} z$ is down-monotone wrt z in any model of Ω. ◁

2.5 Translation from Modal Logic to Ω

Consider a propositional modal logic with the standard language $\mathcal{L}$ over a set of propositional variables, with connectives $\neg$, $\vee$ and modality $\Box$. Other standard propositional connectives and the dual modality $\Diamond$ are defined as usual. Following [4], we define translation tr of modal formulas into terms of Ω inductively, where f is a distinguished variable of Ω:

- $tr(p) \stackrel{\text{def}}{=} x_p$, where p is a variable in $\mathcal{L}$
- $tr(\neg\alpha) \stackrel{\text{def}}{=} f \mathring{-} tr(\alpha)$, where $\alpha \in \mathcal{L}$
- $tr(\alpha \vee \beta) \stackrel{\text{def}}{=} tr(\alpha) \mathring{\cup} tr(\beta)$, where $\alpha, \beta \in \mathcal{L}$
- $tr(\Box\alpha) \stackrel{\text{def}}{=} \mathcal{P}(tr(\alpha))$, where $\alpha \in \mathcal{L}$.

As usual, propositional variables represent sets of states of a modal frame. Along these lines, variable f represents the universe of states of a frame F and, as shown in [4], for any formula $\alpha \in \mathcal{L}$, $F \models \alpha$ iff $\Omega \models f \mathring{\subseteq} tr(\alpha)$. This fact provides us with a starting point for application of the Rasiowa-Sikorski-like deduction system quoted in Section 2.3. Namely, for computing the correspondence for modal axiom, say α, we start with sequent $f \mathring{\subseteq} tr(\alpha)$ and then we apply the suitable rules.

2.6 Fixpoints in Ω

Below we present a definition of fixpoints in Ω, based on the famous Knaster and Tarski fixpoint theorem [14, 35].

Definition 2.4. *Let $F(\bar{x}, z)$ be a term up-monotone wrt z and let us fix a model of Ω, which is a complete partial order, $\mathcal{S} = \langle U, \mathring{\cup}, \mathring{-}, \mathcal{P}, \mathring{\in}, \mathring{\subseteq}, \bot, \top\rangle$. Then the least and the greatest (wrt $\mathring{\subseteq}$) fixpoint of F exist and are denoted by $\mu\, z.F(\bar{x}, z)$ and $\nu\, z.F(\bar{x}, z)$, respectively.*

The least ordinal number α such that, for any up-monotone F, we have that $\mathcal{S} \models F^{\alpha}(\bar{x}, \bot) = F^{\alpha+1}(\bar{x}, \bot)$, is called the closure ordinal *for $\mathcal{S}$. It is well known that such an ordinal exists (see, e.g., [2]).* ◁

For formulas we adopt the standard definitions of fixpoints, where we assume that the corresponding partial order in any model of Ω is defined by means of implication $\rightarrow$.

The following proposition is well-known.

Proposition 2.5. *The closure ordinal for existential up-monotone formulas is ω.* ◁

3 Constraint-Rule Correspondences

In [18] some correspondences between constraints and Rasiowa-Sikorski-like rules have been provided. In the current paper we shall first need correspondences for constraints of the form[1]:

$$\forall \bar{x}\big[(\alpha_1 \wedge \ldots \wedge \alpha_s) \rightarrow \beta\big], \tag{2}$$

where $\bar{x}$ is a non-empty sequence consisting of all variables in literals $\alpha_1, \ldots, \alpha_s$ and the set of variables in β is non-empty and included in the set of variables in $\bar{x}$.

According to [18], constraints of the form (2) correspond to Rasiowa-Sikorski-like rules:

$$\frac{K, \beta, H}{K, \alpha_1, H, \beta \mid \ldots \mid K, \alpha_s, H, \beta} \tag{3}$$

We shall also need a generalization of the above correspondence by allowing infinite number of variables under quantifiers as well as infinite conjunctions:

$$\forall \bar{x}\left[\left(\bigwedge_{i\in\omega} \alpha_i\right) \rightarrow \beta\right], \tag{4}$$

where $\bar{x}$ is a non-empty sequence consisting of all variables in $\{\alpha_i\}_{i\in\omega}$ and the set of variables in β is non-empty and included in the set of variables in $\bar{x}$.

The rule corresponding to (4) is:

$$\frac{K, \beta, H}{K, \alpha_1, H, \beta \mid K, \alpha_2, H, \beta \mid \ldots \mid K, \alpha_s, H, \beta \mid \ldots\ldots} \tag{5}$$

Infinitary rules of that form are needed, for example, in Rasiowa-Sikorski proof system for dynamic logic [22] and for a logic of demonic nondeterministic programs [5].

[1] This is the form (c1) of [18].

4 Ackermann-Like Lemma for Ω

Let us formulate the announced Ackermann-like lemma. This lemma, under additional assumption that the underlying model is a complete partial order, is subsumed by Lemma 6.1, but it is frequently used in applications, so we formulate it separately.

Let $\Psi(x, \bar{y}) \in$ FORMS be a formula and $t \in$ TERMS be a term. Then $\Psi(x, \bar{y})^x_t$ denotes the formula obtained from Ψ by substituting all free occurrences of x by t and leaving all bound occurrences of x unchanged.

Lemma 4.1 (Ackermann-like Lemma). *Let* $\mathcal{S} = \langle U, \mathring{\cup}, \mathring{-}, \mathcal{P}, \mathring{\in}, \mathring{\subseteq} \rangle$ *be a partially ordered model of* Ω. *Let* $F(\bar{z}) \in$ TERMS *be a term not containing* y. *Then:*

1. *if* $\Psi(\bar{x}, y) \in$ FORMS *is a formula up-monotone wrt* y *in* $\mathcal{S}$, *then*

$$\mathcal{S} \models \exists y \big[y \mathring{\subseteq} F(\bar{z}) \wedge \Psi(\bar{x}, y) \big] \equiv \Psi(\bar{x}, y)^y_{F(\bar{z})} \tag{6}$$

2. *if* $\Psi'(y) \in$ FORMS *is a formula down-monotone wrt* y *in* $\mathcal{S}$, *then*

$$\mathcal{S} \models \exists y \big[F(\bar{z}) \mathring{\subseteq} y \wedge \Psi'(y) \big] \equiv \Psi'(y)^y_{F(\bar{z})}. \tag{7}$$

Proof. We prove (6). The proof of (7) is similar.
($\rightarrow$) Let $\mathcal{S}$ be a model of Ω and let v be a valuation in $\mathcal{S}$ such that

$$\mathcal{S}, v \models \exists y \big[y \mathring{\subseteq} F(\bar{z}) \wedge \Psi(\bar{x}, y) \big].$$

Let v' be a valuation extending v by assigning a domain value to y such that

$$\mathcal{S}, v' \models y \mathring{\subseteq} F(\bar{z}) \text{ and } \mathcal{S}, v' \models \Psi(\bar{x}, y).$$

By up-monotonicity of Ψ we have that also $\mathcal{S}, v' \models \Psi(\bar{x}, y) \rightarrow \Psi(\bar{x}, y)^y_{F(\bar{z})}$. Therefore $\mathcal{S}, v' \models \Psi(\bar{x}, y)^y_{F(\bar{z})}$. Observe that $\Psi(\bar{x}, y)^y_{F(\bar{z})}$ does not contain free occurrences of y. Thus we conclude that $\mathcal{S}, v \models \Psi(\bar{x}, y)^y_{F(\bar{z})}$.

($\leftarrow$) Let v be a valuation in $\mathcal{S}$ such that $\mathcal{S}, v \models \Psi(\bar{x}, y)^y_{F(\bar{z})}$. Consider a valuation v' different from v at most on y and such that $v'(y) \stackrel{\text{def}}{=} F(v(\bar{z}))$. Then $\mathcal{S}, v' \models y \mathring{\subseteq} F(\bar{z})$ and $\mathcal{S}, v' \models \Psi(\bar{x}, y)$. Thus we have that $\mathcal{S}, v \models \exists y \big[y \mathring{\subseteq} F(\bar{z}) \wedge \Psi(\bar{x}, y) \big]$. ◁

5 Generation of Rules Using the Ackermann Lemma

Let us illustrate the process of generating rules corresponding to modal axioms in the case of well-known modal axiom 4.

Axiom 4 is formulated as

$$\Box P \rightarrow \Box\Box P. \tag{8}$$

Of course, it is equivalent to $\neg\Box P \vee \Box\Box P$. Thus, according to Section 2.5, axiom (8) translates into

$$f \mathring{\subseteq} (f \mathring{-} \mathcal{P}(y)) \mathring{\cup} \mathcal{P}(\mathcal{P}(y)).$$

The suitable derivation looks as follows (see [21]):

$$
\begin{array}{c|c}
 & f \mathring{\subseteq} (f \mathring{-} \mathcal{P}(y)) \mathring{\cup} \mathcal{P}(\mathcal{P}(y)) \\
\hline
 & z \mathring{\notin} f, z \mathring{\in} (f \mathring{-} \mathcal{P}(y)) \mathring{\cup} \mathcal{P}(\mathcal{P}(y)) \quad (\mathring{\subseteq}) \\
\hline
 & z \mathring{\notin} f, z \mathring{\in} f \mathring{-} \mathcal{P}(y), z \mathring{\in} \mathcal{P}(\mathcal{P}(y)) \quad (\mathring{\in}\mathring{\cup}) \\
\hline
z \mathring{\notin} f, z \mathring{\in} f \ldots & z \mathring{\notin} f, z \mathring{\notin} \mathcal{P}(y), z \mathring{\in} \mathcal{P}(\mathcal{P}(y)) \quad (\mathring{\in}\mathring{-}) \\
\hline
\text{closed} & z \mathring{\notin} f, z \mathring{\notin} \mathcal{P}(y), z \mathring{\subseteq} \mathcal{P}(y) \quad (\mathring{\in}\mathcal{P}) \\
\hline
 & z \mathring{\notin} f, z \mathring{\not\subseteq} y, z \mathring{\subseteq} \mathcal{P}(y) \quad (\mathring{\notin}\mathcal{P}) \\
\hline
 & z \mathring{\notin} f, z \mathring{\not\subseteq} y, w \mathring{\notin} z, w \mathring{\in} \mathcal{P}(y) \quad (\mathring{\subseteq}) \\
\hline
 & z \mathring{\notin} f, z \mathring{\not\subseteq} y, w \mathring{\notin} z, w \mathring{\subseteq} y \quad (\mathring{\in}\mathcal{P}) \\
\hline
 & z \mathring{\notin} f, z \mathring{\not\subseteq} y, w \mathring{\notin} z, u \mathring{\notin} w, u \mathring{\in} y \quad (\mathring{\subseteq})
\end{array}
$$

Now an application of rule $(\mathring{\not\subseteq})$ with u as the required term, results in two sequents. One of them closes a branch and the second one is

$$z \mathring{\notin} f, u \mathring{\in} z, w \mathring{\notin} z, u \mathring{\notin} w, u \mathring{\in} y, z \mathring{\not\subseteq} y \tag{9}$$

Since y is a translation of P, it is the second-order variable to be eliminated from (9). We then eliminate quantifier $\forall y$ from

$$\forall y (u \mathring{\in} y \vee z \mathring{\not\subseteq} y). \tag{10}$$

In order to eliminate $\forall y$, we first negate (10), then we apply (7) and finally we negate the result:

- after negating (10) we obtain $\exists y (u \mathring{\notin} y \wedge z \mathring{\subseteq} y)$, i.e., $\exists y (z \mathring{\subseteq} y \wedge u \mathring{\notin} y)$ — observe that $u \mathring{\notin} y$ is down-monotone wrt y (see Example 2.3)
- after applying (7) with $F(z) \overset{\text{def}}{=} z$ and $\Psi(y) \overset{\text{def}}{\equiv} u \mathring{\notin} y$ we obtain $u \mathring{\notin} z$
- after negating again we get $u \mathring{\in} z$.

Thus formula (10) is equivalent to $u \mathring{\in} z$. In consequence, sequent (9) is equivalent to

$$z \mathring{\notin} f, u \mathring{\in} z, w \mathring{\notin} z, u \mathring{\notin} w. \tag{11}$$

Observe that (11) represents, in fact, the formula

$$\forall u, w, z, f \big[\big(u \mathring{\notin} z \wedge w \mathring{\in} z \wedge u \mathring{\in} w \big) \rightarrow z \mathring{\notin} f \big]. \tag{12}$$

Constraint (12) is in the form (2). Therefore the corresponding rule is of the form (3), i.e., we obtain the following rule corresponding to axiom (8):

$$\frac{K, z \mathring{\notin} f, H}{K, u \mathring{\notin} z, H, z \mathring{\notin} f \mid K, w \mathring{\in} z, H, z \mathring{\notin} f \mid K, u \mathring{\in} w, H, z \mathring{\notin} f}.$$

6 The Fixpoint Lemma

In this section we adapt the fixpoint lemma given in [20] to the framework of elementary set theory. It allows one to deal with sets of higher-order types, too, while the theorem of [20] is formulated for the second-order logic.

Lemma 6.1 (Fixpoint Lemma). *Let $\mathcal{S} = \langle U, \mathring{\cup}, \mathring{-}, \mathcal{P}, \mathring{\in}, \mathring{\subseteq}, \bot, \top \rangle$ be a model of Ω which is a complete partial order. Let $F(\bar{x}, y) \in$ TERMS be a term up-monotone wrt y in $\mathcal{S}$. Then:*

1. *if $\Psi(\bar{z}, y) \in$ FORMS is a formula up-monotone wrt y in $\mathcal{S}$, then*
$$\mathcal{S} \models \exists y \big[y \mathring{\subseteq} F(\bar{x}, y) \wedge \Psi(\bar{z}, y) \big] \equiv \Psi(\bar{z}, y)^{y}_{\nu y.F(\bar{x},y)} \tag{13}$$
2. *if $\Psi'(y) \in$ FORMS is a formula down-monotone wrt y in $\mathcal{S}$, then*
$$\mathcal{S} \models \exists y \big[F(\bar{x}, y) \mathring{\subseteq} y \wedge \Psi'(y) \big] \equiv \Psi'(y)^{y}_{\mu y.F(\bar{x},y)}. \tag{14}$$

Proof. We prove (13). The proof of (14) is similar.
($\rightarrow$) Assume $\mathcal{S}$ and v are such that $\mathcal{S}, v \models \exists y \big[y \mathring{\subseteq} F(\bar{x}, y) \wedge \Psi(\bar{z}, y) \big]$. Since $F(\bar{x}, y)$ is up-monotone wrt y and $\mathcal{S}$ is a complete partial order, the greatest fixpoint of F, $\nu y.F(\bar{x}, y)$, exists and for any v' such that $\mathcal{S}, v' \models y \mathring{\subseteq} F(\bar{x}, y)$ we have that[2] $\mathcal{S}, v' \models y \mathring{\subseteq} \nu y.F(\bar{x}, y)$. By assumption, $\Psi(\bar{z}, y)$ holds, for some y. Thus, by up-monotonicity of Ψ we conclude that $\mathcal{S}, v \models \Psi(\bar{z}, y)^{y}_{\nu y.F(\bar{x},y)}$ holds.

($\leftarrow$) Assume $\mathcal{S}$ and v are such that $\mathcal{S}, v \models \Psi(\bar{z}, y)^{y}_{\nu y.F(\bar{x},y)}$ holds. Then v' which differs from v at most on y and such that $v'(y) \stackrel{\text{def}}{=} \nu y.F(v(\bar{x}), y)$ satisfies

$$\mathcal{S}, v' \models y \mathring{\subseteq} F(\bar{x}, y) \text{ and } \mathcal{S}, v' \models \Psi(\bar{z}, y).$$

Therefore we also have that $\mathcal{S}, v \models \exists y \big[y \mathring{\subseteq} F(\bar{x}, y) \wedge \Psi(\bar{z}, y) \big]$. $\lhd$

Observe that one can easily generalize Lemma 6.1 to allow infinite disjunctions and conjunctions in Ψ and Ψ'.

7 Generation of Rules Using the Fixpoint Lemma

Observe, that a rule might be *recursive* in the sense that the decomposed formula appears also in the consequences of the rule. For example, rule ($\mathring{\not\subseteq}$), quoted in Section 2.3, is recursive. An intuitive meaning of ($\mathring{\not\subseteq}$) is that in order to prove $K, x \mathring{\not\subseteq} y, H$, one has to prove both $K, t \mathring{\in} x, x \mathring{\not\subseteq} y, H$ and $K, t \mathring{\notin} y, x \mathring{\not\subseteq} y, H$, where the set of conditions to be proved is assumed to be a minimal one.

As an example, consider Löb axiom,

$$\Box(\Box p \rightarrow p) \rightarrow \Box p \tag{15}$$

A fixpoint characterization of this axiom has been considered in [20]. Below we show how a fixpoint approach can contribute to automated generation of infinitary proof rule, given in [21], reflecting (15).

According to the method described in Section 2.5, translation of formula (15) leads to the sequent $f \mathring{\subseteq} \overline{\mathcal{P}}(\overline{\mathcal{P}}(y) \mathring{\cup} y) \mathring{\cup} \mathcal{P}(y)$, where $\overline{\mathcal{P}}(t) \stackrel{\text{def}}{\equiv} f \mathring{-} \mathcal{P}(t)$.

The calculations given in [21] lead from the sequent $\big[v_0 \mathring{\notin} f, v_0 \mathring{\not\subseteq} (\overline{\mathcal{P}}(y) \mathring{\cup} y), v_0 \mathring{\subseteq} y\big]$ to the sequent $\big[v_0 \mathring{\not\in} f, v_0 \mathring{\not\subseteq} (\overline{\mathcal{P}}(y) \mathring{\cup} y), v_1 \mathring{\subseteq} y, v_1 \mathring{\not\in} v_0, v_1 \mathring{\in} y\big]$, where both above sequents are implicitly universally quantified over the second-order variable y.

[2] A standard transfinite induction argument applies here.

This observation gives rise to the fixpoint equation

$$v_0 \mathring{\subseteq} y \;\equiv\; \exists v_1[v_1 \mathring{\subseteq} y \vee v_1 \mathring{\notin} v_0 \vee v_1 \mathring{\in} y],$$

where one looks for the least relation " $\mathring{\subseteq}$ " satisfying it. We thus deal with the fixpoint equation of the form

$$\Xi(v_0, y) \;\equiv\; \exists v_1[\Xi(v_1, y) \vee v_1 \mathring{\notin} v_0 \vee v_1 \mathring{\in} y],$$

in which, for readability, Ξ replaces $\mathring{\subseteq}$. The corresponding fixpoint we are looking for is then $\mu\, \Xi(v_0, y).\exists v_1[\Xi(v_1, y) \vee v_1 \mathring{\notin} v_0 \vee v_1 \mathring{\in} y]$. Since formula under the fixpoint operator is existential, by Proposition 2.5 we have that the closure ordinal for this formula is ω and therefore this fixpoint is equivalent to the infinite disjunction

$$\bigvee_{i\in\omega} [v_{i+1} \mathring{\notin} v_i \vee v_{i+1} \in y].$$

In consequence, we end up here with the disjunction

$$v_0 \mathring{\notin} f \vee v_0 \mathring{\nsubseteq} (\overline{\mathcal{P}}(y) \mathring{\cup} y) \vee \bigvee_{i\in\omega} [v_{i+1} \mathring{\notin} v_i \vee v_{i+1} \in y] \tag{16}$$

We still have to eliminate the second-order variable y from (16).

Consider disjunction (16), universally quantified over y, i.e.,

$$\forall y\big[v_0 \mathring{\notin} f \vee v_0 \mathring{\nsubseteq} (\overline{\mathcal{P}}(y) \mathring{\cup} y) \vee \bigvee_{i\in\omega} [v_{i+1} \mathring{\notin} v_i \vee v_{i+1} \in y]\big]. \tag{17}$$

In order to eliminate $\forall y$ from (17), we first negate it,

$$\exists y\big[v_0 \mathring{\in} f \wedge v_0 \mathring{\subseteq} (\overline{\mathcal{P}}(y) \mathring{\cup} y) \wedge \bigwedge_{i\in\omega} [v_{i+1} \mathring{\in} v_i \wedge v_{i+1} \mathring{\notin} y]\big].$$

The negated formula is equivalent to

$$v_0 \mathring{\in} f \wedge \exists y\big[(v_0 \mathring{\cap} \mathcal{P}(y)) \mathring{\subseteq} y) \wedge \bigwedge_{i\in\omega} [v_{i+1} \mathring{\in} v_i \wedge v_{i+1} \mathring{\notin} y]\big] \tag{18}$$

where $v_0 \mathring{\cap} \mathcal{P}(y)$ is an abbreviation for $f \mathring{-} \big((f \mathring{-} v_0) \mathring{\cup} \overline{\mathcal{P}}(y)\big)$.

Observe that $v_0 \mathring{\cap} \mathcal{P}(y)$ is up-monotone wrt y and $\bigwedge_{i\in\omega} [v_{i+1} \mathring{\in} v_i \wedge v_{i+1} \mathring{\notin} y]$ is down-monotone wrt y. Thus we can apply the second part of Lemma 6.1 and obtain the following formula equivalent to (18),

$$v_0 \mathring{\in} f \wedge \bigwedge_{i\in\omega} [v_{i+1} \mathring{\in} v_i \wedge v_{i+1} \mathring{\notin} \mu\, y.(v_0 \mathring{\cap} \mathcal{P}(y))]. \tag{19}$$

By unfolding the least fixpoint operator we easily notice that $\mu\, y.(v_0 \mathring{\cap} \mathcal{P}(y)) \equiv \bot$. Thus (19) reduces to $v_0 \mathring{\in} f \wedge \bigwedge_{i\in\omega} [v_{i+1} \mathring{\in} v_i \wedge v_{i+1} \mathring{\notin} \bot]$, i.e., to $v_0 \mathring{\in} f \wedge \bigwedge_{i\in\omega} [v_{i+1} \mathring{\in} v_i]$. Negating this formula, we obtain

$$v_0 \mathring{\notin} f \vee \bigvee_{i\in\omega} [v_{i+1} \mathring{\notin} v_i] \tag{20}$$

which is equivalent to (17). The same result has been obtained in [21]. However, the above method of calculating (20) can be fully automated, e.g., in the spirit of the algorithm given in [31].

In order to generate the corresponding rule we first present (20) in the form required in (4),

$$\forall f, v_0, \ldots, v_i, v_{i+1}, \ldots \left[\bigwedge_{i \in \omega} [v_{i+1} \mathring{\in} v_i] \rightarrow v_0 \mathring{\notin} f \right].$$

According to the general scheme (5), we obtain the following rule:

$$\frac{K, v_0 \mathring{\notin} f, H}{K, v_1 \mathring{\in} v_0, v_0 \mathring{\notin} f, H \mid \ldots \mid K, v_{i+1} \mathring{\in} v_0, v_0 \mathring{\notin} f, H \mid \ldots},$$

reflecting the Löb axiom (15).

8 Conclusions

In the current paper we have provided two methods for quantifier elimination applicable to a large class of formulas of the elementary set theory. The first one is based on an Ackermann-like lemma (Lemma 4.1) and the second one is based on the fixpoint lemma (Lemma 6.1).

We have shown applications of the proposed techniques in the theory of correspondence between modal logics and the elementary set theory.

We have also applied the proposed techniques as a part of automated generation of proof rules based on the semantic-based translation of axioms of a given logic into the elementary set theory.

The approach presented here can be extended in various directions, including the development of algorithms for transforming terms into the forms required in Lemmas 4.1 and 6.1. One could also try to provide lemmas extending the class of formulas for which second-order quantifier elimination would still be possible.

References

1. W. Ackermann. Untersuchungen über das Eliminationsproblem der Mathematischen Logik. *Mathematische Annalen*, 110:390–413, 1935.
2. A. Arnold and D. Niwinski. *Rudiments of μ-calculus, Studies in Logic and the Foundations of Mathematics, Elsevier, 2001.*
3. R. Berghammer, B. Moeller, and G. Struth, editors. *Proc. 7th Int. Seminar on Relational Methods in Comp. Science (RelMiCS7) and 2nd Int. Workshop on Applications of Kleene Algebra*, volume 3051 of *LNCS*. Springer Verlag, 2004.
4. G. D'Agostino, A. Montanari, and A. Policriti. A set-theoretic translation method for polymodal logics. *Journal of Automated Reasoning*, 3(15):317–337, 1995.
5. S. Demri, E. Orłowska. Logical analysis of demonic nondeterministic programs. *Theoretical Computer Science*, 166:173–202, 1996.
6. P. Doherty, J. Kachniarz, and A. Szałas. Using contextually closed queries for local closed-world reasoning in rough knowledge databases. In S.K. Pal, L. Polkowski, and A. Skowron, editors, *Rough-Neuro Computing: Techniques for Computing with Words*, Cognitive Technologies, pages 219–250. Springer–Verlag, 2003.

7. P. Doherty, W. Łukaszewicz, and A. Szałas. Computing circumscription revisited. *Journal of Automated Reasoning*, 18(3):297–336, 1997.
8. P. Doherty, W. Łukaszewicz, and A. Szałas. Declarative PTIME queries for relational databases using quantifier elimination. *Journal of Logic and Computation*, 9(5):739–761, 1999.
9. P. Doherty, W. Łukaszewicz, and A. Szałas. Computing strongest necessary and weakest sufficient conditions of first-order formulas. *International Joint Conference on AI (IJCAI'2001)*, pages 145 – 151, 2000.
10. I. Düntsch and E. Orłowska. A proof system for contact relation algebras. *Journal of Philosophical Logic*, 29:241–262, 2000.
11. M. Frias and E. Orłowska. A proof system for fork algebras and its applications to reasoning in logics based on intuitionism. *Logique et Analyse*, 150-151-152:239–284, 1995.
12. D. M. Gabbay and H. J. Ohlbach. Quantifier elimination in second-order predicate logic. In B. Nebel, C. Rich, and W. Swartout, editors, *Principles of Knowledge representation and reasoning, KR 92*, pages 425–435. Morgan Kauffman, 1992.
13. J. Kachniarz and A. Szałas. On a static approach to verification of integrity constraints in relational databases. In [29], pages 97–109, 2001.
14. B. Knaster. Un theoreme sur les fonctions d'ensembles. *Ann. Soc. Polon. Math.*, 6:133-134, 1928.
15. B. Konikowska and E. Orłowska. A relational formalisation of a generic many-valued modal logic. In [29], pages 183–202, 2001.
16. V. Lifschitz. Computing circumscription. In *Proc. 9th IJCAI*, pages 229–235. Morgan Kaufmann, 1985.
17. V. Lifschitz. Circumscription. In D. M. Gabbay, C. J. Hogger, and J. A. Robinson, editors, *Handbook of Artificial Intelligence and Logic Programming*, volume 3, pages 297–352. Oxford University Press, 1991.
18. W. MacCaull and E. Orłowska. Correspondence results for relational proof systems with application to the Lambek calculus. *Studia Logica: an International Journal for Symbolic Logic*, 71(3):279–304, 2002.
19. A. Nonnengart, H.J. Ohlbach, and A. Szałas. Elimination of predicate quantifiers. In H.J. Ohlbach and U. Reyle, editors, *Logic, Language and Reasoning. Essays in Honor of Dov Gabbay, Part I*, pages 159–181. Kluwer, 1999.
20. A. Nonnengart and A. Szałas. A fixpoint approach to second-order quantifier elimination with applications to correspondence theory. In E. Orłowska, editor, *Logic at Work: Essays Dedicated to the Memory of Helena Rasiowa*, volume 24 of *Studies in Fuzziness and Soft Computing*, pages 307–328. Springer Physica-Verlag, 1998.
21. E. Omodeo, E. Orłowska, and A. Policriti. Rasiowa-Sikorski style relational elementary set theory. In R. Berghammer, B. Moeller, and G. Struth, editors, [3], pages 213–224, 2004.
22. E. Orłowska. Dynamic logic with program specifications and its relational proof system. *Journal of Applied Non-Classical Logic*, 3:147–171, 1993.
23. E. Orłowska. Relational interpretation of modal logics. In H. Andreka, D. Monk, and I. Nemeti, editors, *Algebraic Logic*, volume 54 of *Colloquia Mathematica Societatis Janos Bolyai*, pages 443–471, 1988.
24. E. Orłowska. Relational proof systems for relevant logics. *Journal of Symbolic Logic*, 57:1425–1440, 1992.
25. E. Orłowska. Relational semantics for non-classical logics: Formulas are relations. In J. Woleński, editor, *Philosophical Logic in Poland*, pages 167–186, 1994.
26. E. Orłowska. Relational proof systems for modal logics. In H. Wansing, editor, *Proof Theory of Modal Logics*, pages 55–77, 1996.
27. E. Orłowska. Relational formalisation of nonclassical logics. In C. Brink, W. Kahl, and G. Schmidt, editors, *Relational Methods in Computer Science*, pages 90–105, 1997.

28. E. Orłowska and W. McCaull. A calculus of typed relations. In [3], pages 152–158, 2004.
29. E. Orłowska and A. Szałas, editors. *Relational Methods for Computer Science Applications*. Springer Physica-Verlag, 2001.
30. H. Simmons. The monotonous elimination of predicate variables. *Journal of Logic and Computation*, 4:23–68, 1994.
31. A. Szałas. On the correspondence between modal and classical logic: An automated approach. *Journal of Logic and Computation*, 3:605–620, 1993.
32. A. Szałas. On an automated translation of modal proof rules into formulas of the classical logic. *Journal of Applied Non-Classical Logics*, 4:119–127, 1994.
33. A Szałas. Second-order quantifier elimination in modal contexts. In M. Flesca, S. Greco, N. Leone, and G. Ianni, editors, *Proceedings of 8th European Conference on Logics in Artificial Intelligence JELIA'2002*, LNAI, pages 223–232. Springer-Verlag, 2002.
34. A Szałas. On a logical approach to estimating computational complexity of potentially intractable problems. In A. Lingas and B.J. Nilsson, editors, *Proceedings of 14th International Symposium on Fundamentals of Computation Theory FCT'2003*, LNCS, pages 423–431. Springer-Verlag, 2003.
35. A. Tarski. A Lattice-theoretical Fixpoint Theorem and its Applications. *Pacific Journal of Mathematics*, 5(2):285-309, 1965.
36. J. van Benthem. *Modal Logic and Classical Logic*. Bibliopolis, Naples, 1983.
37. J. van Benthem. Correspondence theory. In D. Gabbay and F. Guenthner, editors, *Handbook of Philosophical Logic*, volume 2, pages 167–247. D. Reidel Pub. Co., 1984.

Time-Dependent Contact Structures in Goguen Categories

Michael Winter*

Department of Computer Science, Brock University,
St. Catharines, Ontario, Canada L2S 3A1
mwinter@brocku.ca

Abstract. In this paper we focus on a theory of time-extended contact. It turns out that a suitable theory can be defined using an $\mathcal{L}$-valued or $\mathcal{L}$-fuzzy version of a contact relation. We study this structure in the context of Goguen categories - a suitable categorical formalization of $\mathcal{L}$-valued or $\mathcal{L}$-fuzzy relations.

1 Introduction

Contact relations have been studied in the context of qualitative or 'pointless' geometry since the early 1920's and nowadays in qualitative spatial reasoning.

In this paper we want to recall a suitable set of axioms for contact in a static world (c.f. [3]), i.e. in a world without any notion of time. In such a situation the contact relation C is an ordinary (or concrete) binary relation. Such a relation can be represented by a characteristic function of a set of pairs. The range of this function is the set of truth values, i.e. the Boolean lattice with two elements, true (or 1) and false (or 0).

Subsequently we consider regions moving in space. It turns out that a convenient definition of contact uses an $\mathcal{L}$-valued or $\mathcal{L}$-fuzzy relation. The underlying concept is a generalization of fuzzy relations using values of an arbitrary complete Brouwerian lattice $\mathcal{L}$ instead of the unit interval $[0,1]$ of the real numbers [6, 21]. Notice, that, in this paper, Brouwerian lattices and Heyting algebras are interchangeable. In our application the degree of 'x is in contact to y' is given by the time x and y are indeed in contact. Since every complete Brouwerian lattice has a least element 0 and a greatest element 1 a (static) contact relation can be seen as a crisp $\mathcal{L}$-fuzzy relation.

In order to provide a suitable set of axioms in this situation, which coincide with the original axioms in the static case, we use the theories of Dedekind and Goguen categories. A Goguen category is a categorical formalization of $\mathcal{L}$-valued relations as a Dedekind category or an allegory is of ordinary relations [4, 17, 18, 19, 20]. We first translate the original axioms for contact into the language of Dedekind categories. Since the language of Dedekind categories is a

* The author gratefully acknowledges support from the Natural Sciences and Engineering Research Council of Canada.

W. MacCaull et al. (Eds.): RelMiCS 2005, LNCS 3929, pp. 249–262, 2006.

subset of the language of Goguen categories we are able to investigate these axioms in the $\mathcal{L}$-valued world. It turns out that a modification or extension is not necessary.

Last, but not least, we focus on the notion of (relative) movement and universes in such a time-dependent contact structure. Here, the additional operations of a Goguen category are necessary in order to capture those notions.

2 Relations and Dedekind Categories

If R is a concrete relation between two sets A and B, i.e. $R \subseteq A \times B$, we use the notation xRy instead of $(x,y) \in R$ to indicate that x and y are in relation R. A suitable categorical description of relations is given by Dedekind categories [10, 11]. Such categories are called *locally complete division allegories* in [4].

Throughout this paper, we use the following notations. To indicate that a morphism R of a category $\mathcal{R}$ has source A and target B we write $R : A \to B$. Composition of a morphism $R : A \to B$ followed by a morphism $S : B \to C$ is denoted by $R;S$. Finally, the identity morphism on A is denoted by $\mathbb{I}_A$.

Definition 1. *A Dedekind category $\mathcal{R}$ is a category satisfying the following:*

1. *The collection of all morphisms with source A and target B is a complete Brouwerian lattice. The morphisms are also called (abstract) relations. Meet, join, the induced ordering, the least and the greatest element are denoted by $\sqcap, \sqcup, \sqsubseteq, \perp\!\!\!\perp_{AB}, \top\!\!\!\top_{AB}$, respectively.*
2. *There is a monotone operation $\smile$ (called converse) mapping a relation $Q : A \to B$ to $Q^\smile : B \to A$ such that for all relations $Q : A \to B$ and $R : B \to C$ the following holds: $(Q;R)^\smile = R^\smile;Q^\smile$ and $(Q^\smile)^\smile = Q$.*
3. *For all relations $Q : A \to B, R : B \to C$ and $S : A \to C$ the modular law*
$$(Q;R) \sqcap S \sqsubseteq Q;(R \sqcap (Q^\smile;S))$$
holds.
4. *For all relations $R : B \to C$ and $S : A \to C$ there is a relation $S/R : A \to B$ (called the left residual of S and R) such that for all $X : A \to B$ the following holds: $X;R \sqsubseteq S \iff X \sqsubseteq S/R$.*

Notice, that, by convention, composition binds more tightly than meet. Therefore, Axiom 3 may be written as $Q;R \sqcap S \sqsubseteq Q;(R \sqcap Q^\smile;S)$.

Corresponding to the left residual, we define the right residual by $Q\backslash R := (R^\smile/Q^\smile)^\smile$. This relation is characterized by $Q;Y \sqsubseteq R \iff Y \sqsubseteq Q\backslash R$. Using both operation simultaneously we define the symmetric quotient $\mathrm{syQ}(Q,R) : B \to C$ of two relations $Q : A \to B$ and $R : A \to C$ by

$$\mathrm{syQ}(Q,R) = (Q\backslash R) \sqcap (Q^\smile/R^\smile) = (Q\backslash R) \sqcap (R\backslash Q)^\smile.$$

This relation is characterized by

$$Q;X \sqsubseteq R \text{ and } X;R^\smile \sqsubseteq Q^\smile \text{ iff } X \sqsubseteq \mathrm{syQ}(Q,R).$$

Because the so-called Tarski rule

$$R \neq \perp\!\!\!\perp_{AB} \implies \top\!\!\!\top_{CA};R;\top\!\!\!\top_{BD} = \top\!\!\!\top_{CD} \quad \text{for all objects } C \text{ and } D$$

is equivalent to a generalized version of the notion of simplicity known from universal algebra, we call a Dedekind category simple iff the Tarski rule is valid.

Recall that a complete Brouwerian lattice has relative pseudo complements, i.e. for all elements x, y there is an element $x \to y$ so that $x \sqcap z \sqsubseteq y$ iff $z \sqsubseteq x \to y$. Of course, this applies to the sets $\mathcal{R}[A, B]$ for all objects A and B of a Dedekind category. Furthermore, recall that in the case of a Boolean algebra the relative pseudo complement $x \to y$ is given by $\bar{x} \sqcup y$ where $\bar{x}$ denotes the complement of x.

In the Dedekind category of sets and concrete relations, a subset of a set A can be represented by a vector, i.e. a relation $v : A \to A$ with $\top\!\!\!\top_{AA}; v = v$. Consequently, a vector in an arbitrary Dedekind category can be seen as an abstract notion of a subset. Furthermore, a relation $Q : A \to B$ is called *total* iff $Q;\top\!\!\!\top_{BB} = \top\!\!\!\top_{AB}$. Notice that in a simple Dedekind category all vectors $v \neq \perp\!\!\!\perp_{AA}$ are total since $\top\!\!\!\top_{AA} = \top\!\!\!\top_{AA}; v;\top\!\!\!\top_{AA} = v;\top\!\!\!\top_{AA}$.

An important concept in Dedekind categories are relational products, i.e. the abstract version of a Cartesian product. A *relational product* of two objects A and B is an object $A \times B$ together with two relations $\pi : A \times B \to A$ and $\rho : A \times B \to B$ such that

$$\pi^{\smile};\pi \sqsubseteq \mathbb{I}_A, \qquad \rho^{\smile};\rho \sqsubseteq \mathbb{I}_B \qquad \pi^{\smile};\rho = \top\!\!\!\top_{AB}, \qquad \pi;\pi^{\smile} \sqcap \rho;\rho^{\smile} = \mathbb{I}_{A \times B}.$$

Notice, that the relational product of a Dedekind category $\mathcal{R}$ is not a product of $\mathcal{R}$ in the sense of category theory. It constitutes such a product in the subcategory of mappings $\mathrm{Map}(\mathcal{R})$, i.e. the subcategory of all relations $Q : A \to B$ that are univalent, i.e. $Q^{\smile};Q \sqsubseteq \mathbb{I}_B$, and total. However, a product in $\mathrm{Map}(\mathcal{R})$ is not necessarily a relational product in $\mathcal{R}$.

Every concrete relation between sets A and B can be transformed to a mapping between A and the powerset $\mathcal{P}(B)$. Such a powerset is determined by the 'is element relation' $\varepsilon : B \to \mathcal{P}(B)$. The corresponding abstract version is called a *relational power*. A relational power of an object A is an object $\mathcal{P}(A)$ together with a relation $\varepsilon : A \to \mathcal{P}(A)$ so that $\mathrm{syQ}(\varepsilon, \varepsilon) \sqsubseteq \mathbb{I}_{\mathcal{P}(A)}$, and $\mathrm{syQ}(R, \varepsilon)$ is total for every object B and every relation $R : A \to B$. Notice that $\mathrm{syQ}(R, \varepsilon)$ is, in fact, a mapping.

A relation $P : A \to A$ of a Dedekind category is called a *pre-order* iff it is reflexive ($\mathbb{I}_A \sqsubseteq P$) and transitive ($P;P \sqsubseteq P$). If P is, in addition, antisymmetric ($P \sqcap P^{\smile} \sqsubseteq \mathbb{I}_A$) it is called an *ordering*. The upper/lower bounds with respect to a pre-order P of subsets induced by a relation $R : B \to A$ are given by $\mathrm{ub}_P(R) := R^{\smile}\backslash P$ and $\mathrm{lb}_P(R) = \mathrm{ub}_{E^{\smile}}(P) = R^{\smile}\backslash P^{\smile}$, respectively. Consequently, the least upper bound is given by $\mathrm{lub}_P(R) = \mathrm{ub}_P(R) \sqcap \mathrm{lb}_P(\mathrm{ub}_P(R))$. If $\mathrm{lub}_P(R) \neq \perp\!\!\!\perp_{BA}$ this relation computes the equivalence class (with respect to the equivalence relation $P \sqcap P^{\smile}$ induced by P) of least upper bounds. If P is an ordering this class is a singleton. For further details on the relational description of bounds we refer to [13, 14, 16].

A pre-order P is called an *upper semi-prelattice* iff $J_P := \mathrm{lub}_P(\pi \sqcup \rho) : A \times A \to A$ is total, i.e. if there is an equivalence class of least upper bounds for every pair of elements. If P is an ordering we call it an upper semi-lattice. Notice, that it can be shown that $\mathrm{lub}_P(\pi \sqcup \rho) = \pi; P \sqcap \rho; P \sqcap ((\pi; P \sqcap \rho; P)^\smile \backslash P^\smile)$. P is said to provide a least element iff the vector $0_P := \mathrm{lub}_P(\perp\!\!\!\perp_{AA})$ is total. If it is clear from the context we omit the index P.

The following lemma summarizes some properties of residuals and the symmetric quotient used in this paper. A proof may be found in [13, 14, 16].

Lemma 1. *Let $\mathcal{R}$ be a Dedekind category, and be $Q : A \to B$, $R : B \to C, S : A \to C$ and $U : A \to D$. Then we have*

1. $(S/R); R \sqsubseteq S$ *and* $Q; (Q\backslash S) \sqsubseteq S$,
2. *if R is a pre-order, then $R/R = R$ and $R\backslash R = R$,*
3. $\mathrm{syQ}(Q, S)^\smile = \mathrm{syQ}(S, Q)$,
4. *if* $\mathrm{syQ}(Q, S)$ *is total, then* $\mathrm{syQ}(Q, S); \mathrm{syQ}(S, U) = \mathrm{syQ}(Q, U)$.

3 $\mathcal{L}$-Valued/Fuzzy Relations

In fuzzy theory, relations are usually considered to be functions mapping pairs to the unit interval $[0, 1]$ of the real numbers. A more general approach uses $\mathcal{L}$-valued/fuzzy relations, i.e. the interval $[0, 1]$ is replaced by an arbitrary complete Brouwerian lattice $\mathcal{L}$.

Definition 2. *Let $\mathcal{L}$ be a complete Brouwerian lattice. An $\mathcal{L}$-fuzzy Dedekind category $D(\mathcal{L})$ is defined by the following components:*

1. *The objects of $D(\mathcal{L})$ are sets $A, B, \ldots$.*
2. *A morphism of $D(\mathcal{L})$ is a function $Q : A \times B \to \mathcal{L}$.*
3. *For $Q : A \times B \to \mathcal{L}$ and $R : B \times C \to \mathcal{L}$ composition is defined by*
$$(Q; R)(x, z) := \bigsqcup_{y \in B} (Q(x, y) \sqcap R(y, z)).$$
4. *For $Q, R : A \times B \to \mathcal{L}$ the join and meet operations are defined by $(Q \sqcup R)(x, y) := Q(x, y) \sqcup R(x, y)$ and $(Q \sqcap R)(x, y) := Q(x, y) \sqcap R(x, y)$, respectively.*
5. *For $Q : A \times B \to \mathcal{L}$ the converse is defined by $Q^\smile(y, x) := Q(x, y)$.*
6. *The identity, zero and universal elements are defined by*
$$\mathbb{I}_A(x, y) := \begin{cases} 0 : x \neq y \\ 1 : x = y, \end{cases} \qquad \begin{array}{l} \perp\!\!\!\perp_{AB}(x, y) := 0, \\ \top\!\!\!\top_{AB}(x, y) := 1. \end{array}$$

It is easy to verify that the structure of $\mathcal{L}$-fuzzy relations defined above is a Dedekind category. The residual operation is given by

$$(Q\backslash R)(x, y) = \bigsqcap_z (Q(z, x) \to R(z, y))$$

If R is a concrete $\mathcal{L}$-valued/fuzzy relation between two sets A and B we usually write xRy instead of $R(x,y)$. The element $xRy \in \mathcal{L}$ is interpreted as the degree of validity of the property 'x and y are in relation R'. Notice, that ordinary relations are a special case of $\mathcal{L}$-valued/fuzzy relations where $\mathcal{L}$ is the Boolean algebra with two elements.

On the other hand, the set of concrete relations is embedded (up to isomorphism) in the set of $\mathcal{L}$-valued/fuzzy relations. The embedded concrete relations are called crisp relations, i.e. such a relation R is characterized by $xRy \in \{0,1\}$ for all x and y. This substructure induces two additional operations on the set of $\mathcal{L}$-valued/fuzzy relations.

$$yR^{\downarrow}z := \begin{cases} 1 \text{ iff } yRz = 1 \\ 0 \text{ iff } yRz \neq 1 \end{cases} \qquad yR^{\uparrow}z := \begin{cases} 1 \text{ iff } yRz \neq 0 \\ 0 \text{ iff } yRz = 0 \end{cases}$$

The operations above map a relation R to the largest crisp relation contained in R and to the least crisp relation above R, respectively. Notice, that for concrete relations these operations are trivial.

4 Goguen Categories

In this section we want to recall the basic notions of the theory of Goguen categories. For properties, motivations and/or explanations not given in this section we refer to [17, 18, 20, 19].

In some sense, a relation of a Dedekind category may be seen as an $\mathcal{L}$-valued relation. The lattice $\mathcal{L}$ may be characterized by scalar relations, i.e. relations $\alpha : A \to A$ satisfying $\alpha_A \sqsubseteq \mathbb{I}_A$ and $\top\!\!\top_{AA};\alpha_A = \alpha_A;\top\!\!\top_{AA}$. We will denote the set of scalar relations in $\mathcal{R}$ on A by $\mathrm{Sc}_{\mathcal{R}}(A)$. Notice, that, if $\mathcal{R}$ is simple, then $\mathrm{Sc}_{\mathcal{R}}(A) = \{\perp\!\!\perp_{AA}, \mathbb{I}_A\}$, i.e. the relations are based on the Boolean algebra with two elements.

Several notions of crispness within a Dedekind category were introduced and discussed in [5, 7, 8]. In [17] it was shown that the theory of Dedekind categories is too weak to express basic notions of $\mathcal{L}$-fuzzy relations such as crispness. Therefore, an extended categorical structure – a Goguen Category – was introduced. This approach adds abstract versions of the two operations $R^{\downarrow}$ and $R^{\uparrow}$ to Dedekind categories.

Definition 3. *A Goguen category $\mathcal{G}$ is a Dedekind category with $\top\!\!\top_{AB} \neq \perp\!\!\perp_{AB}$ for all objects A and B together with two operations $^{\uparrow}$ and $^{\downarrow}$ satisfying the following:*

1. *$R^{\uparrow}, R^{\downarrow} : A \to B$ for all $R : A \to B$.*
2. *$(^{\uparrow}, ^{\downarrow})$ is a Galois correspondence.*
3. *$(R^{\smile};S^{\downarrow})^{\uparrow} = R^{\uparrow\smile};S^{\downarrow}$ for all $R : B \to A$ and $S : B \to C$.*
4. *If $\alpha_A \neq \perp\!\!\perp_{AA}$ is a non-zero scalar then $\alpha_A^{\uparrow} = \mathbb{I}_A$.*
5. *For all antimorphisms[1] $f : \mathrm{Sc}_{\mathcal{G}}(A) \to \mathcal{G}[A,B]$ such that $f(\alpha_A)^{\uparrow} = f(\alpha_A)$ for all $\alpha_A \in \mathrm{Sc}_{\mathcal{G}}(A)$ and all $R : A \to B$ the following equivalence holds.*

[1] A function $f : \mathcal{L}_1 \to \mathcal{L}_2$ between complete lattices is called an *antimorphism*, iff $f(\bigsqcup M) = \bigsqcap f(M)$ holds for all subsets M of $\mathcal{L}_1$.

$$R \sqsubseteq \bigsqcup_{\alpha_A \in \mathrm{Sc}_{\mathcal{G}}(A)} (\alpha_A; f(\alpha_A)) \iff (\alpha_A \backslash R)^{\downarrow} \sqsubseteq f(\alpha_A) \text{ for all } \alpha_A \in \mathrm{Sc}_{\mathcal{G}}(A).$$

Again, it is not hard to verify that the Dedekind category of $\mathcal{L}$-fuzzy relations with $^{\uparrow}$ and $^{\downarrow}$ establishes a Goguen category.

Notice that we do not need Axiom 5 in this paper so that the weaker theory of arrow categories, which does not require a second order axiom, would be sufficient.

In general, we have $R^{\downarrow} \sqsubseteq R \sqsubseteq R^{\uparrow}$. Consequently, we call a relation $R : A \to B$ of a Goguen category crisp iff $R^{\uparrow} = R$. Notice, that a relation is crisp iff $R^{\downarrow} = R$ iff $R^{\uparrow} = R^{\downarrow}$. The crisp fragment $\mathcal{G}^{\uparrow}$ of $\mathcal{G}$ is defined as the collection of all crisp relations of $\mathcal{G}$. This structure together with the inherited operations and constants is a simple Dedekind category, i.e, an abstract counterpart of concrete binary relations.

Furthermore, in [17] it was shown that the complete Brouwerian lattices $\mathrm{Sc}_{\mathcal{G}}(A)$ in a given Goguen category $\mathcal{G}$ are isomorphic for all objects A, and later in [20], that every Goguen category $\mathcal{G}$ is determined by the crisp relations $\mathcal{G}^{\uparrow}$ and the lattice of scalars $\mathrm{Sc}_{\mathcal{G}}(A)$, i.e. it is an abstract version of $\mathrm{Sc}_{\mathcal{G}}(A)$-fuzzy Dedekind category.

The projections π and ρ of a relational product in a Goguen category need not be crisp. Requiring an additional property of the set of scalar elements (or the underlying lattice $\mathcal{L}$) it can be shown that crisp versions actually exist [19]. In this paper we will assume (or require) that the projections are crisp.

In a Goguen category we may consider two different types of pre-orders. The induced equivalence relation $P \sqcap P^{\smile}$ may relate a pair of different elements with 'full' degree 1. In this situation there are equivalent elements even in a crisp interpretation. On the other hand, each pair of different elements could be related in $P \sqcap P^{\smile}$ with a degree strictly less than 1. A crisp interpretation might conclude that we do not have equivalent elements (in a crisp sense). In the language of Goguen categories this is expressed by validity of $(P \sqcap P^{\smile})^{\downarrow}$ being included in $\mathbb{I}$. We call a pre-order P with $(P \sqcap P^{\smile})^{\downarrow} \sqsubseteq \mathbb{I}$ *weakly antisymmetric.*

Notice that, if the underlying lattice $\mathcal{L}$ is interpreted as time periods, P is weakly antisymmetric if two different elements are not equivalent during the whole time considered.

Last,but not least, in the next lemma we summarize some basic properties of relations in a Goguen category.

Lemma 2. *Let $\mathcal{G}$ be a Goguen category. Then we have for all $Q, R : A \to B$*

1. $Q^{\downarrow\uparrow} = Q^{\downarrow}$ *and* $Q^{\uparrow\downarrow} = Q^{\uparrow}$,
2. $(\bigsqcup_{i\in I} R_i)^{\uparrow} = \bigsqcup_{i\in I} R_i^{\uparrow}$ *and* $(\bigsqcap_{i\in I} R_i)^{\downarrow} = \bigsqcap_{i\in I} R_i^{\downarrow}$,
3. $(Q \sqcap R^{\uparrow})^{\uparrow} = Q^{\uparrow} \sqcap R^{\uparrow}$ *and* $(Q \sqcap R^{\downarrow})^{\uparrow} = Q^{\uparrow} \sqcap R^{\downarrow}$,
4. $Q^{\uparrow\smile} = Q^{\smile\uparrow}$ *and* $Q^{\downarrow\smile} = Q^{\smile\downarrow}$,
5. $R = \bigsqcup_{\alpha_A \in \mathrm{Sc}_{\mathcal{G}}(A)} (\alpha_A; (\alpha_A \backslash R)^{\downarrow})$.

A proof may be found in [17]. Notice, that the last property is an abstract version of the α-cut or decomposition Theorem known from fuzzy theory.

5 Static Contact and Time-Dependent Regions

Several theories of contact have been proposed in the literature [1, 2, 9, 12, 15]. In this paper we want to use a very general and, therefore, basic theory. The order structure that is assumed on the set of regions throughout the different proposals is at least an upper semi-lattice (c.f. [3]).

Let C be a relation on an upper semi-lattice $(L, +, \leq)$ with a least element 0. Then C is called a *contact relation* iff it fulfills the following axioms:

C0: $\forall x : \neg(xC0)$,
C1: $\forall x : x \neq 0 \Rightarrow xCx$,
C2: $\forall x, y : xCy \Rightarrow yCx$,
C3: $\forall x, y, z : xCy$ and $y \leq z \Rightarrow xCz$,
C4: $\forall x, y, z : (x + y)Cz \Rightarrow xCz$ or yCz,
C5: $\forall x, y : (\forall z : xCz \Leftrightarrow yCz) \Rightarrow x = y$.

Notice, that using C4 it can be shown that C5 is equivalent to the so-called compatibility axiom

C5': $\forall x, y : (\forall z : xCz \Rightarrow yCz) \Rightarrow x \leq y$.

Later on, we will generalize this approach to pre-orders instead of orderings. In that case C5' seems to be more suitable since C5 implies that the equivalence relation induced by the pre-order is the identity, i.e. the pre-order is in fact an ordering.

A traditional example of a contact structure is the set of regular closed sets of a connected regular T_0 space together with Whitehead's contact relation xCy iff $x \cap y \neq \emptyset$. This model, and hence the whole theory, is static, i.e. it does not use any notion of time.

By translating the first-order axioms given above into equations of the relational language we are able to define an abstract notion of a contact structure in an arbitrary Dedekind category. As mentioned above we take a more general approach by using a pre-order instead of an ordering.

Definition 4. *Let $\mathcal{R}$ be a Dedekind category, $P : A \to A$ be an upper semi-prelattice with a least element, $C : A \to A$ be a relation, and $(A \times A, \pi, \rho)$ a relational product. Then C is called a contact relation on (A, P) iff*

C0: $C \sqcap 0_P = \perp\!\!\!\perp_{AA}$,
C1: $\mathbb{I}_A \sqsubseteq C \sqcup 0_P$,
C2: $C^{\smile} \sqsubseteq C$,
C3: $C; P \sqsubseteq C$,
C4: $J_P; C \sqsubseteq (\pi \sqcup \rho); C$,
C5: $C \backslash C \sqsubseteq P$.

The triple (A, P, C) is also called a contact structure. Furthermore (A, P, C) is called antisymmetric if P is, i.e. if P is an ordering.

Now, we want to switch to regions parameterized by time. We will establish a notion of a standard model and derive the abstract properties thereof. The standard example will be a subset R of all functions from $\mathbb{R}$ to the regular closed sets of a connected regular T_0 space T, for instance, the Euclidean plane. Most examples will assume additional properties of the elements in R. For example, R might be the set of continuous functions. However, throughout this paper we will just assume that at any point in time every (static) region is covered by a moving region, i.e. that for all $t \in \mathbb{R}$ and regular closed sets r of T there is a function $f \in R$ with $f(t) = r$. This property is, for example, satisfied if the set of constant-valued functions is a subset of R, i.e. if every static region is considered to be a moving region with velocity zero.

Definition 5. *Let X and Y be sets and F be a subset of all functions from X to Y. Then F is called dense iff for all $x \in X$ and $y \in Y$ there is a function $f \in F$ with $f(x) = y$.*

First of all, we want to consider an order structure induced on R. If $\mathcal{P}$ denotes the powerset operation we define an $\mathcal{P}(\mathbb{R})$-valued/fuzzy relation P. The value fPg can be interpreted as the degree of f being less than g and is given by the set of all points in time $t \in \mathbb{R}$ where $f(t)$ is indeed less than $g(t)$.

Proposition 1. *Let X be a set, $P : Y \to Y$ be a pre-order and F be a subset of all functions from X to Y. Then the relation $\hat{P}$ defined by*

$$f\hat{P}g := \{x \in X \mid f(x)Pg(x)\}$$

is an $\mathcal{P}(X)$-valued/fuzzy pre-order on F. If P is an ordering then $\hat{P}$ is weakly antisymmetric.

Proof. Since $f(x)Pf(x)$ for all x we have $f\hat{P}f = X$ and, hence, $\mathbb{I} \sqsubseteq \hat{P}$. Now, suppose $x \in (f\hat{P}g \cap g\hat{P}h)$. Then we have $f(x)Pg(x)$ and $g(x)Ph(x)$, and we conclude $f(x)Ph(x)$ by the transitivity of P. This implies $f\hat{P}g \cap g\hat{P}h \subseteq f\hat{P}h$ and, hence, $f(\hat{P};\hat{P})h = \bigcup_g (f\hat{P}g \cap g\hat{P}h) \subseteq f\hat{P}h$ for all f and h, or, equivalently, $\hat{P};\hat{P} \sqsubseteq \hat{P}$. Finally, suppose P is antisymmetric and $f(\hat{P} \sqcap \hat{P}^\smile)^\downarrow g \neq \emptyset$. Then we have $f(\hat{P} \sqcap \hat{P}^\smile)g = X$, and, hence, $f(x)Pg(x)$ and $g(x)Pf(x)$ for all $x \in X$. This implies $f = g$, and, hence, $(\hat{P} \sqcap \hat{P}^\smile)^\downarrow \sqsubseteq \mathbb{I}$. □

In addition, we want to define a join operation $J_{\hat{P}}$ on R with respect to the pre-ordering $\hat{P}$. $J_{\hat{P}}$ will be a $\mathcal{P}(\mathbb{R})$-valued/fuzzy relation between pairs of elements of R and R. The value $(f, g)J_{\hat{P}}h$ can be interpreted as the degree of h being the join of f and g and is given by the set of all points in time $t \in \mathbb{R}$ where $h(t)$ is indeed the join of $f(t)$ and $g(t)$.

Proposition 2. *Let X be a set, (Y, P) be an upper semi-prelattice and F be a dense subset of functions from X to Y. Then the relation $J_{\hat{P}}$ defined by*

$$(f,g)J_{\hat{P}}h := \{x \in X \mid (f(x),g(x))J_P h(x)\}$$

is total and equal to $\mathrm{lub}_{\hat{P}}(\pi \sqcup \rho)$, *i.e.* $\hat{P}$ *is an upper semi-prelattice on* F.

Proof. Since (Y,P) is an upper semi-prelattice, there is an r with $(f(x),g(x))J_P r$ for all $x \in X$ and every pair $f(x)$ and $g(x)$. By the density of F there is a function $k \in R$ with $k(x) = r$ and, hence, $(f(x),g(x))J_P k(x)$ or, equivalently, $x \in (f,g)J_{\hat{P}}k$. We conclude $x \in \bigcup_k((f,g)J_{\hat{P}}k \cap k\top\!\!\top_{AA}h) = (f,g)(J_{\hat{P}};\top\!\!\top_{AA})h$ for all $f,g,h \in F$ and $x \in X$, so that $J_{\hat{P}};\top\!\!\top_{AA} = \top\!\!\top_{A\times A\,A}$, hence, the totality of $J_{\hat{P}}$ follows. Now, consider the following computation

$$\begin{aligned}
&x \in (f,g)J_{\hat{P}}h\\
&\Leftrightarrow (f(x),g(x))J_P h(x)\\
&\Leftrightarrow f(x)Ph(x) \text{ and } g(x)Ph(x) \text{ and}\\
&\qquad f(x)Pk(x) \text{ and } g(x)Pk(x) \text{ implies } h(x)Pk(x) \text{ for all } k\\
&\Leftrightarrow x \in (f,g)(\pi;\hat{P} \sqcap \rho;\hat{P})h \text{ and}\\
&\qquad x \in (f,g)(\pi;\hat{P} \sqcap \rho;\hat{P})k \text{ implies } x \in h\hat{P}k \text{ for all } k\\
&\Leftrightarrow x \in (f,g)(\pi;\hat{P} \sqcap \rho;\hat{P})h \text{ and}\\
&\qquad x \notin (f,g)(\pi;\hat{P} \sqcap \rho;\hat{P})k \text{ or } x \in h\hat{P}k \text{ for all } k\\
&\Leftrightarrow x \in (f,g)(\pi;\hat{P} \sqcap \rho;\hat{P})h \text{ and}\\
&\qquad x \in (f,g)(\pi;\hat{P} \sqcap \rho;\hat{P})k \to h\hat{P}k \text{ for all } k\\
&\Leftrightarrow x \in (f,g)(\pi;\hat{P} \sqcap \rho;\hat{P})h \text{ and } x \in \bigcap_k((f,g)(\pi;\hat{P} \sqcap \rho;\hat{P})k \to h\hat{P}k)\\
&\Leftrightarrow x \in (f,g)(\pi;\hat{P} \sqcap \rho;\hat{P})h \text{ and } x \in (f,g)((\pi;\hat{P} \sqcap \rho;\hat{P})^{\smile}\backslash\hat{P}^{\smile})h\\
&\Leftrightarrow x \in (f,g)(\pi;\hat{P} \sqcap \rho;\hat{P} \sqcap ((\pi;\hat{P} \sqcap \rho;\hat{P})^{\smile}\backslash\hat{P}^{\smile}))h\\
&\Leftrightarrow x \in (f,g)\mathrm{lub}_{\hat{P}}(\pi \sqcup \rho)h,
\end{aligned}$$

which shows $J_{\hat{P}} = \mathrm{lub}_{\hat{P}}(\pi \sqcup \rho)$. □

If P has a least element we may define a relation $0_{\hat{P}}$ on F by $f0_{\hat{P}}g = \{x \in X \mid g(x) = 0\}$. Similar to the last proposition it can be shown that $0_{\hat{P}}$ is total and $0_{\hat{P}} = \mathrm{lub}_{\hat{P}}(\perp\!\!\perp_{AA})$, i.e. that $0_{\hat{P}}$ is the least element of $\hat{P}$. We omit the proof and just state the following proposition.

Proposition 3. *Let* X *be a set,* (Y,P) *be an upper semi-prelattice with a least element, and* F *be a dense subset of all functions from* X *to* Y. *Then* $\hat{P}$ *is an upper semi-prelattice with a least element. If* P *is an ordering then* $\hat{P}$ *is weakly antisymmetric.*

According to our standard model, a time-dependent contact structure will be based on a weakly antisymmetric upper semi-prelattice with a least element.

We define that two such regions $r_1, r_2 \in R$ are in *contact at a time* $t \in \mathbb{R}$ if $r_1(t)Cr_2(t)$, where C is the (static) Whitehead contact relation on T. As above, this allows us to define a $\mathcal{P}(\mathbb{R})$-valued/fuzzy contact relation on R by $r_1\hat{C}r_2 := \{t \in \mathbb{R} \mid r_1(t)Cr_2(t)\}$. The degree of the property 'r_1 is in contact with r_2' is exactly the set of all points in time when they are actually in (static) contact.

Let $\hat{t}$ be the scalar representing the time point t, i.e. $\hat{t}$ is an atom within the lattice of scalars. Now, using α-cuts (c.f. Lemma 2(5)) the relation $r_1(\hat{t}\backslash C)^{\downarrow}r_2$ is the relational version of the statement 'region r_1 is in contact with region r_2 at time t'.

Proposition 4. *Let X be a set, (Y, P, C) be a (concrete) contact structure on the set Y and F be a dense subset of functions from X to Y. Then the relation $\hat{C}$ defined by*

$$f\hat{C}g := \{x \in X \mid f(x)Cg(x)\}$$

together with $\hat{P}$ is a contact structure on F.

Proof. By Proposition 3 it remains to show that $\hat{C}$ is a contact relation.

C0: Suppose $x \in f(\hat{C} \sqcap 0_P)g$. Then we have $f(x)Cg(x)$ and $x \in f0_{\hat{P}}g$. The latter is equivalent to $g(x) = 0$, a contradiction, and, hence, $f(\hat{C} \sqcap 0_{\hat{P}})g = \emptyset$ for all f and g, or, equivalently, $\hat{C} \sqcap 0_{\hat{P}} = \perp\!\!\!\perp_{AA}$.

C1: Suppose $f(x) = 0$. Then we have $x \in f0_{\hat{P}}f$. If $f(x) \neq 0$ we have $f(x)Cf(x)$ and, hence, $x \in f\hat{C}f$. We conclude $x \in f(\hat{C} \sqcup 0_{\hat{P}})f$ for all $x \in X$ and $f \in F$ which shows C1.

C2: Suppose $x \in f\hat{C}^{\smile}g = g\hat{C}f$. Then we have $g(x)Cf(x)$ so that $f(x)C^{\smile}g(x)$ follows. We conclude $f(x)Cg(x)$ and, hence, $x \in f\hat{C}g$ which shows C2.

C3: Suppose $x \in f(\hat{C};\hat{P})h = \bigcup_g(f\hat{C}g \cap g\hat{P}h)$. Then there is a g so that $f(x)Cg(x)$ and $g(x)Ph(x)$. This implies $f(x)Ch(x)$ and, hence, $x \in f\hat{C}h$.

C4: Suppose $x \in (f,g)(J_{\hat{P}};\hat{C})h = \bigcup_k((f,g)J_{\hat{P}}k \cap k\hat{C}h)$. Then there is a function k so that $(f(x),g(x))J_Pk(x)$ and $k(x)Ch(x)$. This implies $f(x)Ch(x)$ or $g(x)Ch(x)$, which is equivalent to $x \in (f,g)((\pi \sqcup \rho);\hat{C})h$.

C5: Finally, suppose $x \in f(\hat{C}\backslash\hat{C})h = \bigcap_g(g\hat{C}f \to g\hat{C}h)$. Then $x \in g\hat{C}f \to g\hat{C}h$ for all $g \in F$, which is equivalent to $x \notin g\hat{C}f$ or $x \in g\hat{C}h$. We conclude that either not $g(x)Cf(x)$ or $g(x)Ch(x)$ holds, or, equivalently, that $g(x)Cf(x)$ implies $g(x)Ch(x)$ for all $g \in F$. Assume $r \in Y$ is an arbitrary region with $rCf(x)$. By the density of F, there is a function $g_r \in F$ with $g_r(x) = r$. Together, we conclude $r = g_r(x)Ch(x)$, and, hence, $rCf(x)$ implies $rCh(x)$ for all r. Consequently, we have $f(x)(C\backslash C)h(x)$. Since C is a (concrete) contact relation this implies $f(x)Ph(x)$ and, hence, $x \in f\hat{P}h$. □

Notice that the two contact structures (Y, P, C) and $(F, \hat{P}, \hat{C})$, even though they fulfil the same set of axioms, are based on a different notion of a relation and hence the corresponding operations are defined differently.

In the standard example $\hat{P}$ is weakly antisymmetric. Recall that in this case, two different regions cannot be equivalent during the whole time considered.

6 Universes

Contact structures describe regions in a 'pointless' manner. Regions are characterized by their relationship to other regions without referring to their extent in space or time. If we focus on one particular region r, we may recognize that this region is in contact to different sets of regions at different points in time. This situation allows different interpretations. First, we may consider that r is actually moving. All other regions are either moving as well or at rest. Another interpretation might be that r is at rest (not moving at all), and the other regions are moving around r. Since a contact structure does not provide additional information, both interpretations are valid. Consequently, any notion of 'being at rest' depends on the 'beholder'.

In this section we want to introduce the concept of a universe within an arbitrary contact structure. Such a universe is a maximal collection of regions, which can legally be considered to be at rest, i.e. they are not moving. Usually there is more than one candidate for such a collection according to different 'beholders'. First, we want to introduce the notion of two regions being in relative rest to each other.

If two regions r_1 and r_2 are not in contact at any point in time or if they are in contact at all points in time, we can safely assume that they are in relative rest to each other. This is motivated by the 'beholder' focussing on exactly r_1 and r_2. This 'beholder' considers all other regions to be moving around (or within) r_1 and r_2. This interpretation is consistent with the information provided by the contact structure.

The considerations above lead immediately to the following definition.

Definition 6. *Let $\mathcal{G}$ be a Goguen category and (A, P, C) be a contact structure. Then we define the 'relative at rest' relation RR by $RR := C^{\uparrow} \to (C \sqcup \mathbb{I}_A)^{\downarrow}$.*

In the next lemma we summarize some basic properties of the relation RR.

Lemma 3. *1. RR is crisp, reflexive and symmetric.*
2. If C is crisp then $RR = \mathbb{T}_{AA}$.

Proof. 1. In order to prove that RR is crisp, it is sufficient to show $RR^{\uparrow} \sqsubseteq RR$ which follows from $RR^{\uparrow} \sqcap C^{\uparrow} = (RR \sqcap C^{\uparrow})^{\uparrow} \sqsubseteq (C \sqcup \mathbb{I}_A)^{\downarrow\uparrow} = (C \sqcup \mathbb{I}_A)^{\downarrow}$, using Lemma 2(1 & 3). From $\mathbb{I}_A \sqcap C^{\uparrow} \sqsubseteq \mathbb{I}_A \sqsubseteq (C \sqcup \mathbb{I}_A)^{\downarrow}$ we immediately conclude that RR is reflexive. The last assertion follows from

$$
\begin{aligned}
RR^{\smile} \sqcap C^{\uparrow} &= (RR \sqcap C^{\uparrow\smile})^{\smile} & \\
&= (RR \sqcap C^{\smile\uparrow})^{\smile} & \text{Lemma 2(4)} \\
&\sqsubseteq (RR \sqcap C^{\uparrow})^{\smile} & \text{Axiom C2} \\
&\sqsubseteq (C \sqcup \mathbb{I}_A)^{\downarrow\smile} & \\
&= (C^{\smile} \sqcup \mathbb{I}_A)^{\downarrow} & \text{Lemma 2(4)} \\
&\sqsubseteq (C \sqcup \mathbb{I}_A)^{\downarrow} & \text{Axiom C2}
\end{aligned}
$$

2. This follows immediately from $\top\!\!\top_{AA} \sqcap C^{\uparrow} = C^{\uparrow} = C = C^{\downarrow} \sqsubseteq (C \sqcup \mathbb{I}_A)^{\downarrow}$. □

Notice, that 2. of the previous lemma indicates that in a static model all regions are in relative rest to each other.

Using this relation it is now possible to define a universe. It is given by a maximal vector of elements in relative rest to each other.

Definition 7. *Let $\mathcal{G}$ be a Goguen category and (A, P, C) be a contact structure. A crisp vector v is called a universe (for (A, P, C)) iff it is maximal with respect to the property $v^{\smile}; v \sqsubseteq RR$.*

Using the Lemma of Zorn it can be shown that universes exist.

Proposition 5. *Let $\mathcal{G}$ be a Goguen category and (A, P, C) be a contact structure. Then there is a universe for (A, P, C).*

Proof. Consider the set $M := \{v : A \to A \mid v \text{ is a crisp vector and } v^{\smile}; v \sqsubseteq RR\}$. Then M is not empty since $\perp\!\!\perp_{AA} \in M$. Let $v_1 \sqsubseteq v_2 \sqsubseteq \ldots$ be a chain in M. Then $\bigsqcup_i v_i$ is a vector since $\top\!\!\top_{AA}; (\bigsqcup_i v_i) = \bigsqcup_i (\top\!\!\top_{AA}; v_i) = \bigsqcup_i v_i$, and crisp since $(\bigsqcup_i v_i)^{\uparrow} = \bigsqcup_i v_i^{\uparrow} = \bigsqcup_i v_i$. Furthermore, we have

$$(\bigsqcup_i v_i)^{\smile}; (\bigsqcup_i v_i) = \bigsqcup_{i,j} (v_i^{\smile}; v_j) \sqsubseteq \bigsqcup_i (v_i^{\smile}; v_i) \sqsubseteq RR,$$

so that M is closed under unions of chains. The Lemma of Zorn implies that M has a maximal element, which is by definition a universe. □

Now, the question arises whether a region can be uniquely represented by its projection to the universe. At any time we may consider the set of universe regions that include the given region r. In other words, we are interested in the extent of r with respect to the given universe. This projection is given by the function $p = \mathrm{syQ}((\mathbb{I}_A \sqcap v^{\smile}; v); P, \varepsilon)$ that maps each region to the set of universe elements included in. Recall that p is again an $\mathcal{L}$-fuzzy relation so that it is indexed by time. Now, a region can be uniquely represented by its projection if p is injective (up to the equivalence relation induced by the preorder P). Unfortunately, this is not necessarily true. We need the additional property $P \sqsubseteq P; (\mathbb{I}_A \sqcap v^{\smile}; v); P$, i.e. at any time and for every pair of regions there is an element of the universe including the first but not the second region. In that case we say that the universe *has enough elements.*

Proposition 6. *Let $\mathcal{G}$ be a Goguen category, (A, P, C) be a contact structure, and v be a universe. If v has enough elements then $p; p^{\smile} = P \sqcap P^{\smile} = \mathrm{syQ}(C, C)$. If P is, in addition, weakly antisymmetric, then $(p; p^{\smile})^{\downarrow} = \mathbb{I}_A$.*

Proof. The last assertion follows immediately from the first. Suppose v has enough elements, i.e. $P \sqsubseteq P; (\mathbb{I}_A \sqcap v^{\smile}; v); P$ holds. We want to show that $((\mathbb{I}_A \sqcap v^{\smile}; v); P) \backslash ((\mathbb{I}_A \sqcap v^{\smile}; v); P) = P$. This implies the assertion by

$$\begin{aligned} p;p^{\smile} &= \mathrm{syQ}((\mathbb{I}_A \sqcap v^{\smile};v);P,\varepsilon);\mathrm{syQ}(\varepsilon,(\mathbb{I}_A \sqcap v^{\smile};v);P) && \text{Lemma 1(3)} \\ &= \mathrm{syQ}((\mathbb{I}_A \sqcap v^{\smile};v);P,(\mathbb{I}_A \sqcap v^{\smile};v);P) && \text{Lemma 1(4)} \\ &= (((\mathbb{I}_A \sqcap v^{\smile};v);P)\backslash((\mathbb{I}_A \sqcap v^{\smile};v);P)) \\ &\quad \sqcap (((\mathbb{I}_A \sqcap v^{\smile};v);P)\backslash((\mathbb{I}_A \sqcap v^{\smile};v);P))^{\smile} \\ &= P \sqcap P^{\smile}. \end{aligned}$$

The inclusion '$\sqsupseteq$' follows immediately from $(\mathbb{I}_A \sqcap v^{\smile};v);P;P \sqsubseteq (\mathbb{I}_A \sqcap v^{\smile};v);P$. For the other inclusion, we have

$$\begin{aligned} &P;((\mathbb{I}_A \sqcap v^{\smile};v);P)\backslash((\mathbb{I}_A \sqcap v^{\smile};v);P) \\ &\sqsubseteq P;(\mathbb{I}_A \sqcap v^{\smile};v);P;((\mathbb{I}_A \sqcap v^{\smile};v);P)\backslash((\mathbb{I}_A \sqcap v^{\smile};v);P) \\ &\sqsubseteq P;(\mathbb{I}_A \sqcap v^{\smile};v);P && \text{Lemma 1(1)} \\ &\sqsubseteq P;P \\ &\sqsubseteq P \end{aligned}$$

so that $((\mathbb{I}_A \sqcap v^{\smile};v);P)\backslash((\mathbb{I}_A \sqcap v^{\smile};v);P) \sqsubseteq P\backslash P$ follows. Since P is a pre-order Lemma 1(2) implies '$\sqsubseteq$'. □

7 Conclusion

In this paper we have investigated the axioms of a contact relation in the fuzzy world. It turned out that this can be used to model moving regions. In addition, we focused on the notion of being 'at relative rest to each other' and the notion of a universe, which are both trivial in the crisp world.

Further study will concentrate on the expressive power of the language of Goguen categories within this theory. In particular, properties of the lattice $\mathcal{L}$ (and consequently the model of time used) will be considered.

References

1. Biacino L., Gerla G.: Connection Structures. Notre Dame Journal of Formal Logic 32 (1991), 242-247.
2. Clarke B. L.: A calculus of individuals based on 'connenction'. Notre Dame Journal of Formal Logic 22 (1981), 204-218.
3. Düntsch I., Winter M.: Finite Contact Structure. Proceedings of the 8th International Conference on Relational Methods in Computer Science, this volume
4. Freyd P., Scedrov A.: Categories, Allegories. North-Holland (1990).
5. Furusawa H: Algebraic Formalisations of Fuzzy Relations and their Representation Theorems. PhD-Thesis, Department of Informatics, Kyushu University, Japan (1998).
6. Goguen J.A.: L-fuzzy sets. J. Math. Anal. Appl. 18 (1967), 145-157.
7. Kawahara, Y., Furusawa H.: Crispness and Representation Theorems in Dedekind Categories. DOI-TR 143, Kyushu University (1997).

8. Kawahara, Y., Furusawa H.: An Algebraic Formalization of Fuzzy Relations. Fuzzy Sets and Systems 101, 125-135 (1999).
9. Leonard H. S., Goodman N.: The calculus of individuals and its use. Journal of Symbolic Logic 5 (1940), 45-55.
10. Olivier J.P., Serrato D.: Catégories de Dedekind. Morphismes dans les Catégories de Schröder. C.R. Acad. Sci. Paris 290 (1980), 939-941.
11. Olivier J.P., Serrato D.: Squares and Rectangles in Relational Categories - Three Cases: Semilattice, Distributive lattice and Boolean Non-unitary. Fuzzy sets and systems 72 (1995), 167-178.
12. Randell D. A., Cohn A. G., Cui Z.: Computing transitivity tables: A challenge for automated theorem provers. In Kapur D. (ed.): Proceedings of the 11th International Conference on Automated Deduction (CADE-11), LNAI 607 (1992), 786-790.
13. Schmidt G., Ströhlein T.: Relationen und Graphen. Springer (1989); English version: Relations and Graphs. Discrete Mathematics for Computer Scientists, EATCS Monographs on Theoret. Comput. Sci., Springer (1993)
14. Schmidt G., Hattensperger C., Winter M.: Heterogeneous Relation Algebras. *In:* Brink C., Kahl W., Schmidt G. (eds.), Relational Methods in Computer Science, Advances in Computer Science, Springer Vienna (1997).
15. Stell J.: Boolean connection algebras: A new approach to the Region Connection Calculus. Artificial Intellegence 122 (2000), 111-136.
16. Winter M.: Strukturtheorie heterogener Relationenalgebren mit Anwendung auf Nichtdetermismus in Programmiersprachen. Dissertationsverlag NG Kopierladen GmbH, München (1998)
17. Winter M.: A new Algebraic Approach to L-Fuzzy Relations convenient to study Crispness. Information Sciences 139 (2001), 233-252.
18. Winter M.: Derived Operations in Goguen Categories. Theory and Apllications of Categories 10 (2002), 220-247.
19. Winter M.: Relational Constructions in Goguen Categories. Relation Methods in Computer Science. LNCS 2561 (2002), 212-227.
20. Winter M.: Representation Theory of Goguen Categories. Fuzzy Sets and Systems 138 (2003), 85-126.
21. Zadeh L.A.: Fuzzy sets, Information and Control 8 (1965), 338-353.

Author Index

Lecture Notes in Computer Science

For information about Vols. 1–3842

please contact your bookseller or Springer

Vol. 3946: T.R. Roth-Berghofer, S. Schulz, D.B. Leake (Eds.), Modeling and Retrieval of Context. XI, 149 pages. 2006. (Sublibrary LNAI).

Vol. 3945: M. Hagiya, P. Wadler (Eds.), Functional and Logic Programming. X, 295 pages. 2006.

Vol. 3939: C. Priami, L. Cardelli, S. Emmott (Eds.), Transactions on Computational Systems Biology IV. VII, 141 pages. 2006. (Sublibrary LNBI).

Vol. 3936: M. Lalmas, A. MacFarlane, S. Rüger, A. Tombros, T. Tsikrika, A. Yavlinsky (Eds.), Advances in Information Retrieval. XI, 584 pages. 2006.

Vol. 3934: J.A. Clark, R.F. Paige, F.A. C. Polack, P.J. Brooke (Eds.), Security in Pervasive Computing. X, 243 pages. 2006.

Vol. 3933: F. Bonchi, J.-F. Boulicaut (Eds.), Knowledge Discovery in Inductive Databases. VIII, 251 pages. 2006.

Vol. 3931: B. Apolloni, M. Marinaro, G. Nicosia, R. Tagliaferri (Eds.), Neural Nets. XIII, 370 pages. 2006.

Vol. 3930: D.S. Yeung, Z.-Q. Liu, X.-Z. Wang, H. Yan (Eds.), Advances in Machine Learning and Cybernetics. XXI, 1110 pages. 2006. (Sublibrary LNAI).

Vol. 3929: W. MacCaull, M. Winter, I. Düntsch (Eds.), Relational Methods in Computer Science. VIII, 263 pages. 2006.

Vol. 3928: J. Domingo-Ferrer, J. Posegga, D. Schreckling (Eds.), Smart Card Research and Advanced Applications. XI, 359 pages. 2006.

Vol. 3927: J. Hespanha, A. Tiwari (Eds.), Hybrid Systems: Computation and Control. XII, 584 pages. 2006.

Vol. 3925: A. Valmari (Ed.), Model Checking Software. X, 307 pages. 2006.

Vol. 3924: P. Sestoft (Ed.), Programming Languages and Systems. XII, 343 pages. 2006.

Vol. 3923: A. Mycroft, A. Zeller (Eds.), Compiler Construction. XIII, 277 pages. 2006.

Vol. 3922: L. Baresi, R. Heckel (Eds.), Fundamental Approaches to Software Engineering. XIII, 427 pages. 2006.

Vol. 3921: L. Aceto, A. Ingólfsdóttir (Eds.), Foundations of Software Science and Computation Structures. XV, 447 pages. 2006.

Vol. 3920: H. Hermanns, J. Palsberg (Eds.), Tools and Algorithms for the Construction and Analysis of Systems. XIV, 506 pages. 2006.

Vol. 3918: W.K. Ng, M. Kitsuregawa, J. Li, K. Chang (Eds.), Advances in Knowledge Discovery and Data Mining. XXIV, 879 pages. 2006. (Sublibrary LNAI).

Vol. 3917: H. Chen, F.Y. Wang, C.C. Yang, D. Zeng, M. Chau, K. Chang (Eds.), Intelligence and Security Informatics. XII, 186 pages. 2006.

Vol. 3916: J. Li, Q. Yang, A.-H. Tan (Eds.), Data Mining for Biomedical Applications. VIII, 155 pages. 2006. (Sublibrary LNBI).

Vol. 3915: R. Nayak, M.J. Zaki (Eds.), Knowledge Discovery from XML Documents. VIII, 105 pages. 2006.

Vol. 3914: A. Garcia, R. Choren, C. Lucena, P. Giorgini, T. Holvoet, A. Romanovsky (Eds.), Software Engineering for Multi-Agent Systems IV. XIV, 255 pages. 2006.

Vol. 3910: S.A. Brueckner, G.D.M. Serugendo, D. Hales, F. Zambonelli (Eds.), Engineering Self-Organising Systems. XII, 245 pages. 2006. (Sublibrary LNAI).

Vol. 3909: A. Apostolico, C. Guerra, S. Istrail, P. Pevzner, M. Waterman (Eds.), Research in Computational Molecular Biology. XVII, 612 pages. 2006. (Sublibrary LNBI).

Vol. 3907: F. Rothlauf, J. Branke, S. Cagnoni, E. Costa, C. Cotta, R. Drechsler, E. Lutton, P. Machado, J.H. Moore, J. Romero, G.D. Smith, G. Squillero, H. Takagi (Eds.), Applications of Evolutionary Computing. XXIV, 813 pages. 2006.

Vol. 3906: J. Gottlieb, G.R. Raidl (Eds.), Evolutionary Computation in Combinatorial Optimization. XI, 293 pages. 2006.

Vol. 3905: P. Collet, M. Tomassini, M. Ebner, S. Gustafson, A. Ekárt (Eds.), Genetic Programming. XI, 361 pages. 2006.

Vol. 3904: M. Baldoni, U. Endriss, A. Omicini, P. Torroni (Eds.), Declarative Agent Languages and Technologies III. XII, 245 pages. 2006. (Sublibrary LNAI).

Vol. 3903: K. Chen, R. Deng, X. Lai, J. Zhou (Eds.), Information Security Practice and Experience. XIV, 392 pages. 2006.

Vol. 3901: P.M. Hill (Ed.), Logic Based Program Synthesis and Transformation. X, 179 pages. 2006.

Vol. 3899: S. Frintrop, VOCUS: A Visual Attention System for Object Detection and Goal-Directed Search. XIV, 216 pages. 2006. (Sublibrary LNAI).

Vol. 3898: K. Tuyls, P.J. 't Hoen, K. Verbeeck, S. Sen (Eds.), Learning and Adaption in Multi-Agent Systems. X, 217 pages. 2006. (Sublibrary LNAI).

Vol. 3897: B. Preneel, S. Tavares (Eds.), Selected Areas in Cryptography. XI, 371 pages. 2006.

Vol. 3896: Y. Ioannidis, M.H. Scholl, J.W. Schmidt, F. Matthes, M. Hatzopoulos, K. Boehm, A. Kemper, T. Grust, C. Boehm (Eds.), Advances in Database Technology - EDBT 2006. XIV, 1208 pages. 2006.

Vol. 3895: O. Goldreich, A.L. Rosenberg, A.L. Selman (Eds.), Theoretical Computer Science. XII, 399 pages. 2006.

Vol. 3894: W. Grass, B. Sick, K. Waldschmidt (Eds.), Architecture of Computing Systems - ARCS 2006. XII, 496 pages. 2006.

Vol. 3893: L. Atzori, D.D. Giusto, R. Leonardi, F. Pereira (Eds.), Visual Content Processing and Representation. X, 224 pages. 2006.

Vol. 3891: J.S. Sichman, L. Antunes (Eds.), Multi-Agent-Based Simulation VI. X, 191 pages. 2006. (Sublibrary LNAI).

Vol. 3890: S.G. Thompson, R. Ghanea-Hercock (Eds.), Defence Applications of Multi-Agent Systems. XII, 141 pages. 2006. (Sublibrary LNAI).

Vol. 3889: J. Rosca, D. Erdogmus, J.C. Príncipe, S. Haykin (Eds.), Independent Component Analysis and Blind Signal Separation. XXI, 980 pages. 2006.

Vol. 3888: D. Draheim, G. Weber (Eds.), Trends in Enterprise Application Architecture. IX, 145 pages. 2006.

Vol. 3887: J.R. Correa, A. Hevia, M. Kiwi (Eds.), LATIN 2006: Theoretical Informatics. XVI, 814 pages. 2006.

Vol. 3886: E.G. Bremer, J. Hakenberg, E.-H.(S.) Han, D. Berrar, W. Dubitzky (Eds.), Knowledge Discovery in Life Science Literature. XIV, 147 pages. 2006. (Sublibrary LNBI).

Vol. 3885: V. Torra, Y. Narukawa, A. Valls, J. Domingo-Ferrer (Eds.), Modeling Decisions for Artificial Intelligence. XII, 374 pages. 2006. (Sublibrary LNAI).

Vol. 3884: B. Durand, W. Thomas (Eds.), STACS 2006. XIV, 714 pages. 2006.

Vol. 3882: M.L. Lee, K.L. Tan, V. Wuwongse (Eds.), Database Systems for Advanced Applications. XIX, 923 pages. 2006.

Vol. 3881: S. Gibet, N. Courty, J.-F. Kamp (Eds.), Gesture in Human-Computer Interaction and Simulation. XIII, 344 pages. 2006. (Sublibrary LNAI).

Vol. 3880: A. Rashid, M. Aksit (Eds.), Transactions on Aspect-Oriented Software Development I. IX, 335 pages. 2006.

Vol. 3879: T. Erlebach, G. Persinao (Eds.), Approximation and Online Algorithms. X, 349 pages. 2006.

Vol. 3878: A. Gelbukh (Ed.), Computational Linguistics and Intelligent Text Processing. XVII, 589 pages. 2006.

Vol. 3877: M. Detyniecki, J.M. Jose, A. Nürnberger, C. J. '. van Rijsbergen (Eds.), Adaptive Multimedia Retrieval: User, Context, and Feedback. XI, 279 pages. 2006.

Vol. 3876: S. Halevi, T. Rabin (Eds.), Theory of Cryptography. XI, 617 pages. 2006.

Vol. 3875: S. Ur, E. Bin, Y. Wolfsthal (Eds.), Hardware and Software, Verification and Testing. X, 265 pages. 2006.

Vol. 3874: R. Missaoui, J. Schmidt (Eds.), Formal Concept Analysis. X, 309 pages. 2006. (Sublibrary LNAI).

Vol. 3873: L. Maicher, J. Park (Eds.), Charting the Topic Maps Research and Applications Landscape. VIII, 281 pages. 2006. (Sublibrary LNAI).

Vol. 3872: H. Bunke, A. L. Spitz (Eds.), Document Analysis Systems VII. XIII, 630 pages. 2006.

Vol. 3871: E.-G. Talbi, P. Liardet, P. Collet, E. Lutton, M. Schoenauer (Eds.), Artificial Evolution. XI, 310 pages. 2006.

Vol. 3870: S. Spaccapietra, P. Atzeni, W.W. Chu, T. Catarci, K.P. Sycara (Eds.), Journal on Data Semantics V. XIII, 237 pages. 2006.

Vol. 3869: S. Renals, S. Bengio (Eds.), Machine Learning for Multimodal Interaction. XIII, 490 pages. 2006.

Vol. 3868: K. Römer, H. Karl, F. Mattern (Eds.), Wireless Sensor Networks. XI, 342 pages. 2006.

Vol. 3866: T. Dimitrakos, F. Martinelli, P.Y.A. Ryan, S. Schneider (Eds.), Formal Aspects in Security and Trust. X, 259 pages. 2006.

Vol. 3865: W. Shen, K.-M. Chao, Z. Lin, J.-P.A. Barthès, A. James (Eds.), Computer Supported Cooperative Work in Design II. XII, 659 pages. 2006.

Vol. 3863: M. Kohlhase (Ed.), Mathematical Knowledge Management. XI, 405 pages. 2006. (Sublibrary LNAI).

Vol. 3862: R.H. Bordini, M. Dastani, J. Dix, A.E.F. Seghrouchni (Eds.), Programming Multi-Agent Systems. XIV, 267 pages. 2006. (Sublibrary LNAI).

Vol. 3861: J. Dix, S.J. Hegner (Eds.), Foundations of Information and Knowledge Systems. X, 331 pages. 2006.

Vol. 3860: D. Pointcheval (Ed.), Topics in Cryptology – CT-RSA 2006. XI, 365 pages. 2006.

Vol. 3858: A. Valdes, D. Zamboni (Eds.), Recent Advances in Intrusion Detection. X, 351 pages. 2006.

Vol. 3857: M.P.C. Fossorier, H. Imai, S. Lin, A. Poli (Eds.), Applied Algebra, Algebraic Algorithms and Error-Correcting Codes. XI, 350 pages. 2006.

Vol. 3855: E. A. Emerson, K.S. Namjoshi (Eds.), Verification, Model Checking, and Abstract Interpretation. XI, 443 pages. 2005.

Vol. 3854: I. Stavrakakis, M. Smirnov (Eds.), Autonomic Communication. XIII, 303 pages. 2006.

Vol. 3853: A.J. Ijspeert, T. Masuzawa, S. Kusumoto (Eds.), Biologically Inspired Approaches to Advanced Information Technology. XIV, 388 pages. 2006.

Vol. 3852: P.J. Narayanan, S.K. Nayar, H.-Y. Shum (Eds.), Computer Vision – ACCV 2006, Part II. XXXI, 977 pages. 2006.

Vol. 3851: P.J. Narayanan, S.K. Nayar, H.-Y. Shum (Eds.), Computer Vision – ACCV 2006, Part I. XXXI, 973 pages. 2006.

Vol. 3850: R. Freund, G. Păun, G. Rozenberg, A. Salomaa (Eds.), Membrane Computing. IX, 371 pages. 2006.

Vol. 3849: I. Bloch, A. Petrosino, A.G.B. Tettamanzi (Eds.), Fuzzy Logic and Applications. XIV, 438 pages. 2006. (Sublibrary LNAI).

Vol. 3848: J.-F. Boulicaut, L. De Raedt, H. Mannila (Eds.), Constraint-Based Mining and Inductive Databases. X, 401 pages. 2006. (Sublibrary LNAI).

Vol. 3847: K.P. Jantke, A. Lunzer, N. Spyratos, Y. Tanaka (Eds.), Federation over the Web. X, 215 pages. 2006. (Sublibrary LNAI).

Vol. 3846: H. J. van den Herik, Y. Björnsson, N.S. Netanyahu (Eds.), Computers and Games. XIV, 333 pages. 2006.

Vol. 3845: J. Farré, I. Litovsky, S. Schmitz (Eds.), Implementation and Application of Automata. XIII, 360 pages. 2006.

Vol. 3844: J.-M. Bruel (Ed.), Satellite Events at the MoDELS 2005 Conference. XIII, 360 pages. 2006.

Vol. 3843: P. Healy, N.S. Nikolov (Eds.), Graph Drawing. XVII, 536 pages. 2006.